Geodynamics of the Alps 2

SCIENCES

Geoscience, Field Director – Yves Lagabrielle

Dynamics of the Continental Lithosphere,
Subject Head – Sylvie Leroy

Geodynamics of the Alps 2

Pre-collisional Processes

Coordinated by
Claudio L. Rosenberg
Nicolas Bellahsen

ISTE

WILEY

First published 2023 in Great Britain and the United States by ISTE Ltd and John Wiley & Sons, Inc.

ISTE Ltd
27-37 St George's Road
London SW19 4EU
UK

www.iste.co.uk

John Wiley & Sons, Inc.
111 River Street
Hoboken, NJ 07030
USA

www.wiley.com

Library of Congress Control Number: 2022941465

British Library Cataloguing-in-Publication Data
A CIP record for this book is available from the British Library
ISBN 978-1-78945-117-7

ERC code:
PE10 Earth System Science
PE10_4 Terrestrial ecology, land cover change
PE10_5 Geology, tectonics, volcanology
PE10_13 Physical geography

Contents

Introduction

Nicolas BELLAHSEN and Claudio L. ROSENBERG
ISTeP, Sorbonne Université,
UMR 7193, CNRS, Paris, France

The European Alps represent an archetypal mountain belt, mainly because its orogenic wedge is the most studied one. It results from a period of Mesozoic divergence forming the Tethyan Ligurian oceanic realm, linking the Neo-Tethys in the east with the Central Atlantic in the west. This followed the Variscan orogenic event affecting the European lithosphere and preceded a period of convergence starting during upper Cretaceous times with the northward movement of the Adria/Africa plates. This convergence was first accommodated by oceanic subduction, and later followed by continental subduction and eventually collision *sensu stricto*. The transition between oceanic and continental subduction is characterized by exhumation of high-pressure (HP) nappes of oceanic origin during early Eocene times, followed by HP metamorphism and exhumation of continental units during middle to late Eocene (Briançonnais and European distal continental margin). Collision started at the Eocene–Oligocene transition and is characterized by barrovian metamorphism, crustal shortening, high topographic relief and molasse-type peri-orogenic basins. During the whole convergence history, the amount of magmatism was very small. Today, convergence has slowed down, varying from almost zero in the Western Alps to approximately 2 mm/yr in the Eastern Alps, although vertical movements, extensive and compressive earthquakes are recorded, even in the Western Alps.

In spite of the latter observations, the European Alps may be considered as simpler than many other mountain belts that also result from the closure of oceanic domains. Indeed, in many other mountain belts, subduction and collision do not

Geodynamics of the Alps 2,
coordinated by Claudio L. ROSENBERG and Nicolas BELLAHSEN.

follow a simple sequence. For example, in the middle Paleozoic, during the Caledonian cycle and the Iapetus closure and more recently, during Cenozoic times and the Tethys closure between India and Asia, HP metamorphism events occurred long after the onset of collision. This is possibly due to the subduction of larger oceanic domains, and to longer and more vigorous convergence dynamics, which probably did not affect the Alpine realm.

The book *Geodynamics of the Alps* consists of three volumes. The first includes the recent and present-day structure and tectonic setting of the Alpine chain, from lithospheric mantle to brittle crust and surface topography, but also a historical overview of Alpine research and includes two chapters covering specific Alpine regions (Corsica and Eastern Alps), through all phases of Alpine history.

The second volume presents the pre-collisional history of the Alps. It starts with the Variscan orogeny in the Alpine realm, and continues with the inferred paleogeography, structure and extensional processes affecting continental margins, and their mantle structure in the pre-orogenic Alpine realm. This volume finishes with two chapters describing oceanic and continental subduction processes.

The third volume is entirely dedicated to Alpine collision. It describes and interprets the following elements of the Alpine chain: (1) Alpine magmatism, which is mainly contemporaneous to the very first stages of collision; (2) the formation of the external massifs, which represent a peculiarity in the Alpine chain, rarely occurring in other collisional orogens; (3) the foreland (Molasse) basin, in relation to the lithospheric processes that controlled its evolution; (4) exhumation of the internal part of the chain as constrained from the thermo-chronological record of the foreland basin sediments, and finally (5) the northern deformation front of the Alps, the youngest out of the large-scale collisional structures of the orogen.

This book will not provide a single, consistent and unique view on Alpine geodynamics, but rather create a space for experts on Alpine research to present their state of the art of specific subjects, not withdrawing from proposing their own interpretations, whether or not they are entirely consistent with those suggested in other chapters.

Volume 1

The first volume starts with a chapter on the history of Alpine cross-sections. This chapter does not review the historical evolution of tectonic and geodynamic processes in the Alps, but it shows how the style and format of geological cross-sections have evolved through time in relationship and as a consequence of major tectonic discoveries and debates. It also underlines changes through time in the scientific approach of scientists, showing how the very close link to observations in the mid-19th century evolved toward more conceptual interpretations, eventually lacking an observational base at the beginning of Plate Tectonics, when entirely new concepts of Alpine orogeny were formulated. Only in the last decade of the 20th century, have these conceptual models regained a close link to structural and geophysical observable data.

Chapter 2 of Volume 1 describes the present-day deep structure of the Alps, in which several slabs and slab segments can be identified. The dip-direction and the presence or absence of break-offs in the European slab are discussed in this chapter, strongly suggesting continuous, vertical slabs. In addition, the surface of the Alpine Mohos and the first-order structures of the deep crust are reassessed. Finally, based on the integration of receiver function, Vp/Vs seismic tomography and local earthquake tomography, the geometry and orientation of the west-Alpine subduction channel is imaged.

Chapter 3 of Volume 1 provides a quantitative description of the present-day surface topography of the Alps and its inferred uplift history. Several processes affecting the Alps since the Pliocene, such as sedimentation in the foreland basins, cooling ages and geochemically inferred erosion rates are integrated in the discussion of the present-day topographic structure, suggesting that changes in both climatic and tectonic processes must be invoked to explain the uplift history of the Alps.

Chapter 4 of Volume 1 reviews stress patterns inferred by the analysis of the youngest Alpine brittle fault systems, in addition to inferred deformation fields based on seismic and GPS data. These analyses show that the topographically highest portions of the Alpine chain in the Western and Central Alps were affected by orogen-perpendicular extension during Pliocene time, and this state of deformation is still recorded by seismic activity. The areas affected by such orogen-perpendicular extension coincide with high uplift rates in the northern and central parts of the Alpine arc. The possible collisional and post-collisional large-scale geodynamic causes of this deformation and uplift patterns are discussed.

Volume 1 concludes with two chapters which are both regionally focused. Because of their broad approach, providing the geologic al history of the Alpine realm throughout its entire history and up to the present, they are included within the first volume. Chapter 5 of Volume 1 presents the geology of the Eastern Alps discussing its underlying geodynamic processes. It shows that the stacking of nappes within the upper plate of the Alpine orogen can be inferred to represent a syn-collisional Cretaceous event, hence leading to the conclusion that two distinct orogenies, each of them consisting of a subduction/collision process, affected the Eastern Alps. The younger (Cenozoic) history is also described, with its peculiar style of collision, very different from that in the Western and Central Alps. Indeed, syn-collisional extrusion-related faults accommodate a significant part of collisional convergence in the Eastern Alps.

Chapter 6 of Volume 1 gives an overview of the geology of Corsica. Because the Sardo–Corsican block rifted away from the rest of the Alpine chain at the beginning of collision, it is not affected by the latter process, and it is ideally suited to study subduction and Cenozoic extension, and in particular, to relate the Alpine to the Appeninic chain. This chapter describes and illustrates how the Eocene-aged Alpine/Corsican east-directed subduction is followed by slab break-off and the initiation of a new, west-directed subduction system, which eventually generated the Appenninic chain.

Volume 2

Chapter 1 of Volume 2 deals with the Paleozoic evolution of basement rocks in the Alps. The oldest events recorded correspond to the formation of a Neoproterozoic peri-Gondwanian arc before its accretion to Gondwana, presumably during an Ordovician collisional event. The Cambrian–Ordovician period was marked by tectonic subsidence and widespread magmatic activity, prior to the opening of the Rheic Ocean. Convergence between Laurussia and Gondwana started in the Devonian, which marked the closure of Rheic Ocean, followed by the Variscan collision during the Carboniferous. Evidence of this convergence is provided by Devonian HP-LT eclogites preserved in the Austroalpine units, and Devonian–Tournaisian arc-related magmatism in the External Crystalline Massifs (Western Alps). The main period of nappe stacking and crustal thickening took place during the Mississipian and it was followed by strike-slip tectonics and extension from Pennsylvanian to Permian times. The orogen-scale dextral strike-slip east-Variscan shear zone was active from the Late Carboniferous to the Permian. Because of the intense reorganization during the Alpine orogeny, replacing the Alpine basement in the European Variscan puzzle remains a challenge to be solved.

Chapter 2 of Volume 2 provides an overview of the continental margins involved in Alpine orogeny. The reconstructed stratigraphic sequence and structure of both the European and Adriatic margin are presented based on a literature review showing numerous case studies from all over the Alpine chain. The questions on how many ocean basins and continental blocks existed before the onset of Alpine subduction, and their inferred paleo-geographic positions are reviewed, taking into account the nature of the oceanic substratum, that is, exhumed mantle versus oceanic crust. In addition to describing numerous interpretations of this complex paleo-geographic realm, a new Jurassic paleo-geographic map and a Jurassic-to-Cretaceous evolution are proposed. It suggests the existence of a main oceanic axis, which propagated into different parts of adjacent, pre-fractured continental blocks. A similar pattern is inferred to characterize the North Atlantic ocean in the Early Eocene, where numerous oceanization attempts, leading to hyper-extended crust and exhumed mantle, initiated from one major, stable central axis of accretion.

Chapter 3 of Volume 2 summarizes the petrology, the geochemistry and the age data of mantle and oceanic crustal remnants in the Alps and Appennines, and discusses these data in the frame of rifting and seafloor spreading preceding Alpine subduction. The comparison of these data with those of peridotites worldwide, including mantle samples dredged offshore, points to their origin from passive margins and ultra-slow spreading ridges. Their association with lower crustal granulites and mafic intrusions, in addition to their sedimentary cover typical of continental margins, suggest that they originate from the mantle lithosphere beneath a continental crust. The small volumes of gabbroic dikes recording melt extraction and magmatism during formation of the Piemonte–Liguria Ocean, and the similarity between the REE signature of their basalts, and that of MORB from ultra-slow spreading ridges, point to a likely origin from an ultra-slow spreading ridge.

Chapter 4 of Volume 2 is the first of two chapters on Alpine subduction. Based on a large synthesis of P-T conditions and PTt paths determined from HP rocks, this chapter reconstructs the 3D architecture of oceanic subduction in the Alpine realm and draws several conclusions on the dynamics of Alpine subduction. The consistent alignment of compiled P/T data along a P-T gradient corresponding to that of mature subduction zones suggests that Alpine HP rocks record the P-T conditions of subduction without being significantly affected by overpressure. The magmatic age of subducted and exhumed oceanic HP rocks points to the fact that only oceanic domains next to the continental margin and/or OCT were exhumed, whereas the remaining oceanic crust disappeared during the first stages of subduction. Integrating observations on syn-subduction sedimentary deposits all along the Alpine arc to petro-chronological data on HP rocks shows that an along-strike segmentation of the subduction zone both in terms of crustal-lithospheric structure and processes existed, with subduction accretion dominating in the Western Alps and subduction erosion dominating in the Central Alps. These differences in the nature of the subduction processes are discussed in relationship to the inherited structure of the continental margins and the oceanic crust itself. Finally, the Alpine case study is compared with larger scale subduction zones of the Mediterranean and Pacific.

Chapter 5 of Volume 2 reviews the available data about the subduction of continental crust in the European Alps. The type of crustal material that was involved in the subduction process is characterized (lower vs. upper crust, tilted blocks vs. extensional allochthons) and the data on pressure, temperature and age of metamorphic units are summarized. Finally, a discussion of the present geometry of the subducted continental units allows the setting of some constraints on the kinematic evolution of the mountain belt, and on the exhumation of HP-UHP rocks in particular.

Volume 3

Chapter 1 of Volume 3 focuses on Alpine magmatism, reviewing petrological, geochemical and geochronological data of Alpine magmatic bodies in relationship to their distribution all along the Alpine chain. Based on their spatial occurrence, age and geochemical signature, Alpine magmatic bodies are shown to be generated by different geodynamic processes, which are inferred to be (1) subduction of both continental and oceanic crust, whose melting produced magmas that were emplaced in the upper plate in the Middle and Late Eocene; (2) slab break-offs at the transition from subduction to collision, allowing for the rise of asthenosphere and its direct contact to the lower Alpine crust, whose melting and mixing with asthenospheric melts generated the largest volumes of Alpine magmas, forming the Periadriatic Plutons; (3) extensional processes related to mantle dynamics, limited to the marginal areas of the Alpine chain (the Alpine/Apennine junction and the Alps/Dinarides/Pannonian junction); (4) ascent along inherited lithospheric-scale faults, possibly explaining intraplate magmatism as observed in the Veneto magmatic province, which is located relatively far from the Alpine front.

Chapter 2 of Volume 3 summarizes the present-day knowledge on the external crystalline massifs, basement units deformed during the collisional stage of the mountain belt. In these massifs, the amount of shortening increases from south to north, as well as the P-T conditions of barrovian metamorphism. Each massif experienced a complex sequence of shortening: after slicing off the Dauphinois cover, ductile internal shortening affected the basement, before localization of deformation along basement thrusts resulted in "en bloc" exhumation of the massifs. Specificities between different massifs are described in detail and different hypotheses on the driving forces leading to the formation of the massifs are synthetized.

Chapter 3 of Volume 3 reviews the structure, age and stratigraphy of the northern foreland basin of the Alps, and it discusses the large-scale geodynamic processes underlying its stratigraphic architecture, its facies relationships and the provenance of its clastic material. The along-strike similarity versus change of the stratigraphy through time is synthetized and related to the drainage pattern of the basin. The historical evolution of inferred geodynamic processes is reconstructed over the past decades, thus showing that the stratigraphy and structure of the Molasse Basin may not only reflect thickening of the more internal part of the orogen, but rather the slab dynamics of the lower plate. Hence, subduction loads at deeper crustal levels rather than surface loads may exert the first-order control on basin subsidence.

Chapter 4 of Volume 3 reviews the currently available detrital thermochronological record of exhumation for the European Alps. It allows tracing and discussing temporal and spatial patterns of exhumation across the Alpine chain. The higher temperature white mica 40Ar/39Ar ages, if considered as cooling ages, depict phases of very rapid exhumation, on the order of 3 ± 1 km/Myr, in the Western Alps during the early Oligocene, and somewhat slower exhumation of ~1 km/Myr in the Central Alps during the late Oligocene–early Miocene. Detrital zircon fission-track data indicate a relatively steady exhumation signal since the late Oligocene for the Central Alps, whereas detrital apatite fission-track data emphasize regional differences in exhumation rates since at least Miocene times.

Chapter 5 of Volume 3 focuses on the northern deformation front of the Alps, that is, on the youngest, large-scale structure accommodating collisional shortening by the relative displacement of the orogen on its foreland basin. It shows that the structural style of this fault system varies along strike, with triangle zones being replaced by duplexes and basal décollements changing depth and the stratigraphic level. These variations are correlated to lateral changes in the mechanical stratigraphy, to differences in the links between these faults and their roots in the more internal parts of the orogen, or even to differences in the mechanical properties of the entire European Plate.

Acknowledgments

Romain Bousquet generously provided the vectorized version of his Tectonic Map of the Alps, used in most of the chapters in the three volumes of this book.

1

Paleozoic Evolution and Variscan Inheritance in the Alps

Jean-Baptiste JACOB[1], Stéphane GUILLOT[1], Denis THIÉBLEMONT[2], Jonas VANARDOIS[3], Pierre TRAP[3], Michel FAURE[4] and Didier MARQUER[3]

[1] *Université Savoie Mont Blanc, CNRS, IRD, UGE, ISTerre, Université Grenoble Alpes, France*

[2] *Bureau de Recherches Géologiques et Minières, Orléans, France*

[3] *UMR 6249 Chrono-environnement, Université de Bourgogne-Franche-Comté, Besançon, France*

[4] *Institut des Sciences de la Terre d'Orléans, Université d'Orléans, CNRS, France*

Large outcrops of the Paleozoic basement have formed about half of the exposed rock in the Alpine belt. These Paleozoic domains preserve evidence of the long and complex pre-Alpine history, which has spanned more than 300 million years, from the Ediacaran to the Permian, and reflect the general evolution of the European Paleozoic basement: the oldest events recorded to correspond to the formation of a Neoproterozoic peri-Gondwanian arc that remained active during the Cambrian before its accretion into Gondwana, presumably during an Ordovician collision. The Cambrian–Ordovician period was marked all over the Alps by tectonic subsidence and widespread magmatic activity, which reflects the general extension that affected North Gondwana prior to the opening of the Rheic Ocean and the rifting of Avalonia and the

Geodynamics of the Alps 2,
coordinated by Claudio L. ROSENBERG and Nicolas BELLAHSEN.

Hun Terranes during the Ordovician. The following Silurian period was relatively calm, marked by crustal subsidence and extension at the margin, and deposition of clastic sediments. These are mostly preserved in the Eastern Alps, and they announced the future opening of the Paleotethys from the Late Silurian to the Devonian. Then, convergence between Laurussia and Gondwana started in the Devonian, which marked the closure of the Rheic Ocean, followed by the Variscan collision during the Carboniferous. Evidence of this convergence is provided by Devonian HP-LT eclogites preserved in the Austroalpine units, and Devonian–Tournaisian arc-related magmatism in the External Crystalline Massifs (The Western Alps). The Variscan collision is recorded to various degrees across all of the Alpine basement domains. The main period of nappe-stacking and crustal thickening took place in ca. 350–320 Ma, which was followed by strike-slip tectonics and an extension from the Westphalian to the Permian (ca. 315–290 Ma). In the Helvetic domain, the orogen-scale dextral strike-slip East-Variscan shear zone was active from the Late Carboniferous to the Permian and was responsible for large-scale displacements in the Variscan structure at the end of the orogeny. Following the Variscan orogeny, a high-temperature event occurred during the Permian, which preceded the opening of the Tethys Ocean and the initiation of the Alpine cycle. This evolution is in many ways similar to that of other Paleozoic massifs in Europe. However, due to the intense reorganization of the crustal basement during the Alpine orogeny, the Alpine domain does not form a coherent block that can be easily correlated with other Variscan domains in Europe. Replacing the Alpine basement in the European Variscan puzzle remains a challenge to be solved.

1.1. Introduction

Prior to the Alpine cycle, which began with the extension and the opening of the Tethyan oceanic domains during the Lower Jurassic, the Alpine domain underwent a long and complex evolution that can be traced back to the Late Neoproterozoic. In many places, the basement that preserves this pre-Alpine history has been covered by a thick layer of Mesozoic and Cenozoic sediments and is therefore not accessible to geologists. However, areas where the pre-Mesozoic basement is exposed represent about half of the Alpine domain (Von Raumer et al. 2013), which makes this one of the largest Paleozoic massifs in Europe, together with the Bohemian Massif, the Massif Central, the Armorican Massif, and the Iberian Massif. This basement is exposed along all of the domains that form the Alpine belt and was variously overprinted by the Alpine tectonics and metamorphism (Bousquet et al. 2004; Schmid et al. 2004). Like the other Paleozoic massifs of southern and central Europe, the Paleozoic basement of the Alps derives from the northern margin of Gondwana (Stampfli et al. 2011) and has recorded a long and complex history that culminated during the Carboniferous with the Variscan orogeny, during which Gondwana and

Laurussia became amalgamated to form the Pangea supercontinent (Matte 2001). However, this apparently simple history hides great complexity, which is far from being fully understood. This chapter aims to present the current knowledge and outstanding issues concerning the pre-Mesozoic history of the Alpine basement.

1.2. The Paleozoic setting in Europe

1.2.1. *General setting: a crustal basement structured by Paleozoic orogenies*

The pre-Mesozoic domains of the Alps represent only a small part of the European Paleozoic basement and form a small piece of the former Variscan belt, a large orogen that can be followed across all of the different Paleozoic massifs of southern and central Europe (Figure 1.1). The Variscan belt of Europe itself belongs to a larger orogenic system that extended over about 8,000 km, from the Caucasus to the Appalachian and Ouachita mountains of North America (Matte 1986). It arose from the Devonian to the Permian by progressive closure of the oceanic domains between Laurentia–Baltica in the North, and Gondwana in the South (Figures 1.2 and 1.3). Between these two large continental masses, small continental blocks detached from the northern margins of Gondwana during the early Paleozoic, and then from the Late Devonian to the Carboniferous, they underwent accretion to form Laurussia and the Variscan orogen (e.g. Franke 1989; Matte 2001; Stampfli et al. 2013).

There have been major advances during the last few decades in our understanding of the Paleozoic evolution of the European basement, and most studies have agreed on the first-order structure of the Variscan belt (e.g. Kroner and Romer 2013; Ballèvre et al. 2014; Lardeaux et al. 2014; Franke et al. 2017). On a European scale, its strongly arcuate shape was acquired more recently in Variscan history (Upper Carboniferous–Early Permian). It can be subdivided into several lithotectonic domains. On the northern and southern flanks of the Variscan assemblage, the basement is formed by the Avalonia and North Gondwana shelves, respectively (Figure 1.1), which formed the foreland basins of the belt. In between this, the Variscan basement is composed of several lithotectonic domains which represent the assemblage of peri-Gondwana terranes that were squeezed between Gondwana and Laurussia during the Variscan orogeny, and then were variously overprinted by the associated tectono-metamorphic and magmatic events. The "core" of the European Variscan belt is formed by the Moldanubian zone, which includes a large part of the Bohemian Massif and the Massif Central, and extends toward the southern Armorican massif and the northern Iberian massif (Figure 1.1). This represents the exhumed crustal root of the belt, and it contains high-grade migmatites and granulites, with relics of high-pressure rock (eclogites, high-temperature eclogites/ high-pressure granulites), and dismembered ophiolites (Lardeaux et al. 2014).

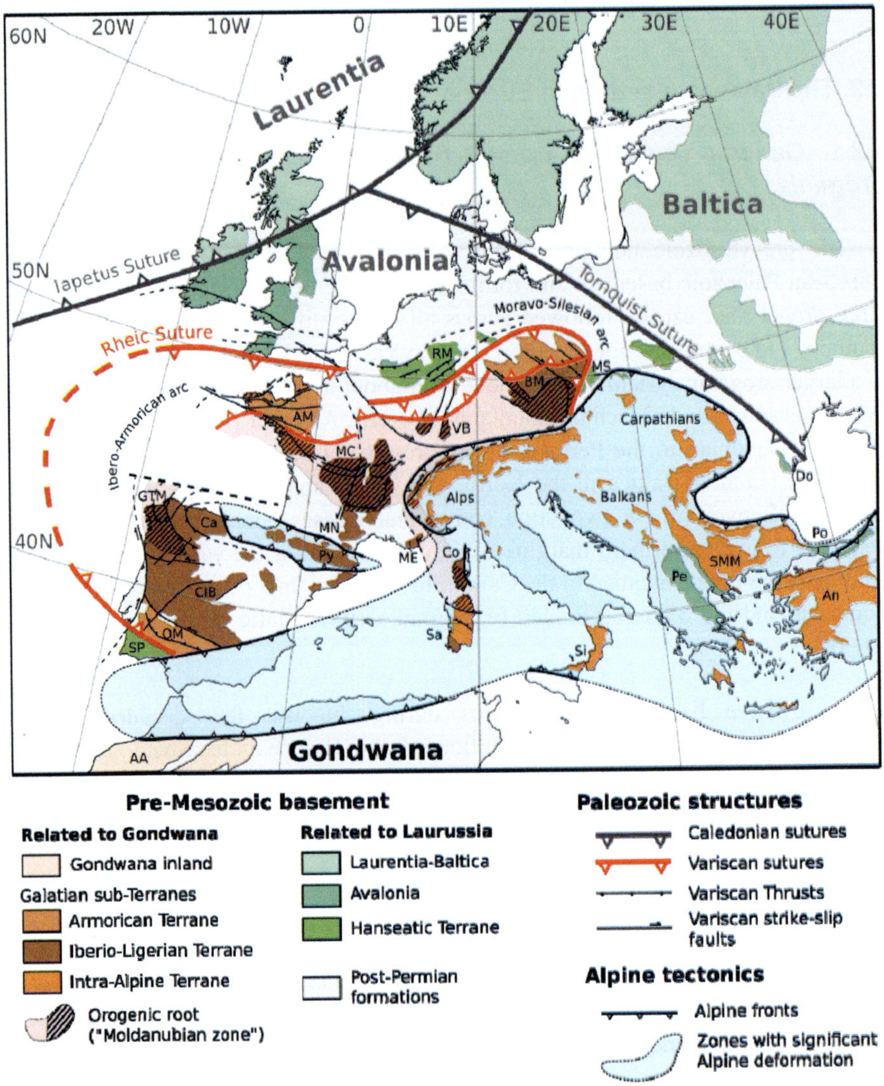

Figure 1.1. *Simplified tectonic map of Europe that shows the exposures of the Paleozoic basement and the Paleozoic sutures*

COMMENTS ON FIGURE 1.1.– *The Avalonian crust in the middle of Gondwana-derived domains in southeastern Europe (the Pelagonian zone) is mostly inferred from the study of sedimentary sequences and detrital zircons (Stephan et al. 2019 and references therein) and is not discussed here. AM, Armorican Massif; An, Anatolia; BM, Bohemian Massif; Ca, Cantabrian zone; CIB, Central Iberian zone; Co, Corsica; Do, Dobrogea; GTM, Galicia–Trás-os-Montes; MC, Massif Central; ME, Maures-Esterel; MN, Montagne Noire; MS, Moravo–Silesian zone; OM, Ossa–Morena zone; Po, Pontides; Pe, Pelagonian zone; Py, Pyrénées; RM, Rhenish Massif; Sa, Sardinia; Si, Sicilia; SMM, Serbo-Macedonian Massif; SP, South Portugese zone; VB, Vosges, Black Forest. Modified from Stephan et al. (2019a), with data from Ballèvre et al. (2014) and Franke et al. (2017).*

1.2.2. *Pre-Variscan history of the European basement*

Prior to the opening of the Rheic Ocean during the Ordovician, the future Variscan domains of Europe were located along the northern margin of Gondwana (Matte 2001; Von Raumer et al. 2002; Stampfli et al. 2011). From the Neoproterozoic to the Early Cambrian, a western-Pacific-type accretion zone was developed along this margin, which resulted in the accretion of various elements to form a consolidated crust (Garfunkel 2015). This period of accretion peaked during the Late Ediacaran (ca. 600–580 Ma) with the accretion of an arc which had previously rifted off Gondwana, during the Lower Ediacaran. This event is known as the Cadomian orogeny, and it has been poorly preserved in the European basement, except in a few places. These include northern Brittany and the region of Caen in northwestern France (Ballèvre et al. 2001; Chantraine et al. 2001), from which the name "Cadomian" originates (as derived from Cadomus, the Latin name of the city of Caen). Elsewhere, the Cadomian orogeny is mainly inferred from the study of zircons in detrital sediments, which has defined a major peak of zircon crystallization at ca. 600 Ma, which corresponds to the Cadomian orogenic events (e.g. Chu et al. (2016); Chelle-Michou et al. (2017)).

The Cambrian–Ordovician period (ca. 540–450 Ma) was characterized by extensional tectonics, with rifting and subsidence over northern Gondwana in a back-arc setting (von Raumer and Stampfli 2008). This eventually led to the opening of the Rheic Ocean (Linnemann et al. 2007; Nance et al. 2010) and the northward drift of the Avalonian microplate (Figure 1.2, sketches 1–4). This widespread extension resulted in the formation of a thinned continental shelf on the northern margin of Gondwana, which was associated with the opening of back-arc basins

(Kroner and Romer 2013; Stampfli et al. 2013) and the limited production of some arc or back-arc oceanic crust (e.g. Chamrousse: Figure 1.2, sketch 1; Pin and Carme 1987; Ménot et al. 1988; Guillot et al. 1992). The Cambrian–Ordovician extension was accompanied by widespread bimodal magmatism (Figure 1.2, sketch 3), which consisted of alkaline and tholeiitic mafic rock that was associated with alkaline and peralkaline rhyolite (e.g. Pin and Marini (1993); Crowley et al. (2000)). A possible orogenic event occurred during the Ordovician (Figure 1.2, sketch 2) is referred to as the Sardic phase or the Cenerian orogeny and was inferred from the widespread production of peraluminous granitoids from ca. 500–450 Ma (Valverde-Vaquero and Dunning 2000; Villaseca et al. 2016; Zurbriggen 2017). However, the geodynamic setting associated with this peraluminous magmatism remains unclear. Different mechanisms have been proposed, which include continental rifting associated with the break-up of northern Gondwana (Montes et al. 2010; Ballèvre et al. 2012), back-arc extension in an active arc setting (Valverde-Vaquero and Dunning 2000; Fernández et al. 2008), thermal relaxation of a thickened arc crust (Villaseca et al. 2016; Soejono et al. 2019), and melting of fore-arc sediments driven by underplating of mafic magma (Zurbriggen 2017).

1.2.3. *The Variscan orogeny in Europe*

From the Late Silurian to the Early Devonian, convergence between Gondwana and Laurussia led to the closure of the Rheic and Saxo-Thuringian Oceans (Figure 1.2, sketches 5 and 6). This then culminated during the Late Devonian–Carboniferous (360-320 Ma) with the collision between Laurussia, Gondwana, and peri-Gondwana fragments (Figure 1.3). During the Middle-to-Late Carboniferous, two successive stages of post-orogenic extension occurred (Faure and Becq-Giraudon 1993; Burg et al. 1994). The first of these was associated with lateral East-West extrusion of material in a still convergent setting and the second was associated with North-South orogenic collapse via the activation of low-angle detachment faults. These were accompanied by the development of migmatite complexes, emplacement of large volumes of peraluminous granitoids, and formation of intra-continental, coal-bearing sedimentary basins (e.g. Ledru et al. 2001; Faure et al. 2009; Lardeaux et al. 2014). This general collapse associated with extensive crustal melting may have been driven by the delamination of the subcontinental mantle lithosphere during the late orogenic stages (e.g. Žák et al. (2018)).

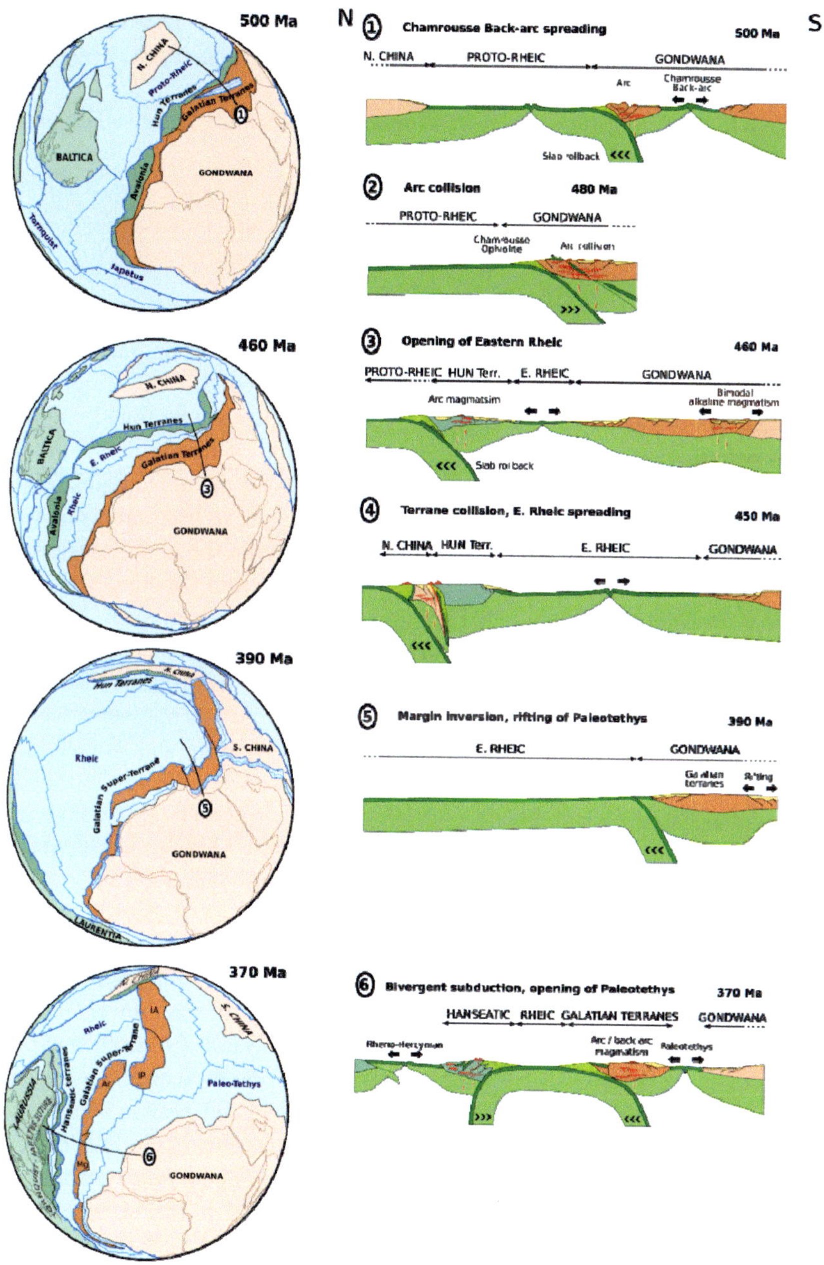

Figure 1.2. *Palinspastic reconstruction that shows the evolution of North Gondwana from the Cambrian to the Devonian, with the associated cross-sections that cut through the northern Gondwana margin*

COMMENTS ON FIGURE 1.2.– *The position of the cross-sections is indicated by the number on the corresponding globe. This period records the progressive opening of the Rheic domain and the drifting of Avalonian and Hun Terrane during the Ordovician, followed by the rifting of Paleo-Tethys during the Silurian and the Devonian, and the progressive closure of the Rheic Ocean. Mg, Meguma; Ib, Iberia; Ar, Armorica; IA, Intra–Alpine. Colors of the continents follow the same code as for Figure 1.1: light green, Laurentia–Baltica; dark green, Avalonia and Hun Terrane; light brown, Gondwana; dark brown, Galatian Terranes. Modified from Stampfli et al. (2011, 2013).*

However, the Variscan orogen has a complex structure that cannot be solely explained by a simple continent-continent collision. In particular, many relics of mafic-ultramafic complexes, mid-ocean ridge basalt (MORB)-type eclogites, and alkalic magmas of the Upper Cambrian–Lower Ordovician can be found within the Variscan domains of western and central Europe, south of the Rheic suture. This disposition suggests that multiple oceanic basins separated the microplates south of the Rheic Ocean (Figure 1.1). Different tectonic models have been proposed which differ in the numbers of plates and oceans involved, and in the initial pre-Variscan positioning of these blocks along the peri-Gondwana margin (Matte 2001; Kroner and Romer 2013; Stampfli et al. 2013; Franke et al. 2017). Moreover, discussion continues as to the size of the basins that separated the different blocks, and the subduction vergencies. Based on the continuity of the benthic fauna and paleomagnetic data, it is now recognized that these continental fragments were not separated from Gondwana by a large oceanic domain (Cocks and Torsvik 2002; Fortey and Cocks 2003). These micro-continents were more likely separated by either small oceanic basins similar to the Mesozoic Alpine Ocean (Franke et al. 2017), or by zones of hyperextended crust that possibly showed continent-ocean transition (Kroner and Romer 2013; Lardeaux et al. 2014).

Finally, the global picture is further complicated by the non-cylindricity of the Variscan belt and its diachronous evolution from West to East, as indicated by palinspastic reconstructions, detrital zircon patterns in the Lower Paleozoic sediments, and lithostratigraphic, tectonic, and magmatic records (Stampfli et al. 2013; Casas and Murphy 2018; Stephan et al. 2019a). Several recent models now separate the Variscan domains south of the Rheic suture into eastern and western segments that underwent contrasting geodynamic evolution (Stampfli et al. 2013; Stephan et al. 2019a, 2019b). In the western domain, the final collision of Gondwana with Laurussia has been well recorded and corresponds to the indentation of Avalonia by the western Gondwana shelf during the Late Devonian–Carboniferous (ca. 360–340 Ma; Faure et al. 2009; Ballèvre et al. 2014). In the eastern domain, the situation was complicated by the opening of Paleo-Tethys during the Devonian (Figure 1.2, sketch 6 and Figure 1.3). The main Tournaisian–Visean tectonothermal event corresponded to the accretion of the terranes into a cordillera-like orogen (Figure 1.3, sketches 7 and 8). The final collision of Gondwana with Laurussia occurred later (330–300 Ma; Figure 1.3, sketches 9 and 10) and was less pronounced than for the western domain (Haas et al. 2020). Eastward, the collisional system evolved toward a purely accretionary orogen along the northern Paleo-Tethys, to form the Altaids in central Asia (Wilhem et al. 2012).

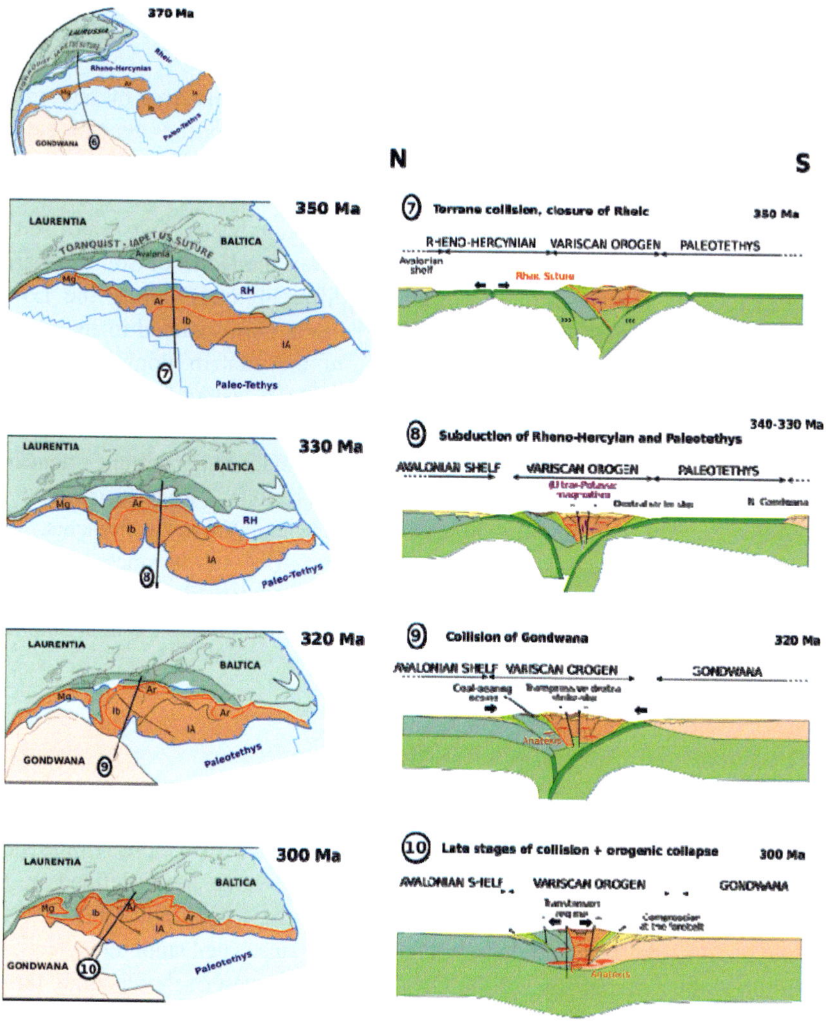

Figure 1.3. *Palinspastic reconstruction of the Variscan domain from the Devonian to the Late Carboniferous, with associated cross-sections through the Variscan collision zone*

COMMENTS ON FIGURE 1.3.– *This period records the closure of the Rheic Ocean and other smaller basins, and the collision between Laurussia, Galatian Terranes, and Gondwana. This collision is polyphasic: a first arc–arc collision occurred at ca. 350 Ma between the two sets of Galatian Terranes; a second collision occurred at ca. 320–300 Ma with indentation of Gondwana in the Galatia–Laurussia terrane assemblage; this was followed by extension and lateral extrusion of material accommodated by the lithospheric-scale strike-slip faults. See Figure 1.2 for labels. Modified from Stampfli et al. (2011, 2013).*

The following sections will present an overview of the Paleozoic massifs in the Alps and discuss their pre-Mesozoic history according to the general setting proposed by Stampfli et al. (2013) for the Paleozoic evolution of northern Gondwana (see Figures 1.2 and 1.3). According to this model, the Variscan belt was formed by the amalgamation of terranes derived from Gondwana. Two major sets of ribbon-like peri-Gondwana terranes that extended from the North of South America to Southern China drifted from Gondwana at different times during the Lower Paleozoic. A first segment known as the Hun Terrane corresponded to the eastern equivalent of Avalonia, and this detached from the eastern domain of North Gondwana during the Ordovician, with the opening of the eastern Rheic Ocean. This segment then drifted to the northeast and collided with northern China during the Silurian. A second segment known as the Galatian super-terrane detached from Gondwana during the mid-Devonian, synchronous with the opening of Paleo-Tethys, and that comprised most of the European Variscan elements. This ribbon-like micro-continent quickly separated into four sub-terranes. The final collision between Gondwana, Laurussia, and the Galatian Terranes during the Middle-to-Upper Carboniferous (ca. 340–300 Ma) created a complex amalgamation that formed the Variscan belt.

1.3. The basement outcrops in the Alps

The Paleozoic basement is exposed discontinuously all along the Alpine arc, and it forms about half of the Alpine domain (Von Raumer et al. 2013). Pre-Alpine rocks are exposed in the four main tectonic domains that have been defined at the lithospheric scale for the Alpine belt (Figure 1.4): the Dauphinois–Helvetic; the Penninic; the Austroalpine; and the south Alpine domains. The Dauphinois–Helvetic domain forms the external zone of the European margin west of the Penninic thrust. The Penninic domain is composed of various nappes that are derived from the distal European margin (e.g. the Tauern Window basement nappes; Schmid et al. 2013), the Valaisan basin, and the Briançonnais microplate in the Western and Central Alps. The Austroalpine domain is composed of allochthonous nappes that form the remnants of the southern margin of Piemonte–Liguria Ocean along with the Apulian plate, under which the Penninic nappes have been thrust. Finally, the south Alpine domain forms the autochthonous part of the Apulia plate south of the Periadriatic fault.

The different Paleozoic domains have been variously overprinted by Alpine tectono-metamorphic events, which depended on their location along the Alpine belt (Bousquet et al. 2004; Schmid et al. 2004). In the more external Helvetic and southern Alpine domains, the Alpine metamorphic overprint never exceeds greenschist facies conditions, which indicates that these domains were not deeply buried and belong entirely to the Alpine suprastructure. Between the Penninic thrust and the Periadriatic fault, the Penninic and Austroalpine domains are composed of allochthonous nappes that show various degrees of metamorphic overprint, which range from lower greenschist in the more external units, to high-pressure–ultrahigh-pressure eclogite in the internal units. Re-equilibration along a Barrovian gradient occurred in the Central and Eastern Alps (Lepontine dome, Tauern Window), and this was characterized by the development of metamorphic assemblages of the amphibolite facies associated with migmatization (Bousquet et al. 2004).

As the Helvetic and south Alpine domains form coherent blocks devoid of pervasive Alpine deformation and metamorphic overprint, they are the ideal places to study pre-Mesozoic evolution in the Alps. In the Penninic and Austroalpine domains, reconstitution of the Paleozoic evolution is more challenging. Difficulties arise from the more pervasive Alpine deformation that has partially or totally obliterated the former structures. Tectonic slicing of these domains into numerous nappes (Figure 1.4) that are disconnected from their lower crustal base makes large-scale correlations more speculative. Moreover, in the areas with Alpine metamorphic overprint that exceeds the upper greenschist facies, the interpretation of the successive metamorphic stages is particularly challenging, because the mineral assemblages cannot be unambiguously attributed to one particular Paleozoic or Alpine stage. This is, for instance, the case for the Gruf complex (Central Alps), where the age of the ultra-high-temperature metamorphism has been a matter of debate for a long time, as it can be attributed to either a Permian (e.g. Galli et al. 2012) or an Eocene (e.g. Nicollet et al. 2018) high-temperature event. Deciphering the metamorphic history in such complex areas requires very careful investigations using modern petrochronological methods (e.g. Sandmann et al. 2014; Giuntoli et al. 2018; Kunz et al. 2018). Despite these difficulties, investigations of the Paleozoic basement in the Austroalpine and Penninic units have provided useful constraints on the pre-Mesozoic history of the Alpine domain, especially for the low-grade Austroalpine nappes and for the external Briançonnais (Zone Houillère, and its basement).

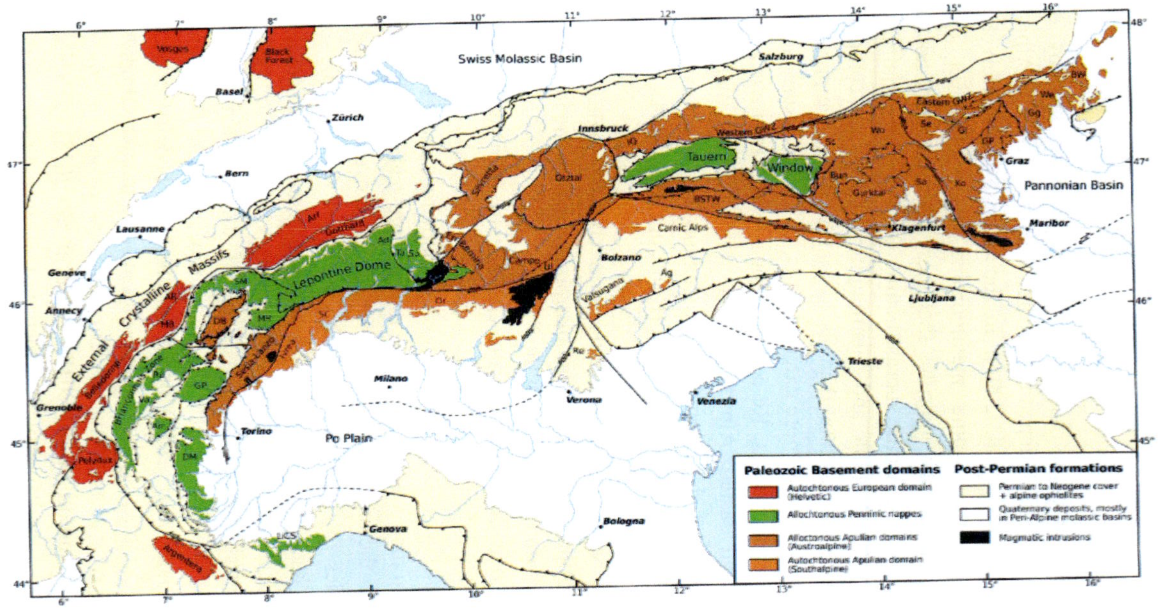

Figure 1.4. *Paleozoic basement units in the Alps in their current tectonic setting and classified according to the Alpine lithotectonic domain to which they belong (i.e. Helvetic, Penninic, Austroalpine, south Alpine). For clarity, no distinction of the lithotectonic domains is made for the post-Permian units, which group the Permian to Neogene sedimentary sequences and the Alpine ophiolites. Dark gray, Cenozoic peri-Adriatic magmatic intrusions. Contours of the massifs were redrawn from Frey et al. (1999). Alpine faults and thrusts are from Bousquet et al. (2004). See comments on Figure 1.5 for abbreviations*

1.4. Paleozoic evolution in the Alpine basement

1.4.1. *Late Proterozoic evolution: an active margin setting along northern Gondwana*

During the Neoproterozoic, the Alpine domains were located on North Gondwana in a Cordillera-like active margin setting, where various terranes underwent accretion to form a consolidated continental crust. This concept of composite terranes was first proposed in the Eastern Alps by Neubauer (1989), to explain the apparently unrelated origins of different crustal fragments. This model was further supported by geochronological studies that identified Ediacaran magmatic protoliths at different places in the Austroalpine nappes. The oldest magmatic protoliths of this period are meta-diorites from the Silvretta nappe that became emplaced at 609 ±3 Ma in an arc-type setting (Schaltegger et al. 1997). Other Neoproterozoic ages have been further identified in the basement of the Tauern (Eichhorn et al. 1995, 2001) and in the basement units south of the Tauern Window, where normal-MORB-type metabasites attributed to subduction-related magmatism were dated at 590 ±4 Ma (Schulz et al. 2004, 2008). According to these Ediacaran magmatic protoliths, Schulz et al. (2008) defined an eastern Cadomian arc in the Eastern Alps which was active from the Late Precambrian to the Cambrian, before undergoing accretion with Gondwana during the Early Ordovician collisional phase (ca. 470 Ma; Figure 1.2, sketch 2). Further south, similar fragments of arc-related Cambrian to Neoproterozoic crust have been reported in the Hellenides (Dörr et al. 2015; Zulauf et al. 2015). No sections of the preserved Neoproterozoic basement have been identified in the Western Alps yet. However, the recent dating of an orthogneiss sampled in crustal fragments embedded within the Mesozoic sediments of the Briançonnais unit close to the city of Briançon yielded an age of 598 ±4 Ma (our unpublished data, Thiéblemont et al.) (Figure 1.6).

Apart from the Cadomian magmatic protoliths, significant information about the pre-Cambrian evolution of the Alpine basement is provided by the inherited zircons in the magmatic rock, and detrital zircons from the Lower Paleozoic sediments. The zircon age spectra here have typically shown a dominant Neoproterozoic age population (ca. 550–650 Ma) for all of the Alpine domains (e.g. Chu et al. 2016; Manzotti et al. 2016; Haas et al. 2020), which indicates that most of the Alpine basement is made of material recycled from the Cadomian arcs. In addition, the detrital zircon spectra have frequently shown age populations as Mesoproterozoic–Neoproterozoic (ca. 800–1100 Ma) and Mesoproterozoic–Archean (2.0–3.0 Ga), with very few Mesoproterozoic zircons for 1.2–1.6 Ga. This distribution is typical for material derived from the West African and northeast African cratons and proterozoic shields, with possible sources coming from the Arabian–Nubian shield and the southern China block (von Raumer and Stampfli 2008; Von Raumer et al. 2013; Stephan et al. 2019a).

1.4.2. *Cambrian–Ordovician extension and opening of eastern Rheic Ocean*

The Cambrian–Ordovician period is marked all across the European Variscides by tectonic subsidence (von Raumer and Stampfli 2008), a widespread magmatic activity that consists mainly of bimodal felsic–mafic igneous associations of alkaline affinity (Pin and Marini 1993; Crowley et al. 2000), and intrusion of mainly peraluminous granitoids (Montero et al. 2007; Bea et al. 2010; Villaseca et al. 2016; Couzinié et al. 2017; Zurbriggen 2017). This intense magmatic activity is generally attributed to a regional extensional event that affected the entire margin of North Gondwana (Pin and Marini 1993; Faure et al. 2009) and that preceded the opening of the Rheic Ocean and the separation of Avalonia from Gondwana during the Mid-Ordovician (Matte 2001; Stampfli et al. 2011). However, the general setting during this period was probably more complex. The evolution along northern Gondwana was dominated by an active margin setting that was accompanied by a contemporaneous intra-continental extension (Figure 1.2, sketches 1–3) and stretching from west to east (von Raumer and Stampfli 2008), which was associated with several magmatic pulses.

The Cambrian–Ordovician magmatic records of the Alpine basement provide a good illustration of this complex evolution. Different sets of intrusive and extrusive magmatic rock were emplaced or erupted from the Cambrian to the Early Silurian (ca. 530–430 Ma; Schaltegger and Gebauer 1999; Bussien et al. 2011; von Raumer et al. 2013), which were attributed to either intra-continental rifting and back-arc extension, or magmatic accretion and orogenic events that occurred in the arc (Figures 1.5 and 1.6).

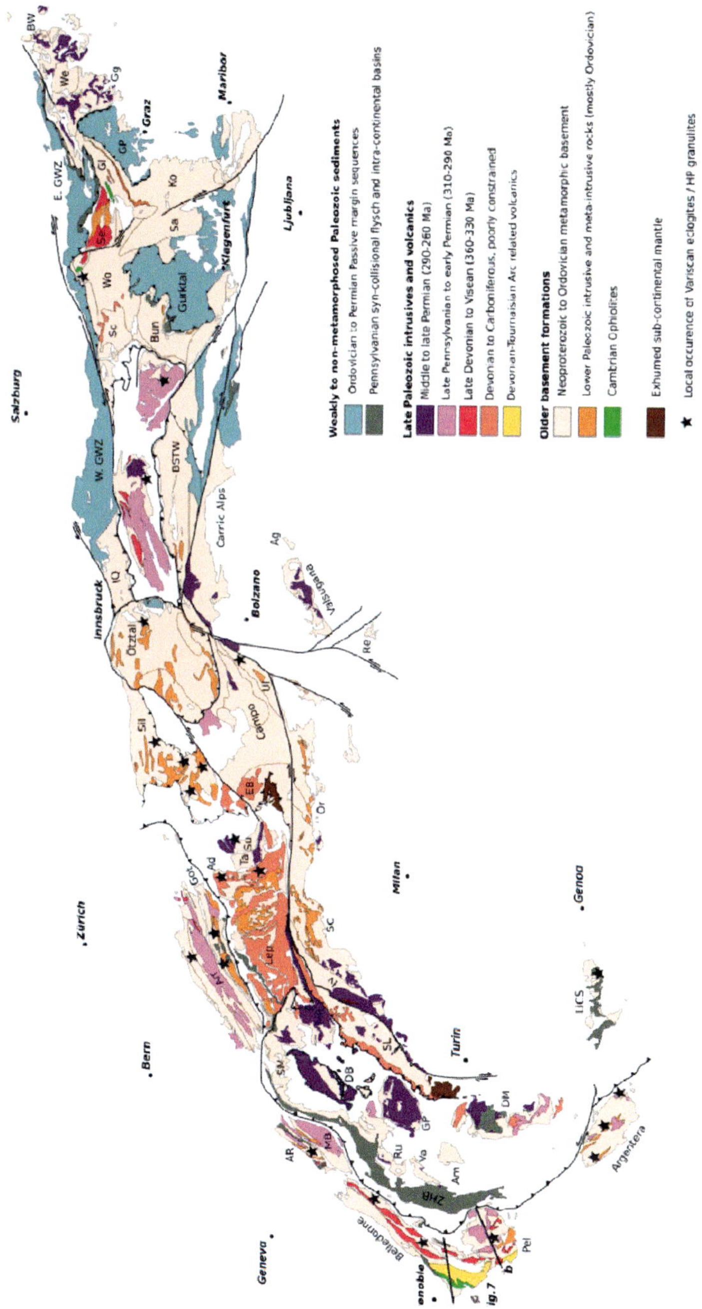

Figure 1.5. Simplified geological map of the lithological units within the pre-Mesozoic basement of the Alps

COMMENTS ON FIGURE 1.5.– *Contours of the massifs were redrawn from Frey et al. (1999). Alpine faults and thrusts are from Bousquet et al. (2004). The classification of the lithologies is based on geological and geochronological data from the Carte Géologique de la France at 1/1,000,000 BRGM (2004); Multithematische Geologische Karte von Österreich, GBA; Guillot and Ménot (2009) (Helvetic zone); Ballèvre et al. (2018), and references therein (Briançonnais); Veselà et al. (2011) (Tauern); Schmidt et al. (2013) (Tauern); Mandl et al. (2018) (Seckau plutonic complex). From southwest to northeast: LiCS, Ligurian, Calizzano–Savona; DM, Dora–Maira; Pel, Pelvoux; Am, Ambin massif; ZHB, Zone Houillère Briançonnaise; Va, Vanoise; GP, Gran Paradiso; SL, Sesia–Lanzo; Ru, Ruitor; AR, Aiguilles Rouges; MB, Mont Blanc; DB, Dent Blanche; SM, Siviez–Mischabel; MR, Monte Rosa; Iv, Ivrea Zone; SC, Strona–Ceneri zone; Lep, Lepontine dome; Ad, Adula nappe; Ta, Tambo nappe; Su, Suretta nappe; Got, Gotthard massif; Or, Orobic Alps; EB, Err–Bernina nappe; Sil, Silvretta nappe; Ul, Ulten zone; Re, Recoaro; Ag, Agordo; IQ, Innsbruck Quartzite; BSTW, Basement units south of the Tauern Window; Sc, Schladming nappe; Bun, Bundschuh nappe; Wo, Wölz nappe; Sa, Saualpe; Ko, Koralpe; Se, Seckau nappe; Gl, Gleinalm; GWZ, Grauwacken Zone; GP, Graz Paleozoic; Gg, Grobgneiss; We, Wechsel; BW, Bucklige Welt.*

1.4.2.1. *Cambro–Ordovician extension magmatism in the Alpine basement*

Alkaline and tholeiitic magmatism has been widely recorded for the Alpine basement (Figure 1.6), and it has been attributed to back-arc extension and continental rifting that affected North Gondwana (Bussien et al. 2011). In the basement of the Vanoise Massif and the Siviez–Mischabel unit (Penninic domain), A-type alkaline metagranitoids have been dated at 507 ±9 Ma (Guillot et al. 2002) and 500 ±4 Ma (Bussy et al. 1996), respectively. In the Silvretta nappe (Austroalpine), two ages at 526 ±7 Ma and 519 ±7 Ma have been reported for an alkaline orthogneiss (Müller et al. 1995). These early stages of continental rifting associated with alkaline magmatism were followed by emplacement of tholeiites in a thinned continental crust, with the local production of oceanic lithosphere in short-lived back-arc basins. This evolution has been well described for the external crystalline massifs of the Western Alps (Helvetic domain), where tholeiitic metabasalts with a normal/enriched-MORB signature were emplaced in a thinned continental crust during the Ordovician (ca. 480-450 Ma; Paquette et al. 1989; Rubatto et al. 2001 and 2010). In the Belledonne Massif, the ophiolitic complex of Chamrousse (Figure 1.5) is a well-preserved kilometer-scale slab of Cambrian oceanic lithosphere (ca. 500 Ma; Pin and Carme 1987; Ménot et al. 1988), and it is regarded as the relic of a small marginal basin that spread in a back-arc setting (Guillot et al. 1992) before it closed, presumably during a mid-Ordovician tectonic phase (Guillot and Ménot 2009; von Raumer et al. 2009). Such tholeiitic suites have also been described for the Penninic and Austroalpine domains. Dismembered relics of mafic and ultramafic rock from Loderio–Biasca (Ticino, Lower Penninic nappes) that have been dated at 518 ±11 Ma might represent the relics of an oceanic domain (Schaltegger et al. 2002). In the Tauern Window, the Habach Terrane corresponds to an active continental margin where back-arc spreading occurred during the Cambrian, which was associated with the emplacement of diorites, gabbros, and trondhjemite, from ca. 550–507 Ma (Neubauer et al. 1989; Eichhorn et al. 2001). Finally, for the Austroalpine domain, oceanic plagiogranites dated at ca. 530 Ma that appear to be emplaced in a back-arc setting were reported by Müller et al. (1996) for the Silvretta nappe, and several dismembered Late Neoproterozoic to Cambrian ophiolites are exposed in different crystalline Austroalpine nappes (Neubauer et al. 1989). This is seen especially in the Speik complex, where a nearly complete ophiolitic sequence dated ca. 550 Ma is preserved (Melcher and Meisel 2004).

1.4.2.2. *Cambro–Ordovician arc and collisional magmatism in the Alpine basement*

Subduction-related magmatism has been recorded contemporaneous to the back-arc rifting. This consists mainly of mafic and felsic meta-igneous rock of calc-alkaline affinity (Figure 1.6). In the Silvretta nappe, eclogitized gabbroic and tonalitic melts were emplaced at ca. 530–520 Ma, and are attributed to arc magmatism (Schaltegger et al. 1997). A similar evolution has been described for the basement south of the Tauern Window, where volcanic arc basaltic suites from 550–530 Ma have been reported (Schulz et al. 2004 and 2008). Additional data that support an Ordovician subduction setting have come from the Aar massif (Helvetic domain), where metagabbros dated at 478 Ma were affected by an Ordovician high-pressure metamorphic event that was followed by a high-temperature overprint accompanied by partial melting at ca. 450 Ma (Schaltegger et al. 2003).

The active margin setting in the Alpine domain has also been characterized by several pulses of peraluminous and calc-alkaline granitoids that were emplaced between 480 and 450 Ma. These metagranitoids are ubiquitous in the basements of all paleogeographic Alpine domains (Figures 1.5 and 1.6), and these attest to the wide extension of this magmatism (e.g. Schaltegger and Gebauer 1999; Bussien et al. 2011; Von Raumer et al. 2013, and references therein). The age of this magmatism is from 470 to 440 Ma in the Helvetic domain (Schaltegger and Gebauer 1999; Bussy et al. 2011), and from 480 to 450 Ma in the Penninic and Austroalpine domains (Guillot et al. 2002; Schulz et al. 2008; Liati et al. 2009). In the south Alpine domain, large outcrops of these Ordovician metagranitoids are exposed in the Strona–Ceneri zone, where they represent around 20% of the exposed outcrops of the Paleozoic basement (Zurbriggen 2015). These consist mainly of peraluminous orthogneisses derived from a greywacke source rock, with subordinate amounts of metaluminous calc-alkaline orthogneisses (ca. 90% peraluminous, 10% metaluminous orthogneisses) and rare mafic end-members (Zurbriggen 2017).

However, the origin of these granitoids has not been well constrained. The predominance of crustal-derived peraluminous granitoids is not expected in a Cordillera-like setting, which typically contains a wide variety of mafic to felsic igneous rock derived from both the fractional crystallization of mantle-derived magma and the partial melting of various igneous and sedimentary crustal components (Barbarin 1999). This is also not expected in an extensional rift setting, and more generally, in all environments in which the bulk of the heat was advected from the mantle by the ascent of mafic melts, which should be dominated by mantle-derived magmatism (Annen et al. 2006). Zurbriggen (2015, 2017) proposed an alternative model of subduction–accretion complexes, in which the dominant peraluminous magmatism results from the melting of flush-type sediments in the accretionary prism of an arc fed by a large sedimentary input from the continent. However, this model still involves heat advection from the mantle as the main driver of crustal melting and attributes the relative scarcity of exposed mafic and intermediate intrusives to exposure bias. Finally, other alternative models proposed for other parts of the European Variscides relate this magmatism to the thermal relaxation of a thickened arc crust (Villaseca et al. 2016; Soejono et al. 2019), which would generate large amounts of peraluminous anatectic granites.

1.4.3. *Ordovician to Silurian crustal extension*

The Cambrian–Ordovician crustal extension led to the detachment of small continental blocks from northern Gondwana during the Ordovician. In the western domain, the Avalonian block detached from Gondwana at ca. 480 Ma, and drifted northward to eventually undergo accretion with Laurentia during the Silurian (Figure 1.2). In the eastern domain, the detachment of the Hun Terranes resulted in the opening of the eastern branch of the Rheic Ocean, slightly later (ca. 460 Ma) than the western branch (Figure 1.2, sketch 3). This opening left a passive margin along North Gondwana (Figure 1.2, sketch 4). From the mid-Ordovician to the Lower Carboniferous, sedimentation occurred along this passive margin. These clastic deposits are mainly exposed in the Austroalpine nappes (Greywacke Zone, Gürktal nappe, and Paleozoic of Graz) and the Carnic Alps (Neubauer et al. 2007), and this provides a record of the crustal subsidence at the margin (Figure 1.5). A few pulses of alkaline magmatism have been recorded for this period; for instance, in basement massifs south of the Tauern Window (Austroalpine), where alkaline within-plate basalts were emplaced at ca. 430 Ma (Schulz et al. 2004). These Late Ordovician–Early Silurian magmatic events might have been related to either late extension after the eastern Rheic opening or an initial rifting stage that preceded the opening of the Paleo-Tethys (Schulz et al. 2004; Stampfli et al. 2011).

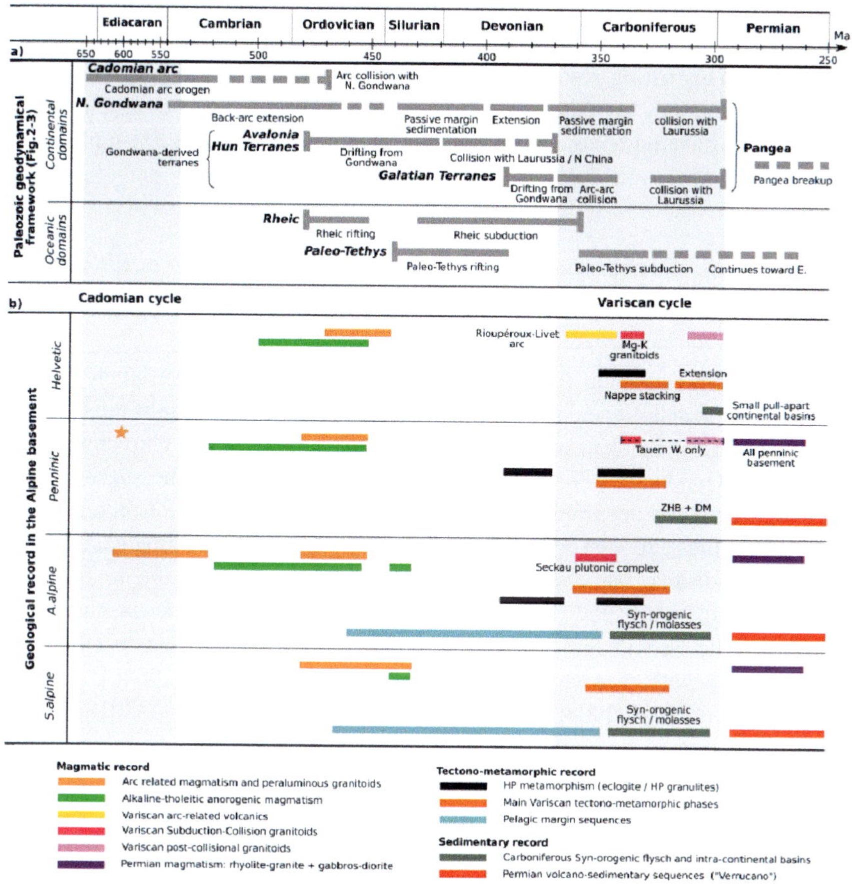

Figure 1.6. *a) Chronological chart for the Neoproterozoic–Paleozoic that summarizes the geodynamical evolution of the European Variscides from the Ediacaran to the Permian. b) Geological record for the Alpine basement for the same period. The four lithotectonic domains in the Alpine belt (i.e. Helvetic, Penninic, Austroalpine, and south Alpine) show contrasted records and are thus treated independently*

The opening of the Paleo-Tethys has not been well recorded for the European domains, as it was largely obliterated by subsequent Variscan and Alpine events, and parts of the former Paleo-Tethyan margin have been covered under Mesozoic sediments (Von Raumer et al. 2013). In Sardinia, the volcano-sedimentary deposits of the Late Ordovician–Early Silurian (ca. 440 Ma) that consist of alkali basalts and trachyandesites might be associated with the initial rifting that preceded the opening of the Paleo-Tethys (Gaggero et al. 2012). This opening has been better documented in Iran (Bagheri and Stampfli 2008), and it resulted from two rifting events that occurred during the Late Ordovician (ca. 440 Ma) and the Middle Devonian (ca. 390 Ma). This rifting resulted in the separation of the Galatian Terranes from Gondwana, where they contained the Alpine domain (Von Raumer et al. 2013) and most of the crustal fragments that formed the European Variscides (Stampfli et al. 2013).

1.4.4. *Devonian–Carboniferous convergence and Variscan collision*

1.4.4.1. *A poly-phased history during the Variscan collision*

From the Mid-Devonian (ca. 390 Ma), the Rheic Ocean progressively closed, simultaneously with the opening of the Paleo-Tethys and the drifting of the Galatian Terranes (Figure 1.2, sketches 5 and 6), which initiated the convergence of Laurussia and Gondwana. The Variscan orogeny corresponds to the succession of collisional events that occurred from the Late Devonian to the Late Carboniferous/Early Permian (ca. 370–295 Ma). This eventually resulted in the amalgamation of Laurussia, Gondwana, and the peri-Gondwana terranes (Figure 1.3), to form the most recent supercontinent, Pangea. In many basement massifs in the Alps, polymetamorphic assemblages prevail and multiple nappe stacking events have been identified (e.g. Schaltegger and Gebauer 1999; Raumer and Bussy 2004; Benciolini et al. 2006; Schulz et al. 2008; Guillot and Ménot 2009; Fréville et al. 2018), which testifies to the polycyclic evolution during the Variscan orogeny. The intensity of the Variscan deformations and metamorphic overprint varies greatly among the basement areas in the Alps, from lower greenschist to granulite and/or eclogite facies (Frey et al. 1999). These different Paleozoic basement areas with various degrees of Variscan overprint might represent either different structural levels in the crust (Schaltegger et al. 2003) or different paleogeographic domains that were variously affected by the Variscan orogeny and subsequently juxtaposed due to the Late Variscan and Alpine tectonics (Guillot and Ménot 2009). Two major Variscan tectono-metamorphic events have been identified for the Alpine domain (Von Raumer et al. 2013). The first occurred during the Early-to-Mid Carboniferous (ca. 350–330 Ma) and corresponds to the closure of the Rheic domain and amalgamation of the Galatian Terranes to Laurussia (Figure 1.3, sketches 7 and 8). The second event occurred during the Late Carboniferous/Early Permian (ca. 320–295 Ma) and corresponds to the final collision of Gondwana and the post-collisional collapse of the Variscan belt (Figure 1.3, sketches 9 and 10).

1.4.4.2. *Devonian–Visean magmatism*

Prior to this collision, early Variscan events can be identified. In the massif of Belledonne (Western Alps), the mafic-felsic bimodal suites of volcanic and plutonic rock which were emplaced at ca. 360–350 Ma (Figures 1.5, 1.6, and 1.7) have been attributed to continental extension in a back-arc setting (Guillot and Ménot 2009; Fréville et al. 2018), similar to the Brévenne unit in the French Massif Central (Faure et al. 2005) and the central part of the Moldanubian zone (Lardeaux et al. 2014). The magmatic evolution suggests a transition from an extensional to a compressional regime, which marks the beginning of the Visean collision. A similar evolution took place along the margin of Laurussia, where the Devonian back-arc extension resulted in the opening of the Rheno-Hercynian domain, which separated ribbon-like terranes from the Laurussian plate (Franke et al. 2017). Volcano–sedimentary deposits of an early Visean age overlie the older gneisses in the basement of the Helvetic domain (Guillot and Ménot 2009; von Raumer et al. 2009). These present an arc-related tholeiitic geochemical signature with contamination from a continental crust component (Carme and Pin 1987). It is therefore believed that these formations were deposited in a continental arc setting, and mark the subduction of the Rheic/Paleo-Tethyan domains during the Lower Carboniferous (Guillot and Ménot 2009; Fréville et al. 2018). In the Seckau nappe (Austroalpine domain), the Hintertal plutonic suite is composed of Late Devonian to Visean (ca. 365–340 Ma) granites and granodiorites (Figures 1.5 and 1.6), which were possibly emplaced in a continental arc setting (Mandl et al. 2018).

1.4.4.3. *Variscan high-pressure metamorphism*

Relics of high-pressure rock, such as eclogites and high-pressure granulites, have been reported for many basement areas in the Alpine domain (Figure 1.5), where they formed small bodies embedded within felsic lithologies (orthogneisses, metasediments). These high-pressure assemblages have been documented in the external crystalline massifs (Liégeois and Duchesne 1981; Paquette et al. 1989; Ferrando et al. 2008; Guillot and Ménot 2009), in the Penninic nappe of Adula in the Lepontine Alps (Liati et al. 2009; Herwartz et al. 2011; Sandmann et al. 2014), and several crystalline nappes of the Austroalpine domain (Thöni 2006 and references therein). For most of these bodies, the age of the high-pressure stage has been estimated at ca. 350–330 Ma (Figure 1.6; Miller and Thöni 1995; Ladenhauf et al. 2001; Tumiati et al. 2003; Thöni, 2006; Liati et al. 2009; Rubatto et al. 2010, Jacob et al. 2021; Jacob et al. 2022; Vanardois et al. 2022). In the eclogites from the Adula nappe (Penninic domain), both U-Pb zircon and garnet Lu-Hf geochronology have indicated a Mid-Carboniferous age for the high-pressure metamorphism (ca. 330–340 Ma). In addition, some zircon rims that were dated by Liati et al. (2009) yielded a Devonian age, at ca. 370 Ma, which might indicate earlier metamorphism.

Finally, an eclogite body from the Hochgrössen Massif (eastern Austroalpine nappes) yielded an age of 397±8 Ma for the high-pressure stage, with peak pressure conditions estimated at ca. 1.8 to 2.2 GPa and 700°C (Faryad et al. 2002). These ages suggest a Mid-to-Late Devonian subduction cycle that preceded the more widely recorded 350 to 330 Ma high-pressure stage. Paquette et al. (1989) reported a U-Pb zircon age of 395±2 Ma in the external crystalline massifs, but this age was obtained about 30 years ago by multigrain isotope-dilution thermal-ionization mass spectrometry, and it more probably reflects mixing between the Ordovician magmatic cores and the metamorphic rims of the zircons (Paquette et al. 2017). More recent dating in the massifs of Argentera (Rubatto et al. 2010), Belledonne (Jacob et al. 2021), Oisans (Jacob et al. 2022) and Aiguilles Rouges (Vanardois et al. 2022) have yielded ages between 340 and 320 Ma, which are more consistent with other U-Pb data from the Alpine basement.

1.4.4.4. *Variscan nappe-stacking events in the Alpine basement*

The Tournaisian–Visean collision was recorded in most basement domains in the Alps by successive episodes of nappe stacking, Barrovian-type metamorphism, emplacement of granitoids, and detrital zircon and mica records in the Carboniferous–Permian basins (Figures 1.5 and 1.6; Schulz et al. 2008; Guillot and Ménot 2009; Von Raumer et al. 2013; Manzotti et al. 2016; Zanoni and Spalla 2018; Haas et al. 2020). Due to the intense post-Permian reworking of the Alpine basement, regional geometric correlations at the scale of the Alpine domain are particularly challenging, in contrast to other Variscan massifs in Europe (e.g. Faure et al. 2009; Ballèvre et al. 2014; Lardeaux et al. 2014). Such correlations have nevertheless been carried out for the external domains of the Western Alps, which underwent very limited reworking by Alpine tectonics (Guillot et al. 2009; Guillot and Ménot 2009; von Raumer et al. 2009). In these domains, two successive Visean nappe-stacking events have been identified (Figure 1.7). The first corresponds to the development of N-N20° trending transpressive shear zones, which led to the thrusting of the internal high-pressure unit towards the northwest. This deformation was associated with the emplacement of granitoids at ca. 340 Ma (Debon et al. 1998; Debon and Lemmet 1999; Bussy et al. 2000; Guillot and Ménot 2009). The second deformation is associated with a top-to-the-east shearing and northeast thrusting of the Chamrousse ophiolitic complex over the Devonian–Tournaisian magmatic complex. This second nappe-stacking event is associated with crustal thickening and is characterized by a Barrovian metamorphic gradient that increases from West to East (from ca. 0.6 GPa and 600°C to ca. 0.8 GPa and 680°C; Guillot and Ménot 2009; Fréville et al. 2018).

1.4.4.5. *The Visean magnesio-potassic magmatism*

The Visean granitoids are of particular significance in the Variscan belt. These form a series of potassic calc-alkaline rock that is enriched in K_2O and MgO. The Visean granitoids are considered to be derived from a mixed source that combined an enriched sub-continental lithospheric mantle and a continental crust (Debon et al. 1998; Debon and Lemmet 1999). This series is documented for the entire Helvetic domain (Debon and Lemmet 1999; Bussy et al. 2000; von Raumer et al. 2014), as well as for the basement of the Tauern Window (Lammerer 1986; Veselà et al. 2011), and it was emplaced at ca. 340–330 Ma (Figures 1.5 and 1.6). Similar rocks of the Visean age can be found in other parts of the European basement in Corsica, French Central Massif, Black Forest, Vosges, and the Bohemian Massif (von Raumer et al. 2014). These granitoids are probably related to collisional processes that involved both mantle and crustal melting (Schaltegger 1997; von Raumer et al. 2013 and 2014), although it has not been excluded that they were emplaced in a Cordillera-like setting. Based on their spatial distribution and their emplacement along sinistral strike-slip shear zones (Debon and Lemmet 1999; Guillot et al. 2009), it has been suggested that the Mg-K granitoids define a major linear structure in the Variscan belt, and possibly mark an orogen-wide strike-slip fault rooted in the lithospheric mantle or a major suture zone (Schaltegger 1997; Corsini and Rolland 2009; Guillot et al. 2009; Veselà et al. 2011; von Raumer et al. 2014).

1.4.5. *Crustal extension and post-orogenic collapse*

The mid-Carboniferous to Permian corresponds to the final collision of Gondwana with the peri-Gondwanian crustal blocks that were previously assembled as Laurussia. This period records a complex tectonic history (Figure 1.6) that was dominated by strike-slip tectonics and extension from the Late Carboniferous and was associated with the opening of intra-continental basins (Ballèvre et al. 2018). This was accompanied by a generalized mid/low-pressure and high-temperature metamorphism, which led to migmatization in the lower-to-mid crustal levels, and the production of sub-alkaline peraluminous granitoids. This series corresponds to the ferriferous low-Mg suite of Debon and Lemmet (1999), which was emplaced at 310–295 Ma (Bonin et al. 1993; Schaltegger 1997; Debon and Lemmet 1999; Schaltegger and Gebauer 1999; Bussy et al. 2000 Veselà et al. 2011; Fréville 2016).

The Late Variscan extension is well documented in the external zone of the Western Alps, where two phases of extension have been identified (Figure 1.7). The first corresponds to the NE-SW extension from the Westphalian to the Mid-Stephanian (310–305 Ma), contemporaneous with migmatization in the deeper crustal levels. The second phase corresponds to a roughly N-S extension with

E-W shortening, accommodated by NE-SW-trending dextral shear zones and brittle strike-slip faults that were active from the Stephanian to the Early Permian (ca. 305–295 Ma; Guillot et al. 2009). This strike-slip system that runs from the Aar massif to the northeast to the Belledonne massif has a possible connection in the Bohemian massif and the Corsica–Sardinia block, and would thus define an orogen-wide structure known as the east Variscan shear zone, with a lateral offset of at least 300 km (Corsini and Rolland 2009; Guillot et al. 2009; Rossi et al. 2009; Duchesne et al. 2013; Ballèvre et al. 2018; Simonetti et al. 2018).

Intra-continental basins with coal-bearing deposits developed during the Late Carboniferous. The largest of these corresponds to the Zone Houillère basin in the Briançonnais domain (Figure 1.5), which extends over at least 150–200 km from the North to the South, and has a width of ca. 20 km. Even without accounting for the Alpine shortening, this is one of the largest Carboniferous basins in Europe (Ballèvre et al. 2018). Sedimentation in the Zone Houillère mainly took place from the Namurian to the Westphallian (ca. 325–315 Ma; Ballèvre et al. 2018 and references therein). There are also smaller Carboniferous basins in other parts of the Alpine belt (Figures 1.5 and 1.6). In the internal zone of the Penninic domain, Carboniferous deposits are exposed for the Dora Maira and Gran Paradiso massifs (Ballèvre et al. 2018). Late Carboniferous molasse basins are also known for the Austroalpine nappes of the Eastern Alps (Neubauer et al. 2007). In the Helvetic domain, rather small, elongated narrow basins of Westphalian–Stephanian age (305–300 Ma) probably opened when they were pulled apart during the dextral strike-slip tectonic phase (Guillot et al. 2009; Ballèvre et al. 2018 and references therein).

This general evolution that consisted of strike-slip tectonics, extension, the opening of the intra-continental basins, and low-pressure–high-temperature metamorphism that was associated with the production of large amounts of peraluminous granitoids has been observed for all of the western European Variscides (Catalán et al. 2007; Faure et al. 2009; Faure 2014), and it attests to the general collapse of the Variscan belt during the Late Carboniferous–Early Permian. For the eastern domain, and particularly for the Austroalpine domain, this late tectonic phase was relatively mild and was not associated with significant crustal thickening and melting (Haas et al. 2020). Following the Variscan orogeny, Mid-to-Late Permian (ca. 280–250 Ma) metamorphic and magmatic activity was recorded for the Alpine basement (Figures 1.5 and 1.6). Traces of this episode are scarce in the basement of the Helvetic domain (Fréville 2016; Ballèvre et al. 2018). In contrast, they are ubiquitous in the other parts of the Alpine basement, where Permian high-temperature metamorphism and the associated magmatism were widespread (Schaltegger and Gebauer 1999; Schulz et al. 2008; Spalla et al. 2014; Ballèvre et al.

2018; Kunz et al. 2018; Manzotti et al. 2018). This high-temperature episode might be related either to the latest stages of the Variscan post-orogenic collapse or to the earliest events of the lithospheric extension and the thinning that preceded the break-up of Pangea (Spalla et al. 2014).

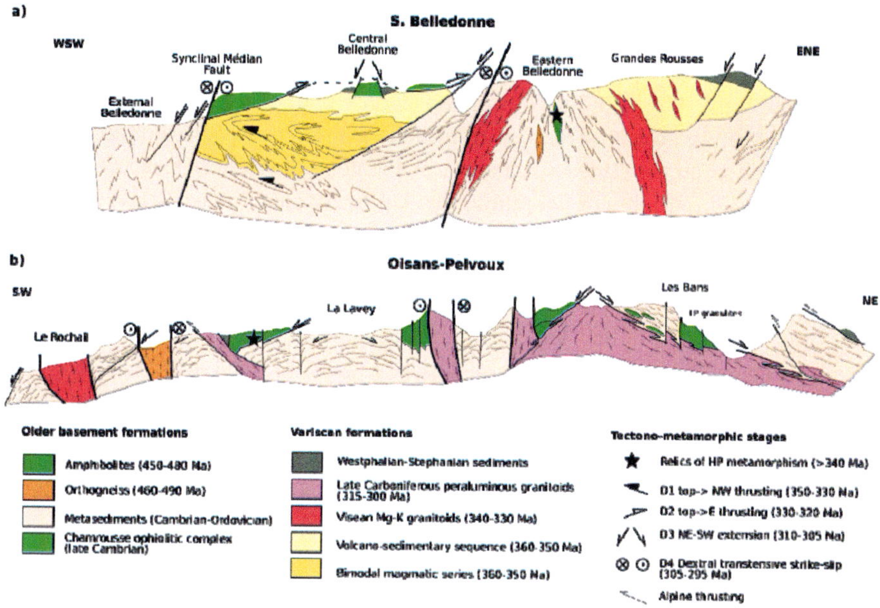

Figure 1.6. *Schematic cross-sections showing the structural relationships for a) the massifs of Belledonne–Grandes Rousses and b) Oisans–Pelvoux. Modified from Guillot and Ménot (2009)*

1.5. Discussion

1.5.1. *Place of the Alpine domain in the Variscan puzzle*

During the last decade, an important effort has been devoted to the correlation between the different Variscan massifs in Europe, and to the reconstruction of the general structure of the belt (Schulmann et al. 2014). The exploitation of large geochronological and geochemical datasets collected from all over the European Paleozoic basement has significantly improved the geodynamic reconstructions, through more precise identification of the different crustal blocks which form the Variscan puzzle, and their initial positions along the northern Gondwana margin, prior to the Ordovician rifting (Blichert-Toft et al. 2016; Stephan et al. 2019a, 2019b).

Although there are large portions of a well-preserved Paleozoic basement in the Alpine domain, this has not been an important focus for the Variscan community, and it has been left out a little from such large-scale correlations. This lack of studies in the Alpine basement is mostly due to the extremely complex tectonic setting in the Mediterranean region, where Alpine tectonics have largely perturbed the original pre-Mesozoic position of the crustal blocks. The Alpine domain is composed of a juxtaposition of crustal blocks with different pre-Mesozoic positions that were stacked during the Alpine orogeny (van Hinsbergen et al. 2019). Only the Helvetic domain has been relatively preserved from this complex tectonic reworking, and it remained located on the European margin of Piemonte–Ligurian Ocean during the Mesozoic. In their current configuration, the Alpine basement massifs can therefore be seen as a mosaic of different pieces of the Paleozoic crust that correspond to different domains of the Variscan belt and that record different tectono-metamorphic and magmatic events. An illustration of this point is given by the Western Alps, where the basement of the Helvetic and Briançonnais domains underwent very contrasted igneous and metamorphic histories (Figures 1.5 and 1.6). The Helvetic basement shows similarities with the Moldanubian core of the Variscan belt (Bohemian Massif, Vosges-Schwarzwald, Massif Central): it has a strong amphibolite/ granulite facies Variscan overprint (Grandjean et al. 1996; Raumer and Bussy 2004; Guillot and Ménot 2009; Jacob et al. 2022), abundant Carboniferous granitic magmatism at ca. 340–330 Ma and 310–300 Ma (Debon and Lemmet, 1999), and very little Permian magmatism. In contrast, with a few exceptions, the basement of the Briançonnais domain has a very weak Variscan metamorphic overprint (Guillot et al. 2002; Ballèvre et al. 2018), and is mainly devoid of Carboniferous granitoids, but is characterized by abundant Permian magmatism (Ballèvre et al. 2018).

This striking difference in terms of pre-Triassic history was regarded by Ballèvre et al. (2018) as the consequence of a major strike-slip displacement along the eastern Variscan shear zone during the Late Carboniferous–Early Permian, that cut through the large-scale zonation of the Variscan belt, and brought internal and external domains into contact. However, this view does not take into account the displacements that occurred during the Alpine cycle. Indeed, it is widely recognized that the Briançonnais domain has a connection with the Corsica–Sardinia block (e.g. Stampfli and Hochard 2009; van Hinsbergen et al. 2019), which was in a more southern position during the Triassic. It is generally assumed that this Corsica–Sardinia–Briançonnais block was positioned close to the Iberian plate during the Triassic, south of the Pyrenees (Stampfli and Hochard 2009), or facing the Gulf of Lion and the Maures-Esterel massif in southern France (van Hinsbergen et al. 2019).

In any case, these reconstructions emphasize the southern origin of the Briançonnais domain. The sharp differences in terms of tectono-metamorphic history with the Helvetic domain may therefore result from an Alpine collage that brought the Briançonnais block (which would belong to the southern foreland of the Variscan belt) into contact with the internal domains that form the Helvetic margin.

Toward the East, the Southalpine and Austroalpine domains were in a southern position during the Triassic, south of the Pyrenees and east of the Iberian plate (van Hinsbergen et al. 2019). This replaces the Austroalpine units close to the southern foreland of the Variscan belt, which is consistent with the weak overprint recorded during the Late Carboniferous for the final stage of the Variscan orogeny (Haas et al. 2020).

Finally, the basement exposed in the Tauern Window is interpreted as derived from the distal European margin (Schmid et al. 2004) and can be seen as the eastward prolongation of the Helvetic basement. This view is supported by the similar Variscan magmatic evolution in these two domains (Eichhorn et al. 2000; Veselà et al. 2011), and by the similarities in the pre-Variscan magmatic and metamorphic records (Eichhorn et al. 2001). However, these domains are not completely identical. The Permian magmatism recorded in the Tauern Window is not seen for the Helvetic basement (Figures 1.5 and 1.6), and the Variscan metamorphic imprint in the Tauern basement reaches a lower grade than in the western domains. Given the relative proximity of the Tauern basement to the Bohemian massif, it is also tempting to see this as the southern continuation of the Moldanubian zone, which forms the core of the Variscan belt. However, there are significant differences between these two domains, in terms of both magmatic and metamorphic history (e.g. Finger et al. 1997). In particular, the Tauern basement completely lacks ultrahigh-temperature granulites that are typically exposed in the adjacent Moldanubian zone. The Tauern basement, therefore, corresponds to a more external domain of the Variscan belt. Eichhorn et al. (2001) came to similar conclusions and suggested that the Tauern basement occupied an intermediate position between the Helvetic basement, which forms parts of the Moldanubian zone, and the more external zones of the Variscan belt, possibly represented by the basement nappes of the Austroalpine domain. In contrast with the Briançonnais zone in the Western Alps, Alpine displacement cannot be invoked here to explain the discontinuity with the Moldanubian zone. This latter more probably results from the Late Variscan strike-slip tectonics, which might be related to the east Variscan shear zone that cuts through the Variscan zonation and put the lower grade Tauern basement in contact with the high-grade Moldanubian core.

1.5.2. *Pre-Variscan position of the Alpine basement units*

Prior to the Variscan cycle, the crustal fragments that formed the Paleozoic basement were located along the northern Gondwana margin. The restoration of the pre-Variscan architecture of this continental margin and the reconstitution of its Late Proterozoic–Early Cambrian evolution is under active research, as it has implications for the crustal growth models and the formation of the European continental crust (von Raumer and Stampfli 2008; Stampfli et al. 2013; von Raumer et al. 2015; Villaseca et al. 2016; Chelle-Michou et al. 2017; Couzinié et al. 2017, 2019). Due to the strong displacement associated with the Late Carboniferous strike-slip tectonics, the replacement of the different crustal blocks in their pre-Ordovician position is not straightforward.

The detrital U-Pb age patterns of zircons from the Paleozoic sediments have proven to be a powerful proxy to unravel their provenance and to infer the relative positions of the crustal fragments in their pre-Variscan setting. Stephan et al. (2019a) used a large compilation of these data from all across Europe and the surrounding cratonic areas to classify the different Paleozoic domains according to their pre-Ordovician positions. Combined with other types of data (e.g. age of magmatism, stratigraphic and paleontological records), this large meta-analysis allowed the identification of three different domains in the Variscan basement, which are related to the Avalonian, the eastern Gondwana, and the western Gondwana shelves. In the main, most of the Moldanubian zone and the Armorican Massif derive from the western Gondwana shelf, while a large part of southeastern Europe, which includes Sardinia and Corsica, derives from the eastern Gondwana shelf. A large part of the Iberian massif also derives from eastern Gondwana, which appears to be counter-intuitive given its westward position in the Variscan assemblage.

The position of the Alpine domains was unfortunately not assessed in the study of Stephan et al. (2019a). However, based on the pre-Alpine southern position of the Austroalpine and south Alpine domains near the Iberian plate (van Hinsbergen et al. 2019), and the connection of the Briançonnais domain with Corsica–Sardinia, it appears reasonable to also attribute an eastern Gondwana origin to these domains. This initial position has been confirmed by recent U-Pb zircon data from metasediments sampled in the Eastern Alps, which attributed these to an origin on the northeastern Gondwana margin for the Austroalpine nappes, next to Anatolia and the Iranian Luth-Tabas blocks (Haas et al. 2020). In contrast, the Helvetic domain shows strong similarities with the Moldanubian zone, which would indicate a more western origin along the northern margin of Gondwana. However, this inference is very speculative because of the lack of detrital zircon data from the Lower Paleozoic sediments in the Western Alps.

1.5.3. *Significance of Variscan high-pressure metamorphism in the Alps*

Eclogites and high-pressure granulites are exposed in different domains of the Alpine basement, mainly in the external crystalline massifs and the Austroalpine nappes (Figure 1.5). Such high-pressure assemblages are not rare for the European Variscides, especially in the internal (Moldanubian) zone of the orogen. Many such occurrences have been reported for Northern Iberian and Southern Armorican massifs, French Massif Central, Vosges-Schwarzwald, Bohemian Massif, Sardinia, and in the Alpine Paleozoic basement (Paquette et al. 2017; Regorda et al. 2019).

The high-pressure metamorphic records in the European Variscides are best understood within a model that involves two distinct high-pressure stages (O'Brien 2000). A Devonian (390–360 Ma) subduction stage produced middle-temperature eclogites ($T < 650°C$) and blueschists by subduction of the oceanic lithosphere and the hyperextended continental margins of the former Ordovician basins, with these now incorporated into gneiss–metabasite complexes (also known as leptyno–amphibolic complexes in the French part of the Variscides). These rocks were already exhumed by the Late Devonian, as shown by the occurrence of the Devonian sediments that contain high-pressure detritus and/or overlie the gneiss–metabasite complexes (e.g. Schäfer et al. 1997). A second mid-Carboniferous (ca. 340 Ma) high-pressure–high-temperature event (1.5–2.2 GPa, 700–1000°C) produced granulitic complexes that contained high-temperature eclogites and pyroxenites, and different types of mantle peridotites, which then underwent very rapid exhumation to the lower crustal levels (e.g. Schantl et al. 2019). This was soon followed by mid-to-low pressure and high-temperature regional metamorphism and granitic magmatism from the Middle-to-Late Carboniferous (335–300 Ma). In contrast with the middle-temperature eclogites, this group rarely preserves prograde metamorphic evolution, and only records exhumation from the deep lower crustal levels, together with the peridotites (O'Brien 2000). These granulitic complexes could represent deep portions of the orogenic lower crust which have been extruded to higher crustal levels during the collision (Schulmann et al. 2005). Alternatively, they might also mark a second subduction stage during the mid-Carboniferous, that might be related to the emplacement of the Mg-K granitoids (Finger et al. 2007).

The Variscan high-pressure rock exposed in the Alps can (with very few exceptions) be related to the mid-Carboniferous group according to the following:

– They mainly consist of small bodies of eclogites and mafic high-pressure granulites embedded within high-grade gneiss (Paquette et al. 1989; Raumer and Bussy 2004; Ferrando et al. 2008), and they are associated with ultramafic bodies (e.g. garnet-spinel-bearing peridotites, pyroxenites, and serpentinites), especially in the external crystalline massifs (Le Fort 1971; Raumer and Bussy 2004) and in the Ulten zone (Godard et al. 1996; Tumiati et al. 2003).

– The high-pressure stage is dated at ca. 350–330 Ma (Miller and Thöni 1995; Ladenhauf et al. 2001; Tumiati et al. 2003; Thöni 2006; Liati et al. 2009; Rubatto et al. 2010).

– Peak pressure conditions occur at relatively high temperatures (>700°C), although no ultra-high-temperature conditions have ever been reported for the Alpine basement. In the Helvetic domain, peak pressure conditions are estimated at ca. 1.3–1.6 GPa and 650–750°C, followed by an amphibolitic overprint at ca. 1.0–1.2 GPa and 750°C (Ferrando et al. 2008, Jouffray et al. 2020; Jacob et al. 2021; Jacob et al. 2022; Vanardois et al. 2022). In the Ulten zone (Austroalpine), eclogites undergo similar equilibration conditions, at ca. 1.4–1.5 GPa and 700°C (Godard et al. 1996; Hauzenberger et al. 1996). Eclogites from the Silvretta and Ötztal nappes (Austroalpine) indicate higher pressure, with peak conditions estimated at ca. 2.7–2.8 GPa and 700–763°C (Miller and Thöni 1995; Ladenhauf et al. 2001; Thöni 2006).

The only reported exception here relates to the Devonian eclogites exposed in the Hochgrössen Massif (eastern Austroalpine nappes), at a metamorphic age of 397 ±8 Ma for the high-pressure stage, with peak pressure conditions estimated at ca.1.8 to 2.2 GPa and 700°C (Faryad et al. 2002). This metamorphism is more consistent with the mid-Devonian subduction cycle recorded elsewhere for the Variscan belt.

The geodynamic significance of the Variscan eclogites in the Alps remains under debate. According to Jouffray et al. (2020), the pressure and temperature conditions estimated for the Variscan eclogites in the external crystalline massifs are consistent with a warm subduction geotherm and might represent a dismembered cryptic suture zone. This view is supported by recent thermomechanical modeling (Regorda et al. 2019), which suggested that the pressure and temperature paths and metamorphic ages of the Variscan high-pressure rocks are better reproduced with a two-stage subduction model. However, a subduction process does not necessarily have to be invoked to explain the formation of some Variscan eclogites. For instance, the Montagne Noire migmatite dome, south of the Massif Central, exposes eclogite bodies at ages and under pressure and temperature conditions very similar to those of the eclogites exposed in the external crystalline massifs (Whitney et al. 2015 and 2020). They, therefore, appear on a warm subduction geotherm, despite being exposed in an external domain of the Variscan belt, far from any recognized paleo-subduction zone. Extrusion of the orogenic lower crust during the formation of the Montagne Noire dome is therefore the favored model to explain the exhumation of these eclogites. Discrimination between the different possible geodynamical scenarios requires the use of thermomechanical models. However, as Regorda et al. (2019) indicated, the large uncertainties on the pressure and temperature data, and even more critically, the age data available for the Variscan eclogites, make it difficult to draw any clear conclusions from these models. Refining these data using all of the modern analytical and modeling tools developed for petrochronology will therefore be essential to building a comprehensive geodynamic model of the Variscan orogeny.

Finally, it can be noted here that two-stage evolution that involves two successive high-pressure stages is also seen for modern orogens, and in particular in the Tibet–Himalaya collisional system that shows some similarities to the European Variscan belt (Maierová et al. 2016). The earliest metamorphism related to the subduction of the Indian plate is dated at ca. 55–45 Ma, and this reached ultra-high-pressure conditions of ca. 2.7 GPa at 700–770°C (Guillot et al. 2008 and references therein). This was followed by the Late Eocene (ca. 37–33 Ma) with high-pressure–high-temperature metamorphism at 2.0–1.4 GPa and 800°C recorded for eclogites and granulites from the south Tibetan Neogene migmatitic domes (Kali et al. 2010; Guilmette et al. 2011; Lanari et al. 2013), which were subsequently re-equilibrated at mid-pressure–high-temperature conditions with the surrounding migmatites at 28–10 Ma. There are also similarities between the Cenozoic Tibet–Himalayas and Paleozoic Variscan orogens relative to the duration of the convergence and the spatial extension of the collisional events (Maierová et al. 2016). Considering all of these similarities, some authors identify the Variscan Belt as an ancient analogue of the Himalaya-Tibet system (e.g. Maierova et al., 2016).

1.6. Conclusion

The Paleozoic domains exposed in the Alps have had a long and complex pre-Alpine history that spanned more than 300 million years, from the Ediacaran to the Permian. This history is in many ways similar to that of other Paleozoic massifs in Europe and reflects the general evolution at the northern margin of Gondwana during the Lower Paleozoic, followed by the Variscan collision during the Devonian and the Carboniferous. However, in contrast to other Paleozoic domains in Europe, the Alpine domain does not form a coherent block. Instead, it represents a collage of several lithotectonic units that were amalgamated during the Alpine orogeny. These blocks are comprised of the different portions of the former Variscan orogen, therefore showing a contrasting Paleozoic evolution. The autochthonous European margin of the Tethys Ocean that currently forms the Helvetic domain has a strong Variscan overprint and exposes deep crustal domains that have been affected by high-grade metamorphism. This probably belongs to the internal Moldanubian zone of the Variscan belt, and possibly corresponds to a domain that was initially located close to the Bohemian massif, that was displaced southward during the Upper Carboniferous–Early Permian by the dextral strike-slip east Variscan shear zone. In contrast, the Briançonnais block and the Apulia-derived units (Austroalpine, south Alpine domains) were located in a more southern position prior to the Alpine tectonic cycle, close to the current Pyrenees and the Iberia plate. They, therefore, represent portions of the southern foreland of the Variscan belt, and have much lower levels of Variscan overprint, although they were more affected by the Permian high-temperature event.

The place of the Alpine Paleozoic domains in the Variscan puzzle is, as of the current state of knowledge, relatively speculative, and still needs to be precisely defined. For this purpose, the succession of pre-Variscan and Variscan tectono-metamorphic, magmatic, and sedimentary events in the Alpine basement need to be better characterized. Correlation with other Paleozoic domains in Europe is not an easy task, given that the Paleozoic massifs in the Alps represent an assemblage of small tectonic blocks that do not preserve a complete section across the Variscan orogen. There remains the need, therefore, to be very careful when considering large-scale Variscan zonation in the Alpine basement massifs. Apart from the Late Variscan strike-slip faults, it appears unlikely that large-scale Variscan zonation or structures can still be unambiguously identified in such a complex assemblage of blocks. In particular, the geodynamic significance of high-pressure relics and their correlation with Variscan suture zones in other parts of the belt remains highly speculative.

1.7. Acknowledgments

This work was supported by the BRGM through the Référentiel Géologique de la France program (RGF). We thank Michel Ballèvre and an anonymous reviewer for their insightful suggestions and comments. We would also like to thank Christopher Berrie for his careful proof reading and correction of the manuscript, which significantly improved the general quality of English.

1.8. References

Bagheri, S. and Stampfli, G.M. (2008). The Anarak, Jandaq and Posht-e-Badam metamorphic complexes in central Iran: New geological data, relationships and tectonic implications. *Tectonophysics*, 451, 123–155.

Ballèvre, M., Le Goff, E., Hébert, R. (2001). The tectonothermal evolution of the Cadomian belt of northern Brittany, France: A neoproterozoic volcanic arc. *Tectonophysics*, 331, 19–43.

Ballèvre, M., Fourcade, S., Capdevila, R., Peucat, J.-J., Cocherie, A., Fanning, C.M. (2012). Geochronology and geochemistry of Ordovician felsic volcanism in the Southern Armorican Massif (Variscan belt, France): Implications for the breakup of Gondwana. *Gondwana Res.*, 21, 1019–1036.

Ballèvre, M., Martínez Catalán, J.R., López-Carmona, A., Pitra, P., Abati, J., Fernández, R.D., Ducassou, C., Arenas, R., Bosse, V., Castiñeiras, P. et al. (2014). Correlation of the nappe stack in the Ibero-Armorican arc across the Bay of Biscay: A joint French–Spanish project. *Geol. Soc. Lond. Spec.*, 405, 77–113.

Ballèvre, M., Manzotti, P., Piaz, G.V.D. (2018). Pre-Alpine (Variscan) inheritance: A key for the location of the future Valaisan basin (Western Alps). *Tectonics*, 37, 786–817.

Barbarin, B. (1999). A review of the relationships between granitoid types, their origins and their geodynamic environments. *Lithos*, 46, 605–626.

Bea, F., Montero, P., Talavera, C., Abu Anbar, M., Scarrow, J.H., Molina, J.F., Moreno, J.A. (2010). The palaeogeographic position of Central Iberia in Gondwana during the Ordovician: Evidence from zircon chronology and Nd isotopes. *Terra Nova*, 22, 341–346.

Benciolini, L., Poli, M.E., Visonà, D., Zanferrari, A. (2006). Looking inside Late Variscan tectonics: Structural and metamorphic heterogeneity of the Eastern Southalpine Basement (NE Italy). *Geodin. Acta*, 19, 17–32.

Blichert-Toft, J., Delile, H., Lee, C.-T., Stos-Gale, Z., Billström, K., Andersen, T., Hannu, H., Albarède, F. (2016). Large-scale tectonic cycles in Europe revealed by distinct Pb isotope provinces: Mapping out Europe with Pb isotopes. *Geochem. Geophys. Geosyst.*, 17, 3854–3864.

Bonin, B., Brändlein, P., Bussy, F., Desmons, J., Eggenberger, U., Finger, F., Graf, K., Marro, C., Mercolli, I., Oberhänsli, R. (1993). Late Variscan magmatic evolution of the Alpine basement. In *Pre-Mesozoic Geology in the Alps*, Von Raumer, J.F., Neubauer, F. (eds). Springer, Berlin, Heidelberg.

Bousquet, R., Engi, M., Gosso, G., Oberhänsli, R., Berger, A., Spalla, M., Zucali, M., Goffé, B. (2004). Explanatory notes to the map: Metamorphic structure of the Alps transition from the western Central Alps. *Mitteilungen Der Österreichischen Mineralogischen Gesellschaft*, 149, 41–51.

BRGM (2004). *Carte géologique de la France*. Éditions BRGM, Orléans.

Burg, J.-P., den Driessche, J.V., Brun, J.-P. (1994). Syn- to post-thickening extension in the Variscan belt of Western Europe: Modes and structural consequences. *Géologie Fr.*, 3, 33–51.

Bussien, D., Bussy, F., Magna, T., Masson, H. (2011). Timing of Palaeozoic magmatism in the Maggia and Sambuco nappes and paleogeographic implications (Central Lepontine Alps). *Swiss J. Geosci.*, 104, 1–29.

Bussy, F., Derron, M.-H., Jacquod, J., Sartori, M., Thelin, P. (1996). The 500 Ma-old Thy on metagranite: A new A-type granite occurrence in the western Penninic Alps (Wallis, Switzerland). *Eur. J. Mineral.*, 8, 565–576.

Bussy, F., Hernandez, J., Von Raumer, J. (2000). Bimodal magmatism as a consequence of the post-collisional readjustment of the thickened Variscan continental lithosphere (Aiguilles Rouges-Mont Blanc Massifs, Western Alps). *Earth Environ. Sci. Trans. R. Soc. Edinb.*, 91, 221–233.

Bussy, F., Péronnet, V., Ulianov, A., Epard, J.L., Raumer, J. (2011). Ordovician magmatism in the external French Alps: Witness of a peri-gondwanan active continental margin. In *International Symposium on the Ordovician System*. Alcalá de Henares, Madrid.

Carme, F. and Pin, C. (1987). Vue d'ensemble sur le magmatisme péri-orogénique et l'évolution métamorphique et tectonique varisque dans le Sud de la chaîne de Belledonne (Massifs cristallins externes, Alpes francaises). *CR Acad. Sci. Paris II*, 304(19), 1177–1181.

Casas, J.M. and Murphy, J.B. (2018). Unfolding the arc: The use of pre-orogenic constraints to assess the evolution of the Variscan belt in Western Europe. *Tectonophysics*, 736, 47–61.

Catalán, J.M., Arenas, R., García, F.D., Cuadra, P.G., Gómez-Barreiro, J., Abati, J., Castiñeiras, P., Fernández-Suárez, J., Martínez, S.S., Andonaegui, P. (2007). Space and time in the tectonic evolution of the northwestern Iberian Massif: Implications for the Variscan belt, In *4-D Framework of Continental Crust*, Hatcher Jr., R.D., Carlson, M.P., McBridge, J.H., Martinez Catalán, J.R. (eds). Geological Society of America Memoir Boulder, Colorado.

Chantraine, J., Egal, E., Thiéblemont, D., Le Goff, E., Guerrot, C., Ballèvre, M., Guennoc, P. (2001). The Cadomian active margin (North Armorican Massif, France): A segment of the North Atlantic Panafrican belt. *Tectonophysics*, 331, 1–18.

Chelle-Michou, C., Laurent, O., Moyen, J.-F., Block, S., Paquette, J.-L., Couzinié, S., Gardien, V., Vanderhaeghe, O., Villaros, A., Zeh, A. (2017). Pre-Cadomian to late-Variscan odyssey of the eastern Massif Central, France: Formation of the West European crust in a nutshell. *Gondwana Res.*, 46, 170–190.

Chu, Y., Lin, W., Faure, M., Wang, Q. (2016). Detrital zircon U-Pb ages and Hf isotopic constraints on the terrigenous sediments of the Western Alps and their paleogeographic implications. *Tectonics*, 35, 2734–2753.

Cocks, L.R.M. and Torsvik, T.H. (2002). Earth geography from 500 to 400 million years ago: A faunal and palaeomagnetic review. *J. Geol. Soc.*, 159, 631–644.

Corsini, M. and Rolland, Y. (2009). Late evolution of the southern European Variscan belt: Exhumation of the lower crust in a context of oblique convergence. *Comptes Rendus Geosci.*, 341, 214–223.

Couzinié, S., Laurent, O., Poujol, M., Mintrone, M., Chelle-Michou, C., Moyen, J.-F., Bouilhol, P., Vezinet, A., Marko, L. (2017). Cadomian S-type granites as basement rocks of the Variscan belt (Massif Central, France): Implications for the crustal evolution of the north Gondwana margin. *Lithos*, 286–287, 16–34.

Couzinié, S., Laurent, O., Chelle-Michou, C., Bouilhol, P., Paquette, J.-L., Gannoun, A.-M., Moyen, J.-F. (2019). Detrital zircon U–Pb–Hf systematics of Ediacaran metasediments from the French Massif Central: Consequences for the crustal evolution of the north Gondwana margin. *Precambrian Res.*, 324, 269–284.

Crowley, Q.G., Floyd, P.A., Winchester, J.A., Franke, W., Holland, J.G. (2000). Early Palaeozoic rift-related magmatism in Variscan Europe: Fragmentation of the Armorican Terrane Assemblage. *Terra Nova*, 12, 171–180.

Debon, F. and Lemmet, M. (1999). Evolution of Mg/Fe ratios in Late Variscan plutonic rocks from the external crystalline massifs of the Alps (France, Italy, Switzerland). *J. Petrol.*, 40, 1151–1185.

Debon, F., Guerrot, C., Ménot, R.-P. (1998). Late Variscan granites of the Belledonne massif (French Western Alps): An Early Visean magnesian plutonism. *Swiss J. Geosci.*, 78, 67–85.

Dörr, W., Zulauf, G., Gerdes, A., Lahaye, Y., Kowalczyk, G. (2015). A hidden Tonian basement in the eastern Mediterranean: Age constraints from U–Pb data of magmatic and detrital zircons of the External Hellenides (Crete and Peloponnesus). *Precambrian Res.*, 258, 83–108.

Eichhorn, R., Schärer, U., Höll, R. (1995). Age and evolution of scheelite-hosting rocks in the Felbertal deposit (Eastern Alps): U-Pb geochronology of zircon and titanite. *Contrib. Mineral. Petrol.*, 119, 377–386.

Eichhorn, R., Loth, G., Kennedy, A. (2001). Unravelling the pre-Variscan evolution of the Habach terrane (Tauern Window, Austria) by U-Pb SHRIMP zircon data. *Contrib. Mineral. Petrol.*, 142, 147–162.

Faryad, S.W., Melcher, F., Hoinkes, G., Puhl, J., Meisel, T., Frank, W. (2002). Relics of eclogite facies metamorphism in the Austroalpine basement, Hochgrössen (Speik complex). *Austria. Mineral. Petrol.*, 74, 49–73.

Faure, M. (2014). Le substratum de la France métropolitaine : de la formation du Gondwana à la constitution de la Pangée, une histoire de 600 Ma. *Géologues*, 180, 13–21.

Faure, M. and Becq-Giraudon, J.-F. (1993). Sur la succession des épisodes extensifs au cours du désépaississement carbonifère du Massif central français. *Comptes Rendus Académie Sci. Sér. 2 Mécanique Phys. Chim. Sci. Univers Sci. Terre*, 316, 967–973.

Faure, M., Bé Mézème, E., Duguet, M., Cartier, C., Talbot, J.-Y. (2005). Paleozoic tectonic evolution of medio-Europa from the example of the French Massif Central and Massif Armoricain. *J. Virtual Explor.*, 19, 1–25. DOI: 10.3809/jvirtex.2005.00120.

Faure, M., Lardeaux, J.-M., Ledru, P. (2009). A review of the pre-Permian geology of the Variscan French Massif Central. *Comptes Rendus Geosci.*, 341, 202–213.

Fernández, C., Becchio, R., Castro, A., Viramonte, J.M., Moreno-Ventas, I., Corretgé, L.G. (2008). Massive generation of atypical ferrosilicic magmas along the Gondwana active margin: Implications for cold plumes and back-arc magma generation. *Gondwana Res.*, 14, 451–473.

Ferrando, S., Lombardo, B., Compagnoni, R. (2008). Metamorphic history of HP mafic granulites from the Gesso-Stura terrain (Argentera Massif, western Alps, Italy). *Eur. J. Mineral.*, 20, 777–790.

Finger, F., Roberts, M.P., Haunschmid, B., Schermaier, A., Steyrer, H.-P. (1997). Variscan granitoids of central Europe: Their typology, potential sources and tectonothermal relations. *Mineral. Petrol.*, 61, 67–96.

Finger, F., Gerdes, A., Janoušek, V., René, M., Riegler, G. (2007). Resolving the Variscan evolution of the Moldanubian sector of the Bohemian Massif: The significance of the Bavarian and the Moravo-Moldanubian tectonometamorphic phases. *J. Geosci.*, 52, 9–28. DOI: 10.3190/jgeosci.005.

Fortey, R.A. and Cocks, L.R.M. (2003). Palaeontological evidence bearing on global Ordovician–Silurian continental reconstructions. *Earth-Sci. Rev.*, 61, 245–307. DOI: 10.1016/S0012-8252(02)00115-0.

Franke, W. (1989). Variscan plate tectonics in Central Europe – Current ideas and open questions. *Tectonophysics*, 169, 221–228.

Franke, W., Cocks, L.R.M., Torsvik, T.H. (2017). The Palaeozoic Variscan oceans revisited. *Gondwana Res.*, 48, 257–284.

Fréville, K. (2016). L'orogenèse varisque dans les massifs cristallins externes de Belledonne et du Pelvoux (Alpes occidentales françaises) : rôle de la fusion partielle et du plutonisme dans la structuration de la croûte continentale. PhD Thesis, Université d'Orléans.

Fréville, K., Trap, P., Faure, M., Melleton, J., Li, X.-H., Lin, W., Blein, O., Bruguier, O., Poujol, M. (2018). Structural, metamorphic and geochronological insights on the Variscan evolution of the Alpine basement in the Belledonne Massif (France). *Tectonophysics*, 726, 14–42.

Frey, M., Desmons, J., Neubauer, F. (1999). The new metamorphic maps of the Alps: Introduction. *Schweiz. Mineral. Petrogr. Mitteilungen.*, 79, 1–4.

Gaggero, L., Oggiano, G., Funedda, A., Buzzi, L. (2012). Rifting and arc-related early paleozoic volcanism along the North Gondwana margin: Geochemical and geological evidence from Sardinia (Italy). *J. Geol.*, 120, 273–292.

Galli, A., Le Bayon, B., Schmidt, M.W., Burg, J.-P., Reusser, É., Sergeev, S.A., Larionov, A. (2012). U–Pb zircon dating of the Gruf Complex: Disclosing the late Variscan granulitic lower crust of Europe stranded in the Central Alps. *Contrib. Mineral. Petrol.*, 163, 353–378.

Garfunkel, Z. (2015). The relations between Gondwana and the adjacent peripheral Cadomian domain – Constraints on the origin, history, and paleogeography of the peripheral domain. *Gondwana Res.*, 28, 1257–1281.

GBA (n.d.). Multithematische geologische Karte von Österreich [Online]. Available at: https://www.geologie.ac.at/services/webapplikationen/multithematische-geologische-karte/.

Giuntoli, F., Lanari, P., Engi, M. (2018). Deeply subducted continental fragments – Part 1: Fracturing, dissolution–precipitation, and diffusion processes recorded by garnet textures of the central Sesia Zone (western Italian Alps). *Solid Earth*, 9, 167–189.

Godard, G., Martin, S., Prosser, G., Kienast, J.R., Morten, L. (1996). Variscan migmatites, eclogites and garnet-peridotites of the Ulten zone, Eastern Austroalpine system. *Tectonophysics*, 259, 313–341.

Grandjean, V., Guillot, S., Pecher, A. (1996). A new record of the LP-HT late-Variscan metamorphism: The Peyre-Arguet unit (Haut-Dauphine). *Comptes Rendus Académie Sci.*, 322, 189–195.

Guillot, S. and Ménot, R.-P. (2009). Paleozoic evolution of the External Crystalline Massifs of the western Alps. *Comptes Rendus Geosci.*, 341, 253–265.

Guillot, S., Ménot, R.-P., Lardeaux, J.-M. (1992). Tectonique intra-océanique distensive dans l'ophiolite paléozoïque de Chamrousse (Alpes occidentales). *Bull. Soc. Géol. Fr.*, 163, 229–240.

Guillot, F., Schaltegger, U., Bertrand, J.-M., Deloule, É., Baudin, T. (2002). Zircon U–Pb geochronology of Ordovician magmatism in the polycyclic Ruitor Massif (Internal W Alps). *Int. J. Earth Sci.*, 91, 964–978.

Guillot, S., Mahéo, G., de Sigoyer, J., Hattori, K.H., Pêcher, A. (2008). Tethyan and Indian subduction viewed from the Himalayan high- to ultra-high-pressure metamorphic rocks. *Tectonophysics*, 451, 225–241.

Guillot, S., di Paola, S., Ménot, R.-P., Ledru, P., Spalla, M.I., Gosso, G., Schwartz, S. (2009). Suture zones and importance of strike-slip faulting for Variscan geodynamic reconstructions of the External Crystalline Massifs of the western Alps. *Bull. Soc. Géol. Fr.*, 180, 483–500.

Guilmette, C., Indares, A., Hébert, R. (2011). High-pressure anatectic paragneisses from the Namche Barwa, Eastern Himalayan Syntaxis: Textural evidence for partial melting, phase equilibria modeling and tectonic implications. *Lithos*, 124, 66–81.

Haas, I., Eichinger, S., Haller, D., Fritz, H., Nievoll, J., Mandl, M., Hippler, D., Hauzenberger, C. (2020). Gondwana fragments in the Eastern Alps: A travel story from U/Pb zircon data. *Gondwana Res.*, 77, 204–222.

Hauzenberger, C.A., Höller, W., Hoinkes, G. (1996). Transition from eclogite to amphibolite-facies metamorphism in the Austroalpine Ulten Zone. *Mineral. Petrol.*, 58, 111–130.

Herwartz, D., Nagel, T.J., Münker, C., Scherer, E.E., Froitzheim, N. (2011). Tracing two orogenic cycles in one eclogite sample by Lu–Hf garnet chronometry. *Nat. Geosci.*, 4, 178–183.

van Hinsbergen, D.J.J., Torsvik, T.H., Schmid, S.M., Maţenco, L.C., Maffione, M., Vissers, R.L.M., Gürer, D., Spakman, W. (2019). Orogenic architecture of the Mediterranean region and kinematic reconstruction of its tectonic evolution since the Triassic. *Gondwana Res.*, 81, 79–229.

Jacob, J.-B., Guillot, S., Rubatto, D., Janots, E., Melleton, J., Faure, M. (2021). Carboniferous high pressure metamorphism and deformation in the Belledonne Massif (Western Alps). *J. Metamorph. Geol.*, 39, 1009–1044.

Jacob, J.-B., Janots, E., Guillot, S., Rubatto, D., Fréville, K., Melleton, J., Faure, M. (2022). HT overprint of HP granulites in the Oisans–Pelvoux massif: Implications for the dynamics of the Variscan collision in the external western Alps. *Lithos*, 106650, 416–417.

Jouffray, F., Spalla, M.I., Lardeaux, J.M., Filippi, M., Rebay, G., Corsini, M., Zanoni, D., Zucali, M., Gosso, G. (2020). Variscan eclogites from the Argentera–Mercantour Massif (External Crystalline Massifs, SW Alps): A dismembered cryptic suture zone. *Int. J. Earth Sci.*, 109, 1273–1294.

Kali, E., Leloup, P.H., Arnaud, N., Mahéo, G., Liu, D., Boutonnet, E., Van der Woerd, J., Liu, X., Liu-Zeng, J., Li, H. (2010). Exhumation history of the deepest central Himalayan rocks, Ama Drime range: Key pressure-temperature-deformation-time constraints on orogenic models. *Tectonics*, 29. DOI: 10.1029/2009TC002551.

Kroner, U. and Romer, R.L. (2013). Two plates – Many subduction zones: The Variscan orogeny reconsidered. *Gondwana Res.*, 24, 298–329.

Kunz, B.E., Manzotti, P., von Niederhäusern, B., Engi, M., Darling, J.R., Giuntoli, F., Lanari, P. (2018). Permian high-temperature metamorphism in the Western Alps (NW Italy). *Int. J. Earth Sci.*, 107, 203–229.

Ladenhauf, C., Armstrong, R., Konzett, J., Miller, C. (2001). The timing of pre-Alpine high pressure metamorphism in the Eastern Alps: Constraints from U–Pb SHRIMP dating of eclogite zircons from the Austroalpine Silvretta nappe. In *Journal of Conference Abstracts*. 6, 600.

Lammerer, B. (1986). Das autochthon im westlichen tauernfenster. *Jahrb. Geol. Bundesanst.*, 129, 51–67.

Lanari, P., Riel, N., Guillot, S., Vidal, O., Schwartz, S., Pêcher, A., Hattori, K.H. (2013). Deciphering high-pressure metamorphism in collisional context using microprobe mapping methods: Application to the Stak eclogitic massif (northwest Himalaya). *Geology*, 41, 111–114.

Lardeaux, J.M., Schulmann, K., Faure, M., Janoušek, V., Lexa, O., Skrzypek, E., Edel, J.B., Štípská, P. (2014). The moldanubian zone in the French Massif Central, Vosges/Schwarzwald and Bohemian Massif revisited: Differences and similarities. *Geol. Soc. Lond. Spec. Publ.*, 405, 7–44.

Le Fort, P. (1971). Géologie du Haut-Dauphiné cristallin (Alpes Françaises) : études pétrologique et structurale de la partie occidentale. PhD Thesis, Université Nancy I.

Ledru, P., Courrioux, G., Dallain, C., Lardeaux, J.M., Montel, J.M., Vanderhaeghe, O., Vitel, G. (2001). The Velay dome (French Massif Central): Melt generation and granite emplacement during orogenic evolution. *Tectonophysics*, 342, 207–237.

Liati, A., Gebauer, D., Fanning, C.M. (2009). Geochronological evolution of HP metamorphic rocks of the Adula nappe, Central Alps, in pre-Alpine and Alpine subduction cycles. *J. Geol. Soc.*, 166, 797–810.

Liégeois, J.-P. and Duchesne, J.-C. (1981). The Lac Cornu retrograded eclogites (Aiguilles Rouges massif, Western Alps, France): Evidence of crustal origin and metasomatic alteration. *Lithos*, 14, 35–48.

Linnemann, U., Gerdes, A., Drost, K., Buschmann, B. (2007). The continuum between Cadomian orogenesis and opening of the Rheic Ocean: Constraints from LA-ICP-MS U-Pb zircon dating and analysis of plate-tectonic setting (Saxo-Thuringian zone, northeastern Bohemian Massif, Germany). *Spec. Pap.-Geol. Soc. Am.*, 423, 61.

Maierová, P., Schulmann, K., Lexa, O., Guillot, S., Štípská, P., Janoušek, V., Čadek, O. (2016). European Variscan orogenic evolution as an analogue of Tibetan-Himalayan orogen: Insights from petrology and numerical modeling: Variscan and Himalayan Orogenies. *Tectonics*, 35, 1760–1780.

Mandl, M., Kurz, W., Hauzenberger, C., Fritz, H., Klötzli, U., Schuster, R. (2018). Pre-Alpine evolution of the Seckau Complex (Austroalpine basement/Eastern Alps): Constraints from in-situ LA-ICP-MS UPb zircon geochronology. *Lithos*, 296–299, 412–430.

Manzotti, P., Ballèvre, M., Poujol, M. (2016). Detrital zircon geochronology in the Dora-Maira and Zone Houillère: A record of sediment travel paths in the Carboniferous. *Terra Nova*, 28, 279–288.

Manzotti, P., Rubatto, D., Zucali, M., Korh, A.E., Cenki-Tok, B., Ballèvre, M., Engi, M. (2018). Permian magmatism and metamorphism in the Dent Blanche nappe: Constraints from field observations and geochronology. *Swiss J. Geosci.*, 111, 79–97.

Matte, P. (1986). Tectonics and plate tectonics model for the Variscan belt of Europe. *Tectonophysics*, 126, 329–374.

Matte, P. (2001). The Variscan collage and orogeny (480–290 Ma) and the tectonic definition of the Armorica microplate: A review. *Terra Nova*, 13, 122–128.

Melcher, F. and Meisel, T. (2004). A metamorphosed early Cambrian crust-mantle transition in the Eastern Alps. *Austria. J. Petrol.*, 45, 1689–1723.

Ménot, R.-P., Peucat, J.J., Scarenzi, D., Piboule, M. (1988). 496 My age of plagiogranites in the Chamrousse ophiolite complex (external crystalline massifs in the French Alps): Evidence of a Lower Paleozoic oceanization. *Earth Planet. Sci. Lett.*, 88, 82–92.

Miller, C. and Thöni, M. (1995). Origin of eclogites from the Austroalpine Ötztal basement (Tirol, Austria): Geochemistry and Sm-Nd vs. Rb-Sr isotope systematics. *Chem. Geol.*, 122, 199–225.

Montero, P., Bea, F., González-Lodeiro, F., Talavera, C., Whitehouse, M.J. (2007). Zircon ages of the metavolcanic rocks and metagranites of the Ollo de Sapo Domain in central Spain: Implications for the Neoproterozoic to Early Palaeozoic evolution of Iberia. *Geol. Mag.*, 144, 963–976.

Montes, A.D., Catalán, J.R.M., Mulas, F.B. (2010). Role of the Ollo de Sapo massive felsic volcanism of NW Iberia in the Early Ordovician dynamics of northern Gondwana. *Gondwana Res.*, 17, 363–376.

Müller, B., Klötzli, U.S., Flisch, M. (1995). U-Pb and Pb-Pb zircon dating of the older orthogneiss suite in the Silvretta nappe, eastern Alps: Cadomian magmatism in the upper Austro-Alpine realm. *Geol. Rundsch.*, 84, 457–465.

Müller, B., Schaltegger, U., Klötzli, U., Flisch, M. (1996). Early Cambrian oceanic plagiogranite in the Silvretta Nappe, eastern Alps: Geochemical, zircon U-Pb and Rb-Sr data from garnet-hornblende-plagioclase gneisses. *Geol. Rundsch.*, 85, 822–831.

Nance, R.D., Gutiérrez-Alonso, G., Keppie, J.D., Linnemann, U., Murphy, J.B., Quesada, C., Strachan, R.A., Woodcock, N.H. (2010). Evolution of the Rheic Ocean. *Gondwana Res.*, 17, 194–222.

Neubauer, F. (1989). Pre-Alpine terranes and tectonic zoning in the eastern Alps. *Terranes Circum-Atl. Paleoz. Orogens.*, 230, 91.

Neubauer, F., Frisch, W., Schmerold, R., Schlöser, H. (1989). Metamorphosed and dismembered ophiolite suites in the basement units of the Eastern Alps. *Tectonophysics*, 164, 49–62.

Neubauer, F., Fried, G., Genser, J., Handler, R., Mader, D., Schneider, D. (2007). Origin and tectonic evolution of the Eastern Alps deduced from dating of detrital white mica: A review. *Austrian J. Earth Sci.*, 100, 8–23.

Nicollet, C., Bosse, V., Spalla, M.I., Schiavi, F. (2018). Eocene ultra-high temperature (UHT) metamorphism in the Gruf complex (Central Alps): Constraints by LA-ICPMS zircon and monazite dating in petrographic context. *J. Geol. Soc.*, 175, 774–787.

O'Brien, P.J. (2000). The fundamental Variscan problem: High-temperature metamorphism at different depths and high-pressure metamorphism at different temperatures. *Geol. Soc. Lond. Spec. Publ.*, 179, 369–386.

Paquette, J.L., Menot, R.P., Peucat, J.J. (1989). REE, Sm-Nd and U-Pb zircon study of eclogites from the Alpine External Massifs (Western Alps): Evidence for crustal contamination. *Earth Planet. Sci. Lett.*, 96, 181–198.

Paquette, J.-L., Ballèvre, M., Peucat, J.-J., Cornen, G. (2017). From opening to subduction of an oceanic domain constrained by LA-ICP-MS U-Pb zircon dating (Variscan belt, Southern Armorican Massif, France). *Lithos*, 294–295, 418–437.

Pin, C. and Carme, F. (1987). A Sm-Nd isotopic study of 500 Ma old oceanic crust in the Variscan belt of Western Europe: The Chamrousse ophiolite complex, Western Alps (France). *Contrib. Mineral. Petrol.*, 96, 406–413.

Pin, C. and Marini, F. (1993). Early Ordovician continental break-up in Variscan Europe: Nd-Sr isotope and trace element evidence from bimodal igneous associations of the Southern Massif Central, France. *Lithos*, 29, 177–196.

Regorda, A., Lardeaux, J.-M., Roda, M., Marotta, A.M., Spalla, M.I. (2019). How many subductions in the Variscan orogeny? Insights from numerical models. *Geosci. Front.*, 11, 1025–1052.

Rubatto, D., Schaltegger, U., Lombardo, D., Colombo, F., Compagnoni, R. (2001). Complex Paleozoic magmatic and metamorphic evolution in the Argentera massif (Western Alps), resolved with U-Pb dating. *Schweiz. Mineral. Petrogr. Mitteilungen.*, 81, 213–228.

Rubatto, D., Ferrando, S., Compagnoni, R., Lombardo, B. (2010). Carboniferous high-pressure metamorphism of Ordovician protoliths in the Argentera Massif (Italy), Southern European Variscan belt. *Lithos*, 116, 65–76.

Sandmann, S., Nagel, T.J., Herwartz, D., Fonseca, R.O.C., Kurzawski, R.M., Münker, C., Froitzheim, N. (2014). Lu-Hf garnet systematics of a polymetamorphic basement unit: New evidence for coherent exhumation of the Adula Nappe (Central Alps) from eclogite-facies conditions. *Contrib. Mineral. Petrol.*, 168, 1075.

Schäfer, J., Neuroth, H., Ahrendt, H., Dörr, W., Franke, W. (1997). Accretion and exhumation at a Variscan active margin, recorded in the Saxothuringian flysch. *Geol. Rundsch.*, 86, 599–611.

Schaltegger, U. (1997). Magma pulses in the Central Variscan Belt: Episodic melt generation and emplacement during lithospheric thinning. *Terra Nova*, 9, 242–245.

Schaltegger, U. and Gebauer, D. (1999). Pre-Alpine geochronology of the central, western and southern Alps. *Schweiz. Mineral. Petrogr. Mitteilungen.*, 79, 79–87.

Schaltegger, U., Nägler, T., Corfu, F., Maggetti, M., Galetti, G., Stosch, H.G. (1997). A Cambrian island arc in the Silvretta nappe: Constraints from geochemistry and geochronology. *Schweiz. Mineral. Petrogr. Mitteilungen.*, 77, 337–350.

Schaltegger, U., Gebauer, D., von Quadt, A. (2002). The mafic-ultramafic rock association of Loderio–Biasca (lower Pennine nappes, Ticino, Switzerland): Cambrian oceanic magmatism and its bearing on early Paleozoic paleogeography. *Chem. Geol.*, 186, 265–279.

Schaltegger, U., Abrecht, J., Corfu, F. (2003). The Ordovician orogeny in the Alpine basement: Constraints from geochronology and geochemistry in the Aar Massif (Central Alps). *Swiss Bull. Mineral. Petrol.*, 83, 183–239.

Schantl, P., Hauzenberger, C., Finger, F., Müller, T., Linner, M. (2019). New evidence for the prograde and retrograde PT-path of high-pressure granulites, Moldanubian Zone, Lower Austria, by Zr-in-rutile thermometry and garnet diffusion modelling. *Lithos*, 342–343, 420–439.

Schmid, S.M., Fügenschuh, B., Kissling, E., Schuster, R. (2004). Tectonic map and overall architecture of the Alpine orogen. *Eclogae Geol. Helvetiae.*, 97, 93–117.

Schmid, S.M., Scharf, A., Handy, M.R., Rosenberg, C.L. (2013). The Tauern Window (Eastern Alps, Austria): A new tectonic map, with cross-sections and a tectonometamorphic synthesis. *Swiss J. Geosci.*, 106, 1–32.

Schulmann, K., Kröner, A., Hegner, E., Wendt, I., Konopásek, J., Lexa, O., Štípská, P. (2005). Chronological constraints on the pre-orogenic history, burial and exhumation of deep-seated rocks along the eastern margin of the Variscan Orogen, Bohemian Massif, Czech Republic. *Am. J. Sci.*, 305, 407–448.

Schulmann, K., Catalán, J.R.M., Lardeaux, J.M., Janoušek, V., Oggiano, G. (2014). The Variscan orogeny: Extent, timescale and the formation of the European crust. *Geol. Soc. Lond. Spec. Publ.*, 405, 1–6.

Schulz, B., Bombach, K., Pawlig, S., Brätz, H. (2004). Neoproterozoic to Early-Palaeozoic magmatic evolution in the Gondwana-derived Austroalpine basement to the south of the Tauern Window (Eastern Alps). *Int. J. Earth Sci.*, 93, 824–843.

Schulz, B., Steenken, A., Siegesmund, S. (2008). Geodynamic evolution of an Alpine terrane – The Austroalpine basement to the south of the Tauern window as a part of the Adriatic Plate (eastern Alps). *Geol. Soc. Lond. Spec. Publ.*, 298, 5–44.

Simonetti, M., Carosi, R., Montomoli, C., Langone, A., D'Addario, E., Mammoliti, E. (2018). Kinematic and geochronological constraints on shear deformation in the Ferriere-Mollières shear zone (Argentera-Mercantour Massif, Western Alps): Implications for the evolution of the Southern European Variscan Belt. *Int. J. Earth Sci.*, 107, 2163–2189.

Soejono, I., Machek, M., Sláma, J., Janoušek, V., Kohút, M. (2019). Cambro-Ordovician anatexis and magmatic recycling at the thinned Gondwana margin: New constraints from the Kouřim Unit, Bohemian Massif. *J. Geol. Soc.*, 177, 325–341.

Spalla, M.I., Zanoni, D., Marotta, A.M., Rebay, G., Roda, M., Zucali, M., Gosso, G. (2014). The transition from Variscan collision to continental break-up in the Alps: Insights from the comparison between natural data and numerical model predictions. *Geol. Soc. Lond. Spec. Publ.*, 405, 363–400.

Stampfli, G.M. and Hochard, C. (2009). Plate tectonics of the Alpine realm. *Geol. Soc. Lond. Spec. Publ.*, 327, 89–111.

Stampfli, G.M., Von Raumer, J., Wilhem, C. (2011). The distribution of Gondwana-derived terranes in the Early Palaeozoic. In *Ordovician of the World Conference*. Instituto Geológico y Minero de España, Madrid.

Stampfli, G.M., Hochard, C., Vérard, C., Wilhem, C. (2013). The formation of Pangea. *Tectonophysics*, 593, 1–19.

Stephan, T., Kroner, U., Romer, R.L. (2019a). The pre-orogenic detrital zircon record of the Peri-Gondwanan crust. *Geol. Mag.*, 156, 281–307.

Stephan, T., Kroner, U., Romer, R.L., Rösel, D. (2019b). From a bipartite Gondwanan shelf to an arcuate Variscan belt: The early Paleozoic evolution of northern Peri-Gondwana. *Earth-Sci. Rev.*, 192, 491–512.

Thöni, M. (2006). Dating eclogite-facies metamorphism in the Eastern Alps – Approaches, results, interpretations: A review. *Mineral. Petrol.*, 88, 123–148.

Tumiati, S., Thöni, M., Nimis, P., Martin, S., Mair, V. (2003). Mantle–crust interactions during Variscan subduction in the Eastern Alps (Nonsberg–Ulten zone): Geochronology and new petrological constraints. *Earth Planet. Sci. Lett.*, 210, 509–526.

Valverde-Vaquero, P. and Dunning, G.R. (2000). New U–Pb ages for Early Ordovician magmatism in Central Spain. *J. Geol. Soc.*, 157, 15–26.

Vanardois, J., Roger, F., Trap, P., Goncalves, P., Lanari, P., Paquette, J.-L., Marquer, D., Cagnard, F., Le Bayon, B., Melleton, J. (2022). Exhumation of deep continental crust in a transpressive regime: The example of Variscan eclogites from the Aiguilles-Rouges Massif (Western Alps). *Journal of Metamorphic Geology*, 40(6), 1087–1120.

Veselà, P., Söllner, F., Finger, F., Gerdes, A. (2011). Magmato-sedimentary Carboniferous to Jurassic evolution of the western Tauern Window, Eastern Alps (constraints from U-Pb zircon dating and geochemistry). *Int. J. Earth Sci.*, 100, 993–1027.

Villaseca, C., Merino Martínez, E., Orejana, D., Andersen, T., Belousova, E. (2016). Zircon Hf signatures from granitic orthogneisses of the Spanish Central System: Significance and sources of the Cambro-Ordovician magmatism in the Iberian Variscan Belt. *Gondwana Res.*, 34, 60–83.

Von Raumer, J.F. and Bussy, F. (2004). Mont Blanc and Aiguilles Rouges geology of their polymetamorphic basement (external massifs, Westerns Alps, France-Switzerland). In *Mémoires de Géologie*. Lausanne.

Von Raumer, J.F. and Stampfli, G.M. (2008). The birth of the Rheic Ocean–Early Palaeozoic subsidence patterns and subsequent tectonic plate scenarios. *Tectonophysics*, 461, 9–20.

Von Raumer, J.F., Stampfli, G., Borel, G., Bussy, F. (2002). Organization of pre-Variscan basement areas at the north-Gondwanan margin. *Int. J. Earth Sci.*, 91, 35–52.

Von Raumer, J.F., Bussy, F., Stampfli, G.M. (2009). The Variscan evolution in the external massifs of the Alps and place in their Variscan framework. *Comptes Rendus Geosci.*, 341, 239–252.

Von Raumer, J.F., Bussy, F., Schaltegger, U., Schulz, B., Stampfli, G.M. (2013). Pre-Mesozoic Alpine basements – Their place in the European Paleozoic framework. *Bulletin*, 125, 89–108.

Von Raumer, J.F., Finger, F., Veselá, P., Stampfli, G. (2014). Durbachites–Vaugnerites – A geodynamic marker in the central European Variscan orogen. *Terra Nova*, 26, 85–95.

Von Raumer, J.F., Stampfli, G.M., Arenas, R., Martínez, S.S. (2015). Ediacaran to Cambrian oceanic rocks of the Gondwana margin and their tectonic interpretation. *Int. J. Earth Sci.*, 104, 1107–1121.

Whitney, D.L., Roger, F., Teyssier, C., Rey, P.F., Respaut, J.-P. (2015). Syn-collapse eclogite metamorphism and exhumation of deep crust in a migmatite dome: The P–T–t record of the youngest Variscan eclogite (Montagne Noire, French Massif Central). *Earth Planet. Sci. Lett.*, 430, 224–234.

Whitney, D.L., Hamelin, C., Teyssier, C., Raia, N.H., Korchinski, M.S., Seaton, N.C.A., Bagley, B.C., Von der Handt, A., Roger, F., Rey, P.F. (2020). Deep crustal source of gneiss dome revealed by eclogite in migmatite (Montagne Noire, French Massif Central). *J. Metamorph. Geol.*, 38, 297–327.

Wilhem, C., Windley, B.F., Stampfli, G.M. (2012). The Altaids of Central Asia: A tectonic and evolutionary innovative review. *Earth-Sci. Rev.*, 113, 303–341.

Žák, J., Svojtka, M., Opluštil, S. (2018). Topographic inversion and changes in the sediment routing systems in the Variscan orogenic belt as revealed by detrital zircon and monazite U-Pb geochronology in post-collisional continental basins. *Sediment. Geol.*, 377, 63–81.

Zurbriggen, R. (2017). The Cenerian orogeny (early Paleozoic) from the perspective of the Alpine region. *International Journal of Earth Sciences*, 106(2), 517–529.

2

Paleogeography and Architecture of the Alpine Tethys Margins

Yves LAGABRIELLE

CNRS, Géosciences Rennes, Université de Rennes 1, France

2.1. Introduction: conjugate passive margins

In the domain of the future Western European Alps (Figures 2.1 and 2.2), the initiation of ocean floor spreading (oceanization) occurred during the Late Jurassic after a long period of rifting during which the continental lithosphere was stretched and fractured over considerable areas. As illustrated in this chapter, this large rifting episode started at the very beginning of the Liassic times with the dislocation of a carbonate marine platform built during the Late Triassic under a tropical climate. Continental extension was active from the Lias (locally as early as the Late Triassic) to the end of the Late Jurassic (Malm), i.e. from 200 to 145 Ma, that is over about 55 Ma. Rifting resulted in the formation of two passive margins that slowly moved away from each other, framing the oceanic domain of the Alpine Tethys, or Piemont-Liguria Ocean, whose main characteristics are discussed in Chapter 4 of this book and a future volume devoted to ophiolites (Lagabrielle and Ceuleneer, in progress).

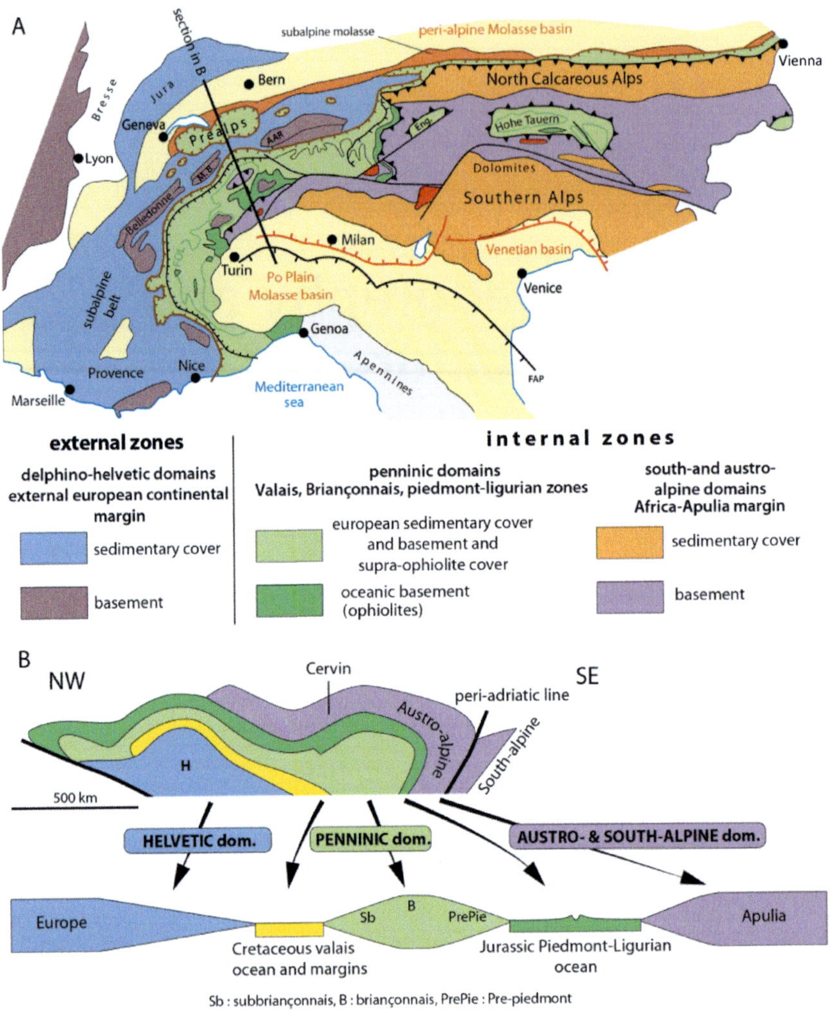

Figure 2.1. *Simplified structural map of the European Alps and definition of the major continental and oceanic domains involved in the Alpine orogeny (from Lemoine et al. 2000)*

To the west and to the north of the Piemont-Liguria Ocean was the margin of the European plate, comprising the future Delphino-Helvetic, Valais, Briançonnais, and Pre-Piedmont domains of the Western Alps. To the east and to the south was the margin of the small Adria plate comprising the future so-called Austroalpine and Southalpine domains, outcropping mainly in the Eastern and Southern Alps in the territories of northern Italy, eastern Switzerland, Liechtenstein and Austria. The aim of this chapter is to outline the stratigraphy and paleogeography of these margins and then to discuss their architecture in the light of our recent knowledge of the present-day passive continental margins, mainly within the Atlantic Ocean realm.

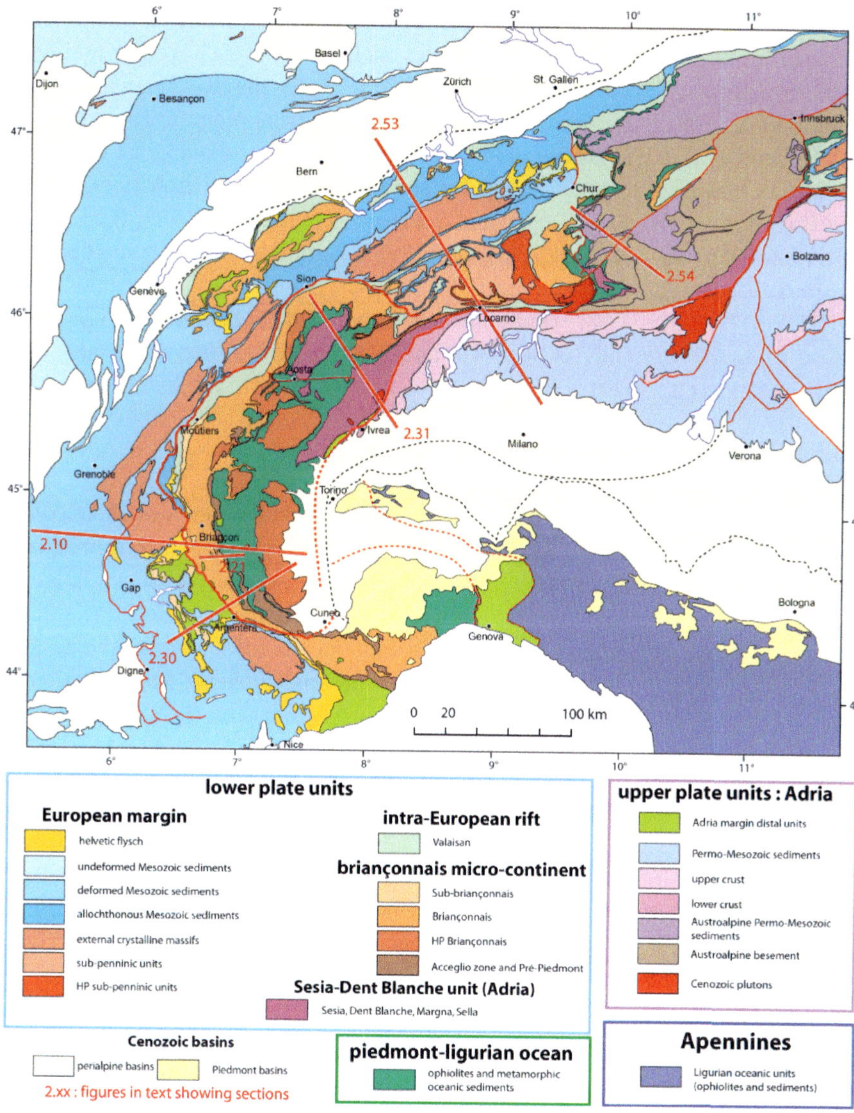

Figure 2.2. *Map of the major western and central Alpine units declined according to their position in the current orogenic prism (modified from Schmid et al. 2017)*

2.2. The contours and the internal division of the Adria plate

Since the end of the 1970s, several European medium-scale research projects, as well as large national or international programs, have focused on the kinematic reconstructions of the Alpine and Mediterranean regions. Numerous individual or collective works have attempted to describe the paleogeography of the regions concerned by the Alpine collision. Examples include the French Tethys and peri-Tethys programs (Dercourt et al. 1986, 1993, 2000), the IGCP 369 program (Stampfli et al. 2001a; Ziegler et al. 2001), and the Darius program (Sosson et al. 2010; Rolland et al. 2012; Hassig et al. 2015). Also worth mentioning are the works of authors or groups of authors (Laubscher and Bernoulli 1977; Frisch 1979; Sengör 1979; Bassoulet et al. 1980; Sengör and Yilmaz 1981; Lemoine 1985; Ricou et al. 1985; 1986; Dercourt et al. 1986; Marcoux 1987; Andrieux et al. 1989; Robertson et al. 1991; Ricou 1988; 1994; Olivet 1996; Dercourt et al. 2000; Ziegler et al. 2001; Ziegler and Stampfli 2001; Stampli and Borel 2002; Schmid et al. 2004; Cavazza et al. 2004; Handy et al. 2010; and more recently, Angrand et al. 2020; Van Hinsbergen et al. 2020).

Many efforts have been devoted to defining the paleo-contours of the Adria plate boundaries. The complexity of the structural pattern in which the units of the Adria lithosphere are involved today allows for numerous and sometimes inconsistent interpretations. Unfortunately, without reliable points of reference such as magnetic anomalies within possible oceans preserved from the Alpine collision[1], it is not possible to accurately reconstruct the outlines of this plate. Despite these difficulties, many authors have ventured to propose kinematic reconstructions that give an idea of how Adria may have behaved. We will see that a pre-orogenic fragmentation of Adria into several subplates is often used.

2.2.1. *Situation in the Triassic (250–200 Ma)*

In all published kinematic reconstructions, the eastern edge of the Adria plate existed since the Paleozoic. During the Carboniferous and the Permian, this border was generally an active margin under which the pre-Mesozoic lithosphere of the Paleotethys was subducting (Figure 2.3). From 260 Ma to the end of the Permian,

1. It is important to remember that only oceanic magnetic anomalies and fracture zone lines can be used to accurately reconstruct the relative kinematics of two plates separated by a divergent oceanic boundary. This is the case for the European and North American plates, separated by the North Atlantic Ridge. However, in the case of the Eurasia and Adria plates, the Piemont-Liguria Ocean separating them – and the anomalies it carried – were entirely swallowed up during the Alpine collision.

the Neotethys started opening in the extreme north of the African plate and spread westward. It progressively cut a long and narrow continental belt whose contours roughly followed the edge of the northern margin of the present African plate. To the east, this continental strip included blocks that formed the core of the Iranian, Taurid and Hellenid chains. As this Neotethyan oceanic opening progressed, the first blocks moved northward while the Paleotethys were closing in front of them. The westward progression of the Neotethys continued until the beginning of the Upper Triassic. It then reached the south of the current Italian boot where it bifurcated toward the north. The small Adria plate thus began to break away from Africa. For many authors, the propagation of the Neotethys stopped there, so that Adria never completely detached from Africa. Some reconstructions, however, suggest an established Africa-Adria separation (this question is discussed in section 2.5).

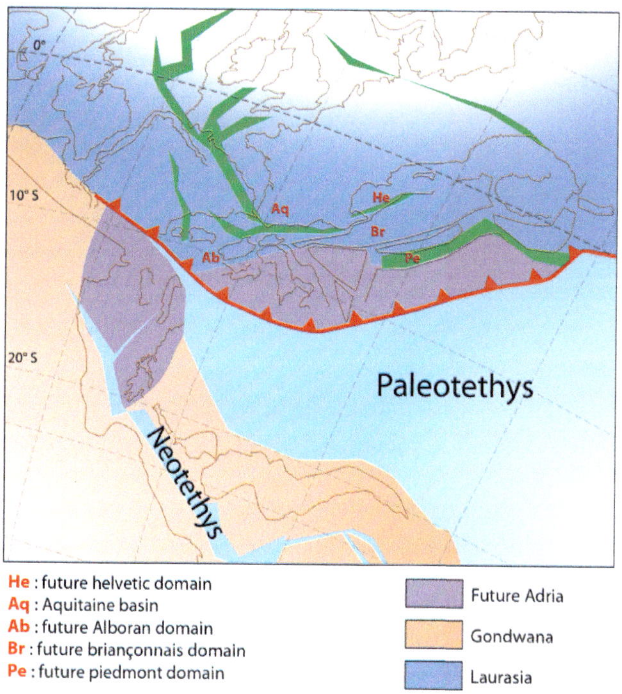

Figure 2.3. *In the Early Permian (290 Ma), the Paleotethys Ocean terminated in a region of Pangea that was fractured from the Triassic onward into a multitude of small blocks whose movements controlled the paleogeography of the Alpine regions. At this time, the Adria plate was not individualized, but the future Neotethys was visible in the south and was starting to open (according to Stampfli et al. 2001)*

Let us now detail the outline of the Adria plate. (1) In some interpretations, the Adria Plate was bounded to the north by the Triassic Meliata Ocean (Kosur 1991; Beltrando et al. 2015). For others, the Meliata Ocean was rather central defining a southern- and a northern-Meliata part (Schmid et al. 2004). (2) To the south, as seen above, it was incompletely (or completely?) separated from Africa by a Neotethyan branch. This branch persisted throughout the Alpine-Mediterranean evolution without completely closing, and is thought to form the bottom of the current Ionian Sea in the eastern Mediterranean. (3) In the west, during the Triassic, the Adria plate was still not separated from Europe, but the regions that correspond to the future Betic, Corsican, and Briançon margins were more or less concerned by continental fracturation in response to the Triassic rifting. (4) The eastern edge of the Adria is now represented by the tectonic units of the Pelagonian margin in Greece and of the Hallstatt margin in the Eastern Alps.

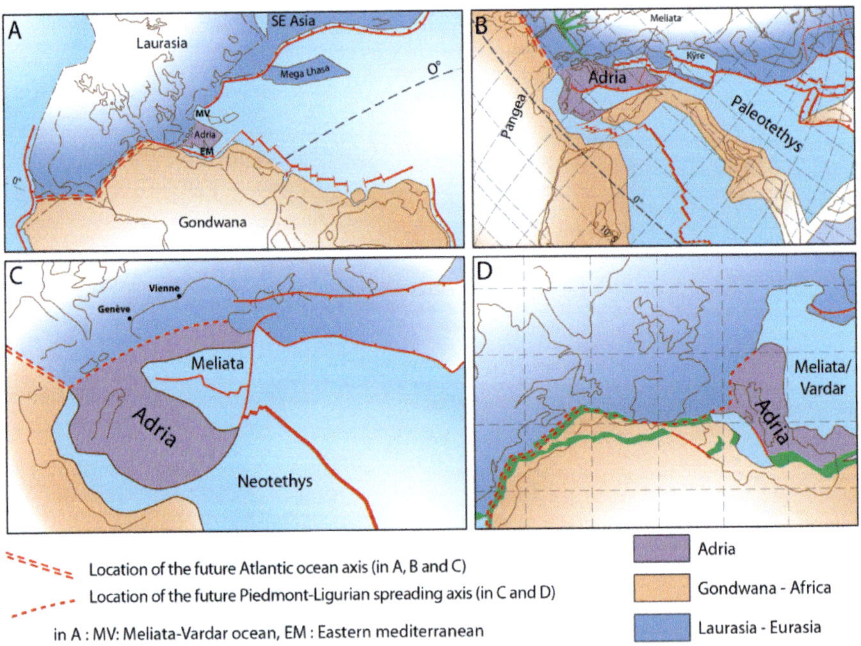

Figure 2.4. *Reconstructions of the Alpine domain and the Adria plate in the Triassic according to A) Ricou (1994), B) Stampfli and Borel (2001) (at 230 Ma, Late Ladinian), C) Schmid et al. (2004), and D) Beltrando et al. (2015). Note the different shapes given to the Adria Plate by each author. MV: Meliata-Vardar, EM: Eastern Mediterranean, Ap: Apennines, NA: North Adria, SA: South Adria*

At the end of the Triassic, the future Adria micro-plate was thus a more or less quadrangular-shaped continental promontory within the Neotethys Ocean. The contours of this overhang and in particular the outline of its eastern edge vary according to the numerous available reconstructions. For Ricou, the Adria Plate was bounded to the north by a single dependency of the Paleotethys, the Maliac-Vardar Ocean (MV in Figure 2.4(A)). According to Stampfli and Borel (2012) (Figure 2.4(B)), Paleotethys subduction continued beneath the eastern Adria edge, while narrow Triassic oceans, isolated in the heart of the continent, opened in the position of back-arc basins of this subduction. This is the case of the Meliata or Meliata-Hallstatt Ocean and its two margins are now closed within the Internal Carpathian chain. Oceanic accretion of the Meliata Ocean ended in the Carnian. Its subduction under Adria initiated in the Late Triassic and ended in the Oxfordian (Ivan 2002). For Schmid et al. (2004), the tip of the Meliata (Vardar) Ocean has a V-shape outline, closing to the west, so that in some reconstructions, the Adria Plate exhibits a semi-circular banana shape (Figures 2.4(C) and (D)).

Other reconstructions, not shown here[2], show triangular-shaped oceans (propagators) located east of the Adria plate, more or less connected to the Neotethys (Maliac-Vardar or Vardar-Pinde according to authors). These oceans propagated from east to west. They were partly overthrust (obducted) during the Upper Jurassic and finally incorporated into the Hellenides-Dinarides orogen. For Ivan (2002), the continuity of the Meliata-Hallstatt Ocean into the Vardar Ocean was not possible because the Vardar Ocean was of Jurassic age (see section 2.4.3).

From the end of the Triassic to the Late Jurassic, a new period of rifting affected the future Alpine regions; Adria then broke away from Europe due to the opening of a new oceanic domain called the “Piemont-Liguria Ocean” or the “Alpine Tethys”.

2.2.2. *Opening of the Piemont-Liguria Ocean (200–145 Ma)*

The opening of a new Jurassic Piemont-Liguria Ocean is the major element in the history of the Alps in Western Europe (de Graciansky and Lemoine 1988). Its relicts, in the form of ophiolites, are found in all the branches of the Alpine chain: in Corsica, in Liguria, in the Western Alps, in the Central Alps, and the Tauern window, in the heart of the Eastern Alps.

The most extreme remains of the Piemont-Liguria Ocean are found in the west in the Betic Cordilleras (small ophiolite bodies within the Mulhacen unit: Puga et al.

2. See the reconstructions for the Triassic proposed in Chapter 2 of the volume devoted to ophiolites (Lagabrielle and Ceuleneer, in progress).

2011) and in the east, at the front of the Eastern Alps (Ybbsitz ophiolite: Lemoine 2003). The presence of Jurassic ophiolites in Liguria, where they are remarkably well preserved, and their abundance in the Piedmont units of the French-Italian Alps have led authors to name this Jurassic ocean, the *Piedmont-Ligurian* or *Piedmont-Ligure Ocean*). The terms "Alpine Ocean", "Alpine Tethys", "Ligurian Tethys" or "Ligure-Piedmont Tethys" are also used.

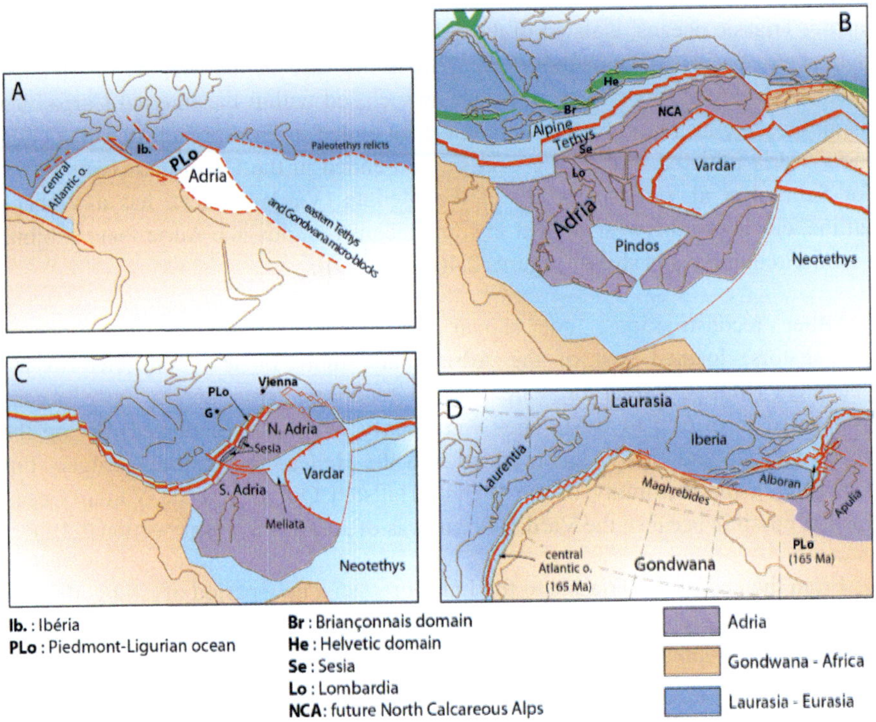

Figure 2.5. *The Piemont-Liguria Ocean: an Atlantic dependency. A) The relations between the Ligurian Tethys and the Central Atlantic during the Jurassic according to Lemoine (1983). In this very simple drawing, Adria remains a promontory of Africa. B) Paleogeography of the Alpine domain in the Late Oxfordian (155 Ma). The Adria plate was completely isolated due to the opening of the Alpine Tethys that formed its northwestern edge. In contrast to Figure A, the connection of the Ligurian Tethys with the central Atlantic was established without major transform faults (from Stampfli and Borel 2001). (C and D) Reconstructions by Schmid et al. (2004) and Cordey and Bailly (2007) showing two other ways of considering the Atlantic-Alpine Tethys connections*

In the case of the Piemont-Liguria Ocean, the continental tear and the oceanic opening did not propagate from the east, that is from the Neotethyan regions, but from the west, that is from the central Atlantic Ocean. The Piemont-Liguria Ocean was therefore not a branch of the Neotethys; it was mainly an Atlantic dependency. For this reason, it is recommended to name it without referring to the Neotethys.

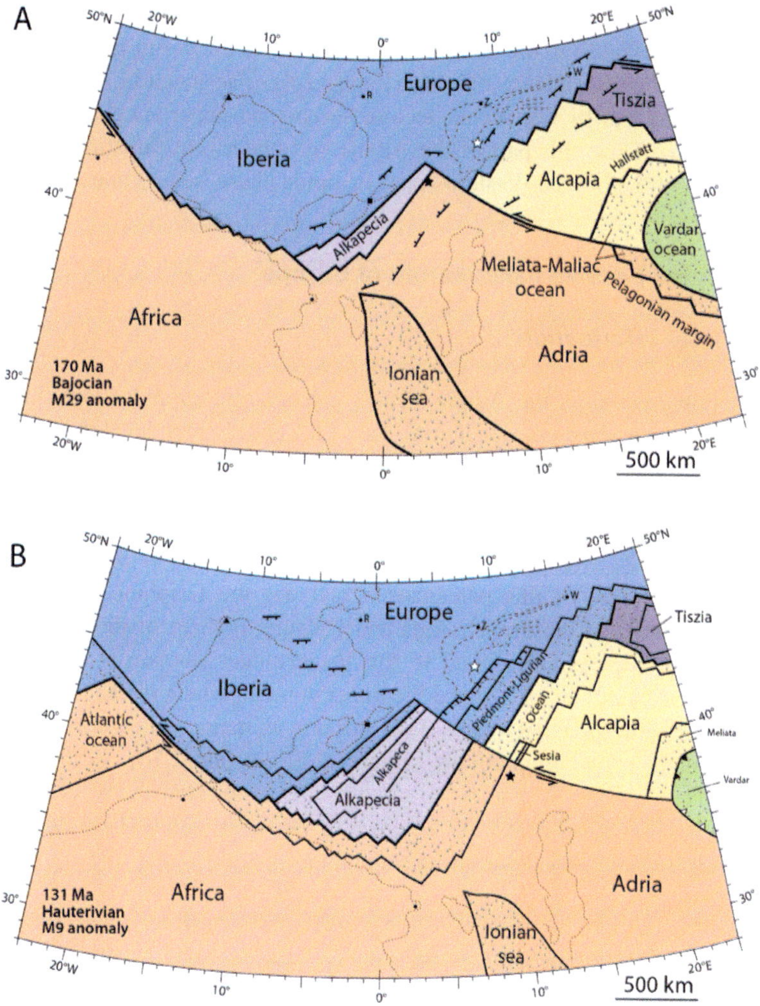

Figure 2.6. *Fragmentation of the Adria plate into three subplates at the beginning of the opening of the Piemont-Liguria Ocean (modified from Handy et al. 2010). Situations at 170 Ma (A) and 130 Ma (B). Portions of the plates with oceanic lithosphere are shown with dotted lines*

2.2.3. *Fragmentation of the Adria plate*

For Schmid et al. (2004), the Adria plate was cut into two regions. In the north, a part was concerned by the closure of the Triassic Meliata Ocean in the Carpathians (Figure 2.5). This took place at the end of the Triassic and during the Jurassic with a very early eclogitization. The division of Adria is more complex in the diagrams of Handy et al. (2010). For these authors, three micro-plates were present: Adria to the south, Alcapia and Tiszia to the north of a proto-Adriatic transform fault. The Tiszia plate detached from Alcapia at 131 Ma (Figure 2.6). The junction between Adria and Alcapia is marked by Cretaceous tectonic events, an important characteristic of the Eastern Alps. Thus, this part of the Alpine belt underwent thrusting and nappe displacement well before the Western Alps (Eoalpine phase, see section 2.4.3).

2.3. The Alpine passive paleomargin of Europe

2.3.1. *General information*

On its European side, the Alpine Tethys paleomargin extends from the Betic Cordilleras (at the eastern edge of Iberia), through Corsica, now isolated in the heart of the Mediterranean, to the eastern end of Austria at the foot of the Eastern Alps (Figure 2.2). It includes all the regions that have undergone marked distension and then subsidence related to the Tethyan Liassic-Jurassic rifting.

The proximal edge of this paleomargin includes the Delphino-Helvetic units (Dauphinois domain of the French Alps and Helvetic domain of the Swiss Alps). The more distal regions correspond to the Briançonnais domain (so-called from Corsica to Switzerland) and to the pre-Piedmont domain, which marks the transition to the oceanic lithosphere domain. This oceanic domain, called the Piedmont-Ligurian domain, contains ophiolites (Figures 2.7, 2.9 and 2.10). In the Western French-Italian Alps, the Alpine paleo-margin is thus divided today between the external Alps (Dauphinois) and the internal or Penninic Alps (Briançonnais and Piedmontese), separated by the frontal Penninic thrust or Penninic front.

We will not discuss the geology of the pre-Mesozoic basement here. We simply recall that Variscan and especially the Late-Variscan tectonic events, through the development of intra-crustal post-orogenic extension zones and Permian basins, exerted a significant control on the development of structures related to the Alpine rifting (see on this subject, Ballèvre et al. 2018). Let us add that the term basement in its classical acceptance referring to a rigid mechanical behavior is adapted to the case of the least metamorphic Alpine zones. It becomes totally inappropriate in the case of internal margin units that have undergone Alpine metamorphism. Indeed, the mechanical behavior of such buried basement units has been ductile during the orogenic compression. This is the case for the Briançonnais zones belonging to the so-called Pennic nappes, characterized by large basement folds found in the western internal Alps (Grand St. Bernard nappes and Gran Paradiso and Monte Rosa nappes) and in the middle Penninic units of the Central Alps (Tambo and Suretta nappes).

Figure 2.7. *In good weather, from the peaks of the Piedmont zone, a large part of the external and internal zones of the Western Alps can be observed. View toward the west from the ophiolitic massif of Rocca Bianca, above Saint Véran at the French-Italian border (photo Lagabrielle)*

If we take as a reference an east-west cross-section of the Western Alps passing through the town of Briançon (Figure 2.10), we can identify the most basic stratigraphic and lithological characteristics of the European margin of the Franco-Italian Alps. Exposures of this margin extend from the Ardèche region to the deepest units of the internal crystalline massif of the Dora Maira. This represents a paleo-domain of considerable extension (Figure 2.8). Today, shortened, it is still about 200 km wide. During subduction of the margin, the deepest units buried to the coesite domain at a depth of about 70 km and were then exhumed during the Paleogene.

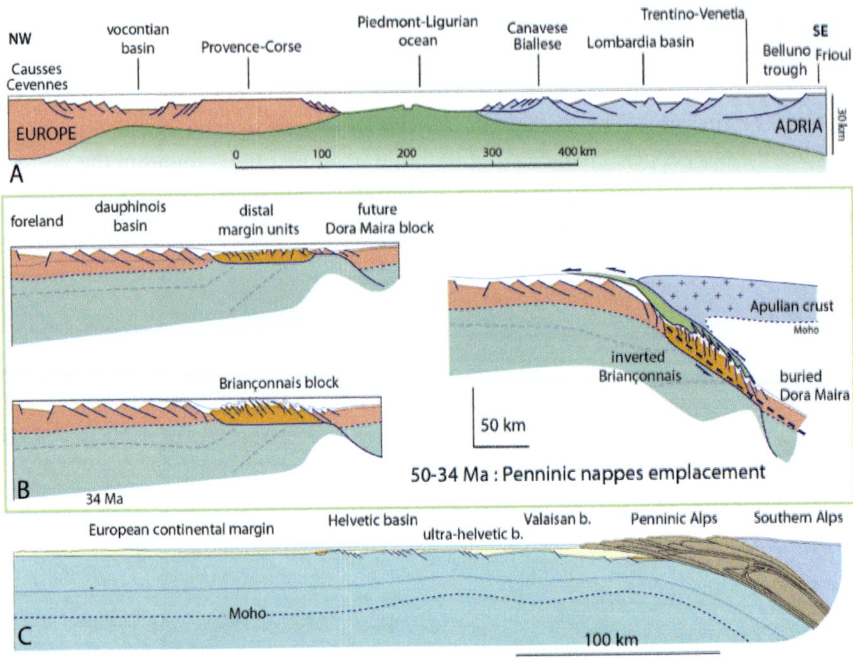

Figure 2.8. *Alpine paleomargins. A) Simplifying diagram from Lemoine (2006) showing the conjugate margins on a section from the Causses to Friuli and representations of the European paleomargin involved in the Alpine subduction according to two authors: Bell and Buttler (2017) (B) and Steck (2008) (C). Note the scale: in these interpretations, each of the two thinned-crust continental margins is over 300 km wide. This is indeed what has to be called a very wide rift*

Obviously, the innermost margin units were flattened, stretched, laminated by the subduction-exhumation process and stacked. They were considerably shortened during the formation of the Alpine orogenic wedge. It is therefore not possible to restore with certainty the architecture and dimensions of the European margin, especially in its innermost part. However, it can be safely assumed that this domain was between 200 and 400 km wide. The remnants of the European passive margin present a succession of stratigraphic series displaying a remarkable diversity of paleoenvironments. Such diversity implies a mosaic of rather independent continental blocks undergoing variable rates of subsidence.

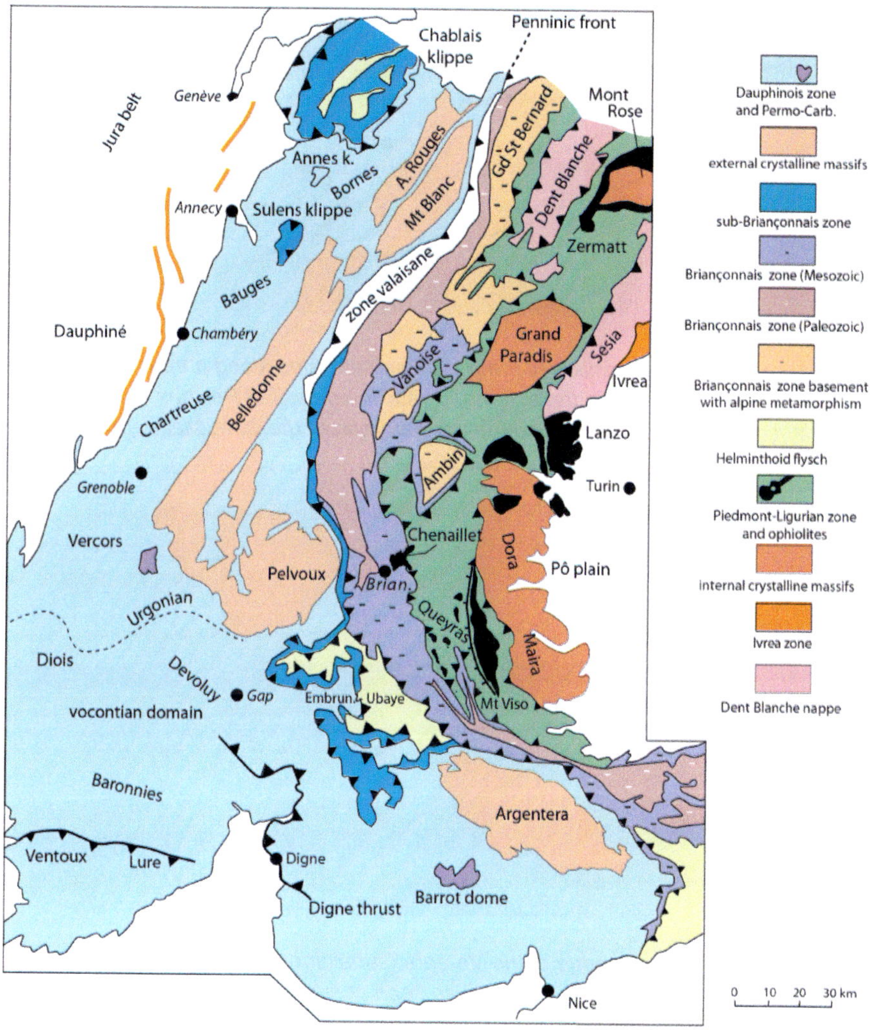

Figure 2.9. *Map of the Western Alps and location of the major units cited in this chapter*

Despite the diversity in their structural evolution, all domains of the European margin can be placed within one of the very classical domains mentioned above: from the outside to the inside of the belt, the Dauphinois, Briançonnais and Pre-Piedmont domains. Transitional subdomains between the main domains display some stratigraphic characters from neighboring domains. They are then designated by using the prefixes sub-, ultra- or pre-.

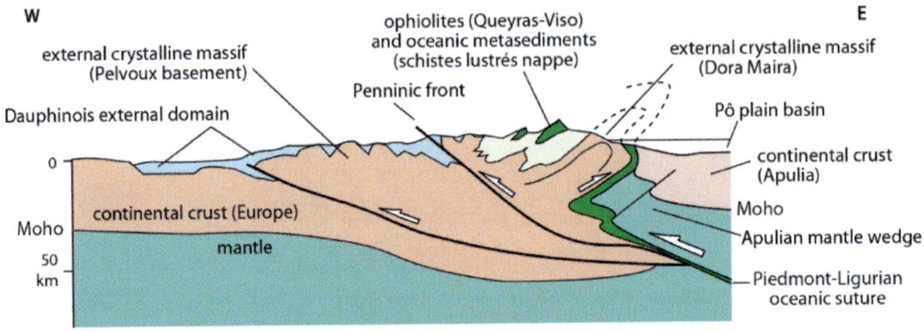

Figure 2.10. *Cross-section of the South-western Alps along a transversal line passing through the external crystalline massif of Pelvoux, the Briançonnais, the Queyras and the northern part of the Dora-Maira massif*

Figure 2.11. *The eastern edge of the Vercors (subalpine zone) and the cliff formed by the Urgonian facies of the Barremian seen at sunrise from the Belledonne basement at Chamrousse. On the left, Mont Aiguille (photo Lagabrielle)*

2.3.2. *The Delphino-Helvetic domain: the external (or proximal) margin*

To the west, the external paleo-margins (also called proximal) form the reliefs of the subalpine zone represented by the Chartreuse and Vercors massifs along the cross-section considered (Figure 2.11). The sedimentary cover corresponds to the Meso-Cenozoic series of the Dauphinois zone with a total thickness of about 4 km (Figures 2.12 and 2.13). On the Paleozoic basement outcropping in the external crystalline massifs, the Triassic was thin. It very often displayed the Germanic trilogy *Bundsandstein*, *Mushelkalk* and *Keuper*, indicating deposition in a very shallow zone, always close to the emersion, as shown by the Keuper evaporite terms. Spilitic magmatic events were known in the Upper Triassic-Hettangian around the external Pelvoux crystalline massif (Adline 1982) and within it (Vatin-Pérignon et al. 1972). They were equivalent to the ophites emplaced at the same time in the future Pyrenees related to the major event of the CAMP (Central Atlantic Magmatic Province) (Azambre et al. 1992).

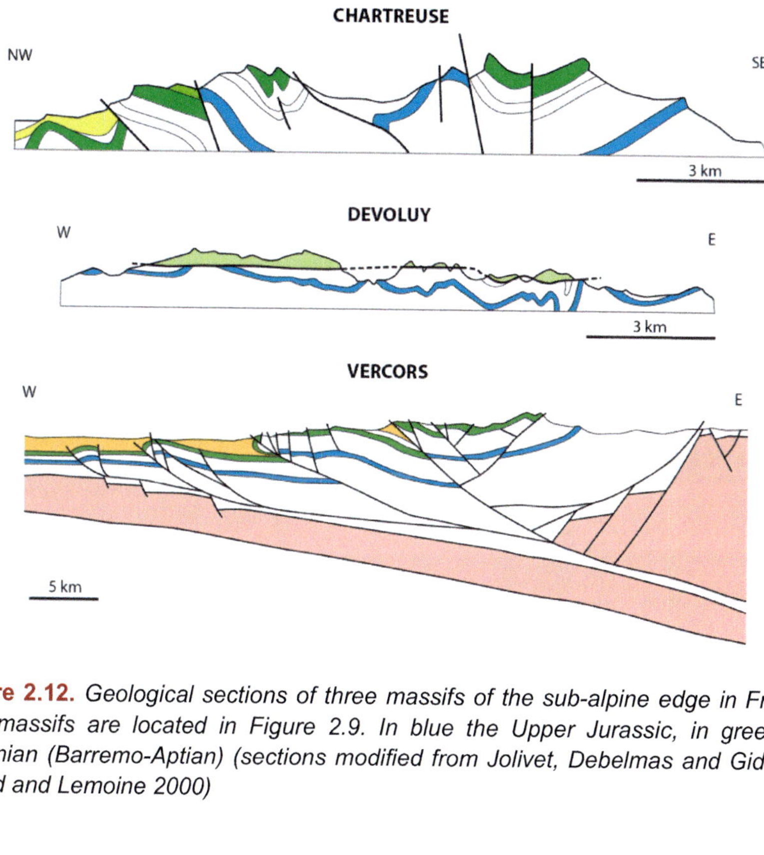

Figure 2.12. *Geological sections of three massifs of the sub-alpine edge in France. The massifs are located in Figure 2.9. In blue the Upper Jurassic, in green the Urgonian (Barremo-Aptian) (sections modified from Jolivet, Debelmas and Gidon, in Agard and Lemoine 2000)*

The carbonate and marl facies of the Lias marine platform are lying over the Triassic layers. The Liassic facies clearly show the invasion of the Jurassic sea in relation to marked subsidence, but the external margin remains rather shallow. In the Upper Lias strata, signs of collapse tectonics marked the beginnings of frank Alpine rifting. These signs are recognized in the subalpine chains but also on the eastern edges of the Massif Central as far as the Languedoc (Dumont 1998).

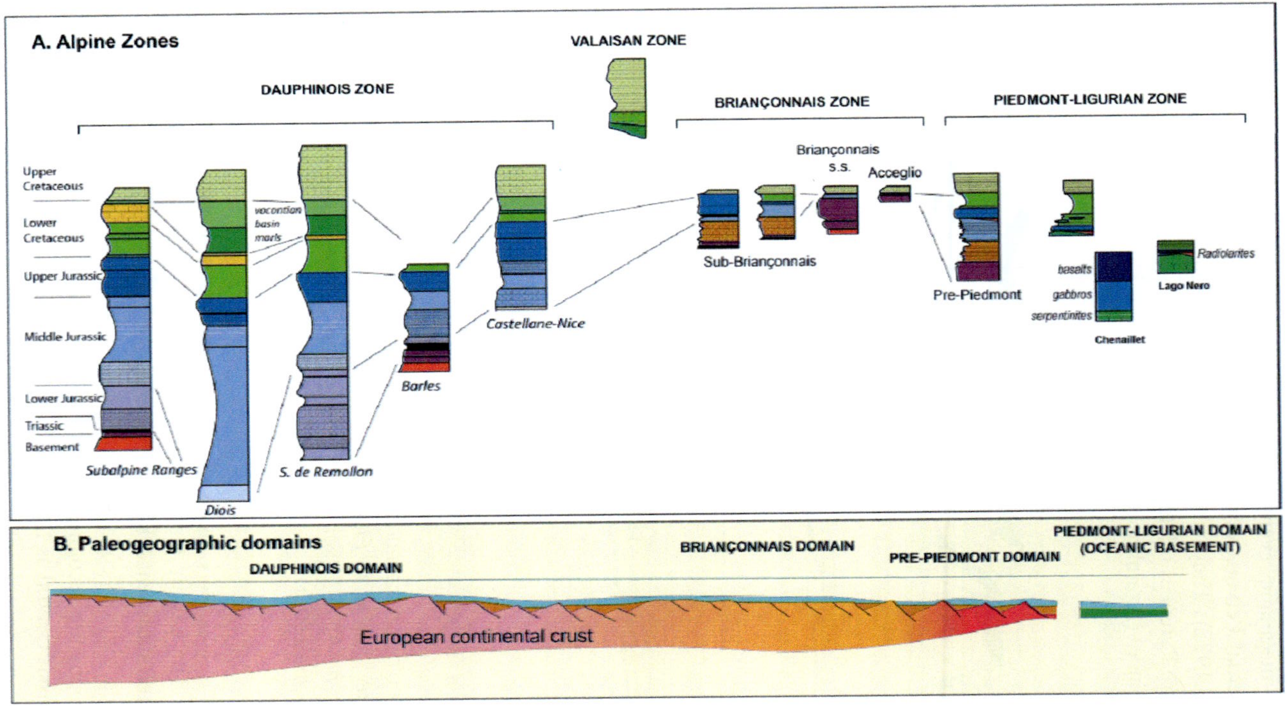

Figure 2.13. *Comparison of sedimentary series from different domains of the European Alpine margin (modified from Jolivet, in Agard and Lemoine 2000)*

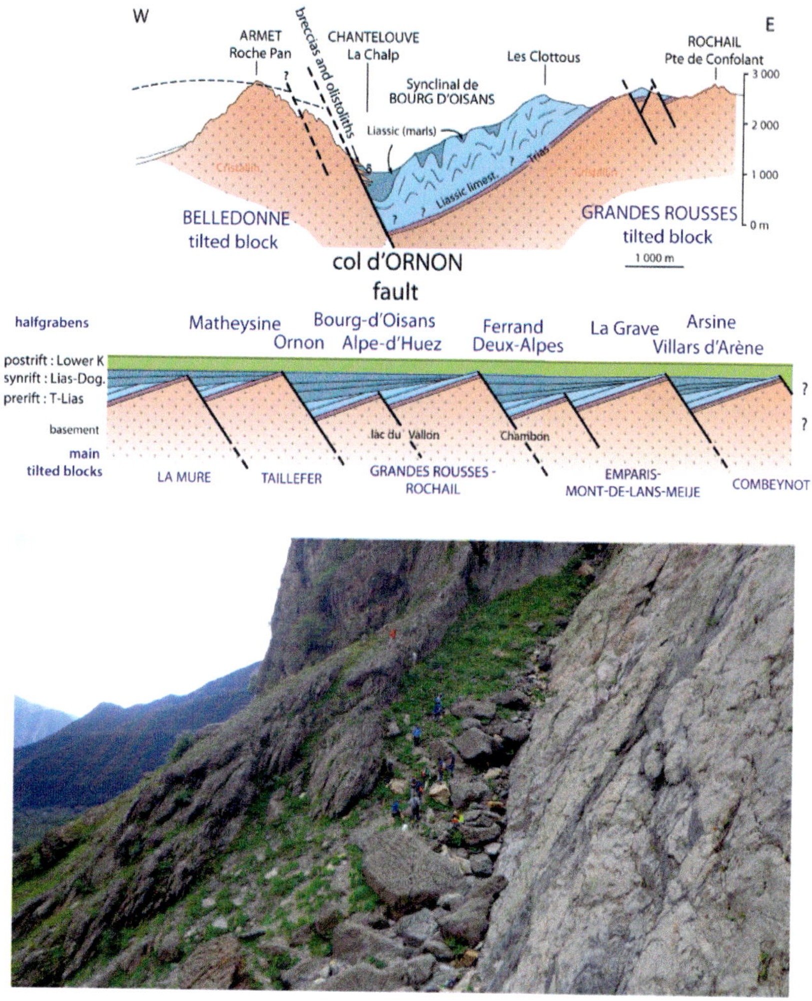

Figure 2.14. *The Col d'Ornon Jurassic fault. Above, the external crystalline massifs are structured in tilted blocks by Jurassic rifting. The former normal fault of the Col d'Ornon is a famous example of the faults controlling the displacements of these blocks. It has not been reversed during alpine compression. The Taillefer basement is on the right and the Liassic sediments filling the Bourg-d'Oisans basin on the left (diagrams from Lemoine et al. 1981, 1989). Below, a view of the fossil escarpment of the Ornon fault (photo Corre)*

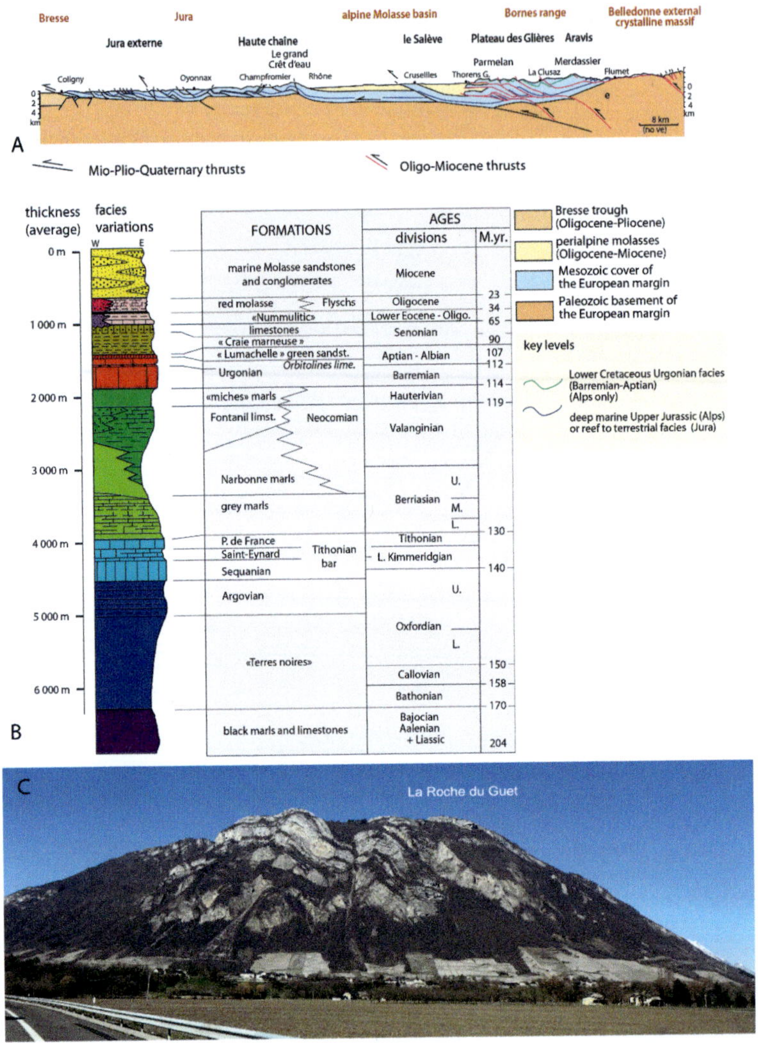

Figure 2.15. *The subalpine belt in France. Top, cross-section from the Bresse to the Belledonne massif, passing through the Bornes massif. The detached cover is stacked in large scales that overthrust the perialpine molasse. The transition to the western Jura series occurred in the Salève Mountain where the pelagic marine Alpine Upper Jurassic layers change from marine to lacustrine facies. Center, stratigraphic column of the Dauphinois series of the Bauges and Bornes' areas (from Gidon 1996, GéolAlpes website, available at http://www.geol-alp.com). Bottom, the faulting and folding of the Tithonian limestones above the town of Montmélian, at the junction between the Combe de Savoie (left) and the Albertville valley (right), looking north (photo Lagabrielle)*

Immediately south of the considered cross-section, in the southern part of the external Belledone-Grandes Rousses-Pelvoux crystalline massif, numerous tilted blocks resulting from the Liassic extension are well preserved due to a relatively moderate tectonic inversion (Barféty et al. 1979; Barféty and Gidon 1983, 1984; Lemoine et al. 1981, 1989; Dumont et al. 2008; Boutoux et al. 2014). The half-grabens are well preserved. They structured the passive margin in a succession of steps facing the more distal margin. Their sedimentary infill form eastward-thinning wedges. Basement olistoliths are present in the Liassic sediments against major fault scarps attesting to their activity during rifting (e.g. Ornon fault; Figures 2.14 and 2.16).

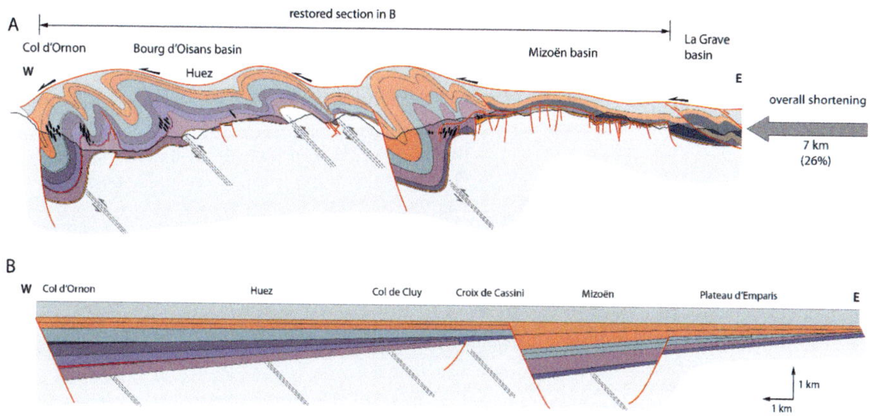

Figure 2.16. *Tilted blocks of the European external margin in the Bourg-d'Oisans region. A) Cross-section of the present structures affecting the Liassic sediments and their basement. B) Reconstruction of the Bourg d'Oisans and Misoën tilted blocks (modified from Boutoux et al. 2014). The Col d'Ornon fault shown in Figure 2.14 is on the left in the section*

Further north from France, in the Delphino-Helvetic zone of Switzerland, the Liassic tilted blocks have been recognized, but the graben sedimentary infills have often been expelled to form detached cover slices. This is found west of the Mont Blanc and Aiguilles Rouges Massifs in the three superimposed helvetic nappes of Morcles, Diablerets, and Wildhorn (Ayrton 1980; Masson et al. 1980) (Figure 2.16). Even further north, in the helvetic domain of the Glarus Alps (Central Alps), listric paleo-faults have been reconstructed. They offset Triassic layers and occurred due to the collapse of submarine reliefs, reflecting extensive tectonics during the Liassic and Dogger periods (Lemoine and Trümpy 1987). Note that a new interpretation of the structure of the Delphino-Helvetic margin is based on recent findings on the architecture of magma-poor passive margins and involves an extensive detachment located at the site of the Mont Blanc block (Ribes et al. 2021). This hypothesis accounts for the origin of the "Grès singuliers" (singular sandstones), a coarse detrital formation whose significance was poorly understood before (Figure 2.17).

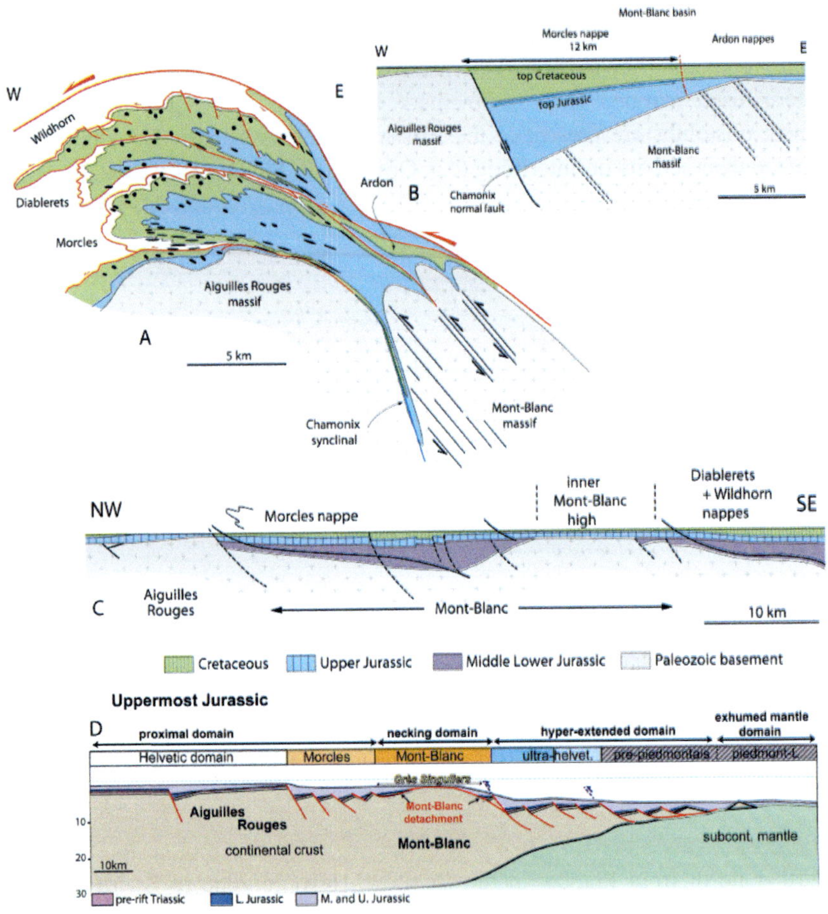

Figure 2.17. *Structure of the external European margin in the Mont Blanc region. Top, cross-section of the Morcles and Diablerets nappes and reconstruction of the Mont Blanc tilted block (modified from Boutoux et al. 2014). Center, balanced section of the helvetic domain according to Lemoine and Trümpy (1987). Note how the tilting of the series on the Mont Blanc block changed between representations from 1987 (eastward slope of the basement) and 2014 (westward slope of the basement). Below, new interpretation of the structure of the delphino-helvetic margin involving an extensive detachment at the location of the future Mont Blanc rather than a simple tilted block (modified from Ribes et al. 2021)*

The Middle Jurassic (Dogger) limestones of the Dauphinois zone formed thick banks, followed by an Upper Jurassic pelites that reflect the subsidence of the whole domain (Arnaud 1966a, 1996b; Arnaud et al. 1987). In the Late Jurassic (Kimmeridgian-Tithonian), fine limestones of the Tithonian facies were deposited in the subalpine, Vocontian and all internal zones. They are characterized by pelagic microfauna such as Aptychus and Calpionellas (see Figure 2.24). The carbonate platforms were then restricted to the Jura and Provence regions. These limestones formed a first benchmark, the "Barre Tithonique" (*Tithonian bar*), which is clearly visible in the landscapes of the Vercors, and further north, in the Chartreuse, Bornes and Bauges (Figure 2.12). The Lower Cretaceous facies remained marine; they correspond to interbeddings of small calcareous and marly bands (Berriasian-Valanginian-Hauterivian). During the Barremian-Aptian times, the carbonate platforms progressed toward the basin and deposited thick peri-reef formations characterized by rudists and neritic microfauna, the "Urgonian" facies. These limestones are resistant to erosion and therefore very visible in the landscape. They constitute the "Barre Urgonienne" (*Urgonian bar*), the second morphological landmark of the subalpine belt in the Vercors region and further north (Figure 2.12). The edge of the platform can still be seen to the south of the Vercors. There, the Urgonian limestones disappear and pass laterally to a domain of marls where the Barremian can reach 1000 m in thickness. This peculiar stratigraphy, which reflects subsidence controlled by faults, characterizes the Diois and Dévoluy region and corresponds to the Vocotian basin (named after the Voconces, Gallic inhabitants of this region) extending between the present-day Drôme and Durance rivers (Figure 2.17).

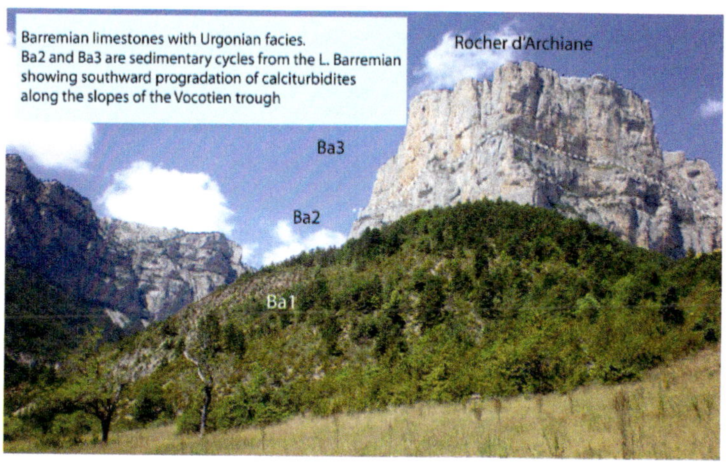

Figure 2.18. *The transition from the Urgonian to the Vocontian domains in the southern Vercors massif at the Archiane rocks. Stratigraphy according to Arnaud (1981). Ba2 and Ba3 rest on Ba1 also of Lower Barremian age (photo Meibom)*

The southern subalpine region of France is affected by early folds, sealed by the Senonian strata (Coniacian to Maastrichtian), related to strike-slip tectonics and/or gravity destabilization of the margin (Michard et al. 2020) (Figure 2.18). Note that at the same time, before the Senonian, a major distension event is recorded in the future Pyrenees that leads to the exhumation of the mantle in the North Pyrenean Zone during the Albian. We shall see that these events were contemporary with the opening of the Valais domain further north, within the lithosphere of Europe. It is possible that the region south of the Vercors constituted an element of a transform corridor, belonging to a series of tectonic connections between the extensional structures affecting both Iberia and Europe during the Middle Cretaceous.

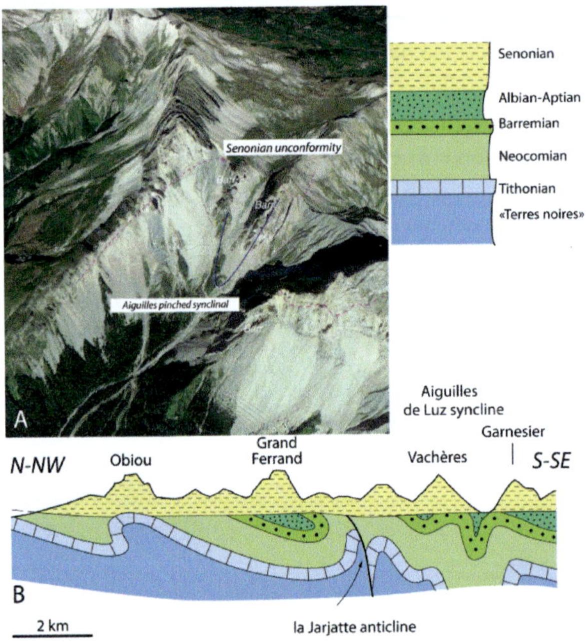

Figure 2.19. *The Aiguilles de Luz and the ante-Senonian unconformity, evidence of an early deformation of the Alpine margin in the Late Cretaceous, possibly related to the Alps-Pyrenees connection active at that time (section from Gidon 1996, GéolAlpes site, accessible at: http://www.geol-alp.com)*

The Upper Cretaceous facies of the Dauphinois domain corresponds in general to pelagic marine sediments rich in foraminifera, reflecting the presence of a deep marine platform. At the beginning of the Tertiary, the margin first emerged but a new transgression occurred in the Eocene times with the deposition of large nummulites-bearing neritic sediments, followed by the deposition of marls and finally by flysch-type sediments (e.g. Annot sandstones), the entirety composing the Priabonian trilogy. This reflects a progressive deepening in the Paleogene by lithospheric flexure, announcing the generalized tectonic inversion of the chain and the entry into the orogenic period.

1) The Triassic succession was of Germanic type as in the rest of France, Iberia and Germany. The Dauphinois zone belonged to the external zone of the margin of Europe, we will see that by contrast, the more internal margin presented a marine Triassic succession of kilometric thickness.

2) The Late Jurassic strata were marine and pelagic (Tithonian facies), in contrast to the Jura region where the sediments of the same age were continental (Purbeckian facies), or to the Provence region which also experienced Purbeckian emersion. The Tithonian facies were deposited after a long period of subsidence of the entire Alpine margin, in contrast to the Jura and Provence regions.

3) We will see that the Upper Jurassic pelagic limestones also constitute the cover of the Briançonnais series and the ophiolites. Thus, they represent a good marker of the beginning of the post-rift period.

Box 2.1. *Three characteristics of the Dauphinois zone*

2.3.3. *The Briançonnais domain: the internal margin*

The internal zones include the Briançonnais, Valais and Piedmont units, all originating from the Penninic domain (Figure 2.9), most of which have been subjected to Alpine metamorphism. The paleo-margins visible in the Briançonnais units are remarkably different from the Dauphinois paleo-margins by their stratigraphic characteristics, which recorded an evolution controlled by extensional rifting tectonics and marked by uplift during the Dogger times, followed by subsidence during the Malm times.

2.3.3.1. *Pre-Mesozoic terrains*

In some units, the pre-Mesozoic terrains of the Briançonnais zone include Carboniferous, coal-rich layers, often devoid of their Mesozoic cover. Therefore, the term "Zone Houillère" or "Houillère Briançonnaise" is frequently used to designate these units with a dominant Paleozoic basement. Under the coal layers, the basement is affected by Variscan metamorphism. The exposures of the variscan basement increase from the south to the north of the Western Alps and then to the Central Alps where outcrops of Mesozoic cover are very scarce (Goffé 1975; Steck 2008). The Briançonnais of the Western and Central Alps has undergone an evolution under low- to medium-temperature high-pressure facies, indicating burial in the Alpine subduction of several tens of kilometers (Platt and Lister 1985; Ballèvre 1988; Fudral 1998; Schwartz 2000; Agard et al. 2001; Ganne et al. 2003; Ganne 2003; Bousquet et al. 2008). The terrains forming the internal crystalline massifs (Dora-Maira, Gran Paradiso, Monte Rosa) and representing the very distal European margin are attached to the Briançonnais basement.

The diversity of the Briançonnais basement units and their contrasted evolution in the P-T paths reflect the variety of subdomains of the Briançonnais margin involved in the Alpine subduction. By itself, such diversity thus reflects the large extent of the passive margin domain involved in the subduction. This is well illustrated in some of the sketches depicting continental subduction, such as the one in Figure 2.19 showing the burial of the Gran Paradiso, Ruitor and Houillère Zone units under the Dent Blanche basement. An additional argument demonstrating the large width of the original Briançonnais domain is the diachronism of the high-pressure peak in the corresponding units, which spanned a large part of the Eocene (see synthesis in Dumont et al. 2012). Given the speed of convergence, the estimation of the width of the absorbed domain can be greater than 200 km.

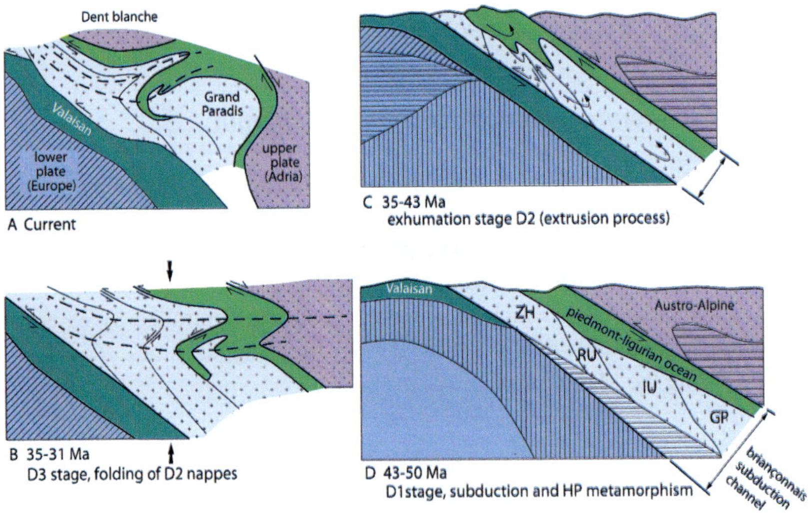

Figure 2.20. *Burial of the Briançonnais basement units within the Alpine subduction (Briançonnais subduction channel). The margin was composed of a succession of domains with contrasting P-T evolution (ZH: Zone Houillère; RU: Ruitor; IU: Internal Briançonnais units; GP: Gran Paradiso; LP: Ligure-Piemontese units). This implies a large initial margin. See Figure 2.8, which shows a very comparable wide margin concept (modified from Bucher et al. 2003)*

2.3.3.2. *The Mesozoic series*

In the Briançonnais zone, an exceptional set of nappes from the internal Alpine margin, predominantly sedimentary in composition, is exposed, allowing the characteristics of the eponymous zone to be established (Debelmas 1955; Debelmas and Lemoine 1970; Bourbon 1980). In the following, we present the main stratigraphic features of these units in the Briançonnais zone. They contain the record of tectonic-sedimentary events related to the development of the more internal Alpine margin than the Dauphinois (Figure 2.20).

The Alpine-type Briançonnais Triassic series strongly contrast with the Germanic-type Dauphinois Triassic. It was much thicker and always of marine origin (Mégard-Galli and Baud 1977). The basal layers are the marine Werfenian sandstones and quartzites (Lower Triassic) following the continental Permian sandstones, some of which are rich in rounded debris of volcanic rocks (Verrucano facies). The quartzites were followed by dolomites, schists and gypsum of the Upper Werfenian. The Middle and Upper Triassic terms (Anisian-Ladinian-Carnian-Norian) were frankly marine. They include dolomites and shallow-water limestones of the lagoon type, deposited over great thicknesses (300–1000 m), without significant facies variation, and thus recording long-term subsidence announcing the birth of the Alpine margin (Dumont 1988, Figure 2.21).

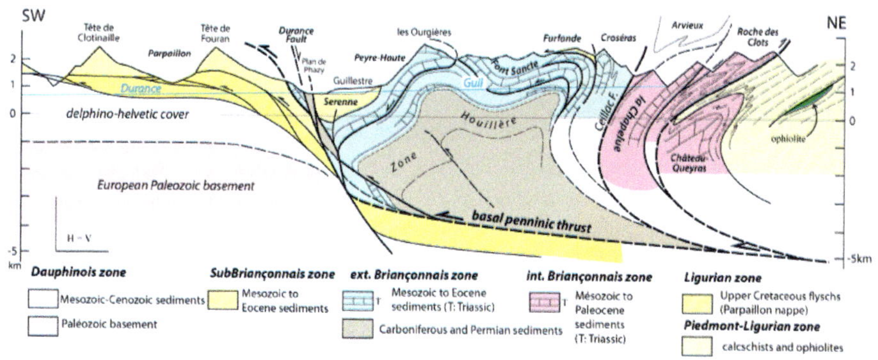

Figure 2.21. *Section of the Briançonnais domain along the Guil valley (modified from Ballèvre et al. 2020)*

The dolomitic series of the Briançonnais Triassic (Figure 2.22) was reminiscent of the *Hauptdolomit* of the Austroalpine margin, forming the bulk of the relief of the Dolomites and the North Calcareous Alps (see section 2.4.4). The latter was equivalent to the Pre-Piedmont Norian. On the European margin, deeper marine facies of the Upper Triassic are not known. In contrast, on the Austroalpine and

South Alpine margins, the *Hauptdolomit* passed laterally eastward to the reef and pelagic limestones (Dachstein and Hallstatt limestones), which in turn passed to the deeper facies of the Triassic Meliata Ocean of the Carpathians (see section 2.4.3).

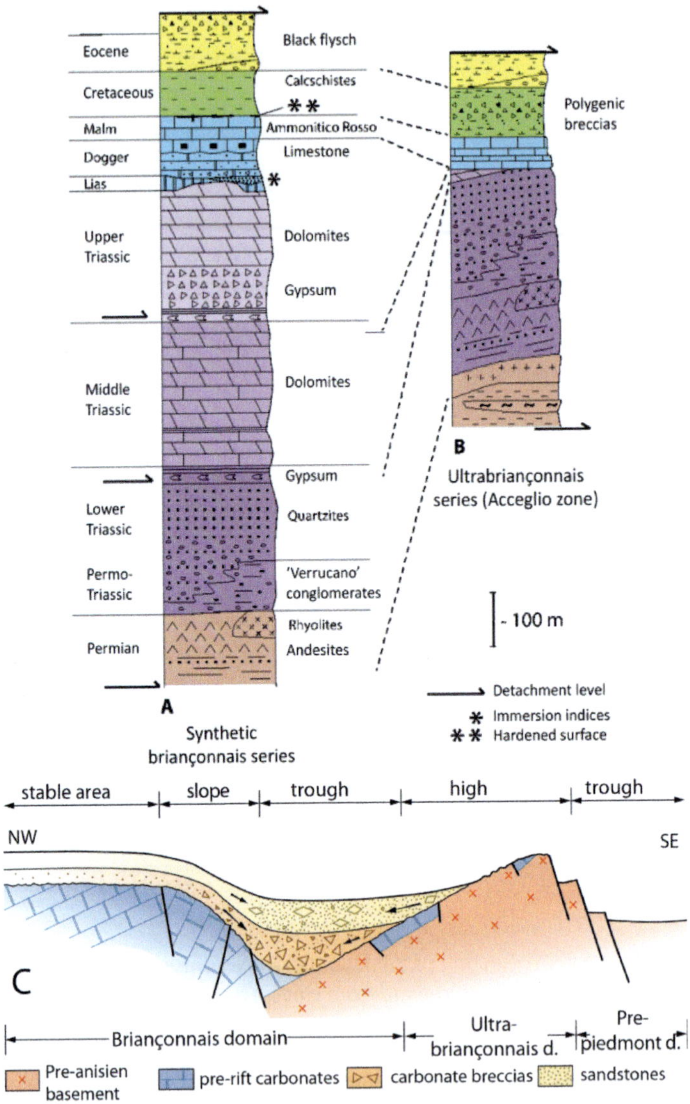

Figure 2.22. *The Brianconnais and Ultrabrianconnais series. Top, stratigraphic columns south of Briançon (modified from Michard et al. 2004). Bottom, reconstruction of the Vanoise margin in the Late Jurassic (modified from Jaillard 1989)*

The influence of a more open sea is well recorded during the Rhetian (208–201 Ma) which constitutes an important turning point marking the beginning of Alpine rifting. The Lias (201–180 Ma) was not continuous and was exceptionally preserved from syn-rift erosion in half-grabens, showing that syn-sedimentary distensional activity (Lower Lias) was rapidly followed by uplift, emersion and erosion throughout the Briançonnais (Upper Lias and Early Middle Jurassic). The Bathonian (167–164 Ma) covered this continental erosion surface and often sealed the last activity of the normal faults of the tilted blocks.

The pre-Bathonian discontinuity, not observed in the Dauphinois domain since sedimentation is continuous, can be called "post-rift unconformity" or "breakup unconformity", and is often identified in geophysics on present-day margins because of its clear geometrical character. It is considered the mark on the margin of the onset of the first ocean-floor spreading that started around 165 Ma (Lemoine et al. 1970, 1987). We will see that the various rifting phases established in the Briançonnais domain are well aligned with the events inferred from the study of the Pre-Piedmont series located more distally in the paleo-margins.

Figure 2.23. *The Briançonnais nappes seen from the Chenaillet ophiolites (looking west). The relief is outlined by the Triassic dolomites (photo Lagabrielle)*

All stratigraphic data in the Briançonnais show that Alpine rifting activity extended from the Latest Triassic to the end of the Dogger (Bourbon et al. 1976; Bourbon 1980). For Lemoine et al. (2000) rifting occurred in successive stages, showing several tectonic phases followed by subsidence, from the Hettangian-Sinemurian (J1 phase), into the Toarcian (J2 phase) and finally into the Bajocian-Bathonian boundary (J3 phase). A general uplift of the margin with emersion and karstification was recorded in the Briançonnais domain after the J2 phase. Thus, for 10 Ma, along a domain extending from Nice to the Grisons, a Briançonnais island or peninsula, several hundreds of kilometers long and 100 km wide persisted.

In the geosynclinal concept[3], this well-known emersion of the Briançonnais domain made it the perfect example of a *geanticlinal* ridge. We shall see that the conjugate Lombardy margin did not experience this surrection, which led to the development of asymmetric rifting models in the 1980s (see section 2.3).

Following the rifting period, known as the syn-rift period, the European margin was affected by generalized subsidence in the Upper Jurassic times (de Graciansky et al. 1998). It sunk in a homogeneous way as evidenced by a radical change in the fauna that became pelagic both in the Dauphinois and Briançonnais domains. In the Briançonnais zone, nodular, red facies with ammonites (*ammonitico rosso*, Guillestre marble) were deposited (Figure 2.23), showing condensed levels and hard-ground surfaces with iron and manganese oxides (recording gaps in sedimentation or submarine erosion). They are covered by white limestones with ammonite opercules (Aptychus) and pelagic microfossils (Calpionella, ciliated Protozoa of the Jurassic-Cretaceous in the form of small amphorae) (Figure 2.23). This shows the extremely condensed character of the Briançonnais sedimentary record, which can be explained by a contrasted bathymetry with highs swept by deep currents, an inheritance of the syn-rift continental dislocation. The Lower Cretaceous sediments are often missing due to continued tectonic activity and lacunar conditions. Pelagic limestones of the Lower Cretaceous, rich in foraminifera (Globotruncanidae), were then deposited in the external Briançonnais domain (Figure 2.24).

North and east of Briançon, the Briançonnais units of the French-Italian Alps are well exposed in the Ambin and Vanoise Massifs. The term "internal Briançonnais" is classically used to designate the easternmost units, more or less detached from the basement of these massifs, and having undergone high-grade Alpine metamorphism (Debelmas and Lemoine 1957; Bocquet-Desmons 1976; Lefèvre and Michard 1976; Chopin 1979; Lefèvre 1982; Jaillard 1985a; 1985b; 1988; 1989; Platt and Lister 1985a,b; Platt et al. 1989).

In Savoie, the internal Briançonnais units are grouped in the Vanoise-Ambin zone of Ellenberger (1958) (Caron and Gay 1977; Fudral et al. 1987). Unlike the external Briançonnais of Savoie, which lacks its Mesozoic cover, Meso-Cenozoic carbonate sediments are well represented in the Vanoise-Ambin zone. This cover (generally detached at its base, above the Lower Triassic quartzites) starts with a thick siliceous and calcareous-dolomitic series of the Lower to Upper Triassic,

3. In geosynclinal theory, the geanticlinal rift separates two trenches. The Alpine eugenosynclinal trench, the innermost and deepest, is filled by sediments of the Piedmont zone and hosts ophiolites. The mid-geosynclinal trench, less significant and more external, corresponds to the Dauphinois domain. The Alpine geanticlinal ridge corresponds to the Briançonnaise zone (see complete history in Lemoine 2006).

deposited in a lagoonal to a neritic environment, and passes into Liassic detrital sediments (Ellenberger 1963). As we have already seen in the more southerly series, an emersion phase affected this domain during the Upper Lias – Lower Dogger period. Emersion is documented locally by karst and bauxites. Then a general transgression occurred in the Callovo-Oxfordian, preceded in some parts of the domain by the Bathonian Mytilus layers. The deposition of a thin and discontinuous pelagic layer then continued until the Late Cretaceous times.

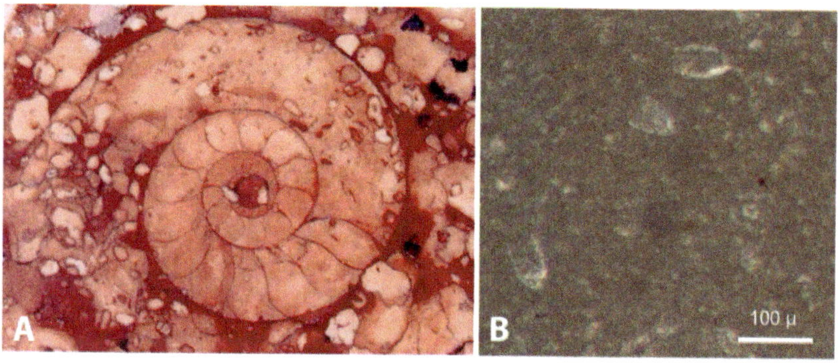

Figure 2.24. *The Briançonnais Upper Jurassic. On the left, the so-called "Guillestre marble" "ammonitico rosso" is a red nodular limestone of the Oxfordian age, rich in ammonites and debris of ammonites exploited in a quarry at the entrance of the Guil Valley (polished sample, a gift from Maison Bérard, La Maison du Roy; photo Lagabrielle). Deposition of the Guillestre nodular limestone preceded white Kimmeridgian pelagic carbonates rich in Aptychus (ammonite opercules) and Calpionella (microphotograph on the right)*

The Vanoise-Ambin zone extends northward into the Mont Pourri, the Ruitor and the Grand Saint-Bernard nappes in the Valais region of Switzerland (Caby 1968) (Figure 2.25). It includes three nappes: the Pontis, Siviez-Mischabel and Mont-Fort nappes. The first two nappes, the most voluminous, comprise an ancient Paleozoic basement associated with Houllier and Permian deposits and with thin scales of Mesozoic metasediments. The case of the Mont-Fort nappe will be detailed further (Figure 2.26). Most of the Mesozoic cover of these basements has been detached and formed the current Prealps klippe, in front of the Penninic front. However, some pellicular layers of Briançonnais metasediments are considered by some authors to remain in an original position over the basement. This is the case of the Barrhorn series lying over the basement of the Siviez-Mischabel nappe. The nappes of the Grand Saint-Bernard occupy a structural position quite close to that of the internal crystalline massif of Monte Rosa, and for this reason, they are ofter described together (Figure 2.28).

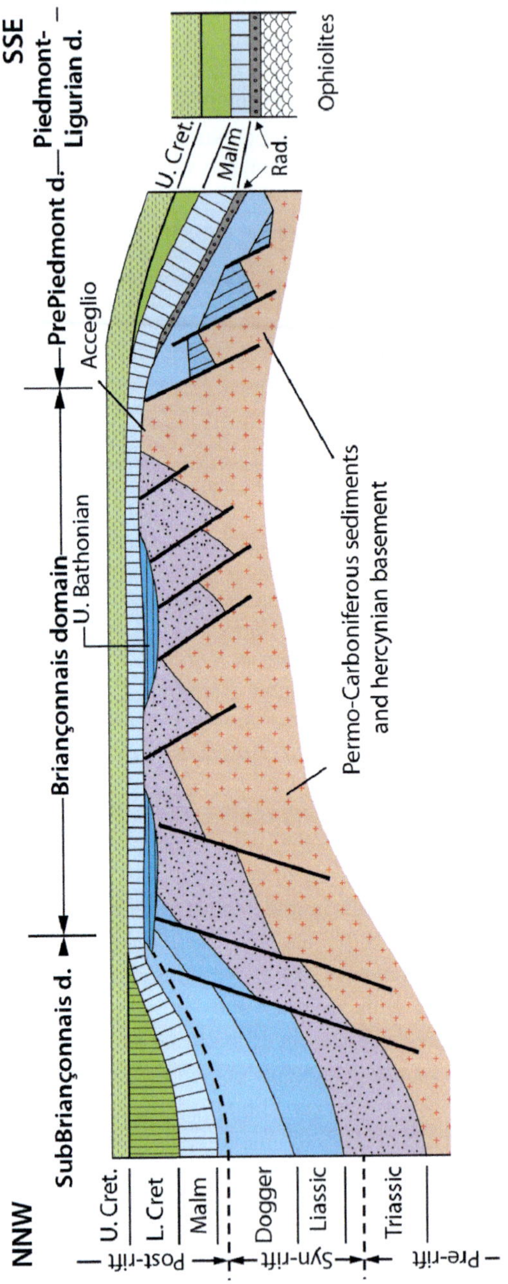

Figure 2.25. *Schematic reconstruction of the Briançonnais, Subbriançonnais, Ultrabriançonnais and Piedmont series along the Briançon traverse (from Lemoine et al. 2000)*

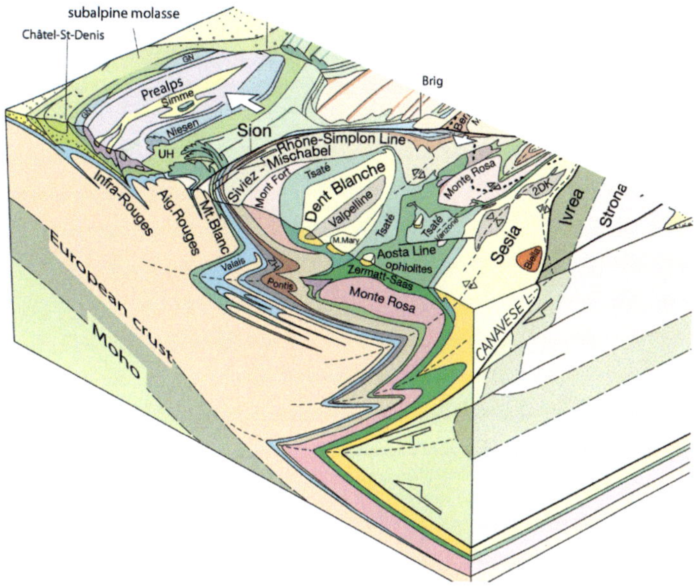

Figure 2.26. *Block diagram of the Western Alps between the Sesia zone and the Prealps (modified from Steck 2008). This diagram shows the complex architecture of nappe stacking, disrupted by the collision of the subadriatic mantle at depth which triggers generalized eastward retro-thrusting*

1) In contrast to the Dauphinois zone, the Briançonnais Triassic series show thick alpine-type successions. They must be considered as external facies of an open margin that develops toward the Triassic Meliata Ocean of the Neotethys and therefore that did not prefigure the Jurassic rifting that followed.

2) The Lower and Middle Jurassic layers are still reduced and patchy, with traces of emersion and dissolution (karst and bauxites) that affected the carbonate terms of the underlying Triassic. Coastal facies (Dogger with Mytilus) were also present, indicating episodic submergence by a very shallow sea. This shows that during Jurassic rifting, as the margin was breaking up, the Briançonnais domain underwent local surrections.

3) Subsidence occurred, triggering the return of the deep sea conditions in the Late Jurassic times. The ammonitic-rosso facies of the Oxfordian and the pelagic limestones with Aptychus and Calpionella of the Terminal Jurassic are the markers of the initiation of the post-rift period.

Box 2.2. *Three main characteristics of the Briançonnais domain and main differences with the Dauphinois domain*

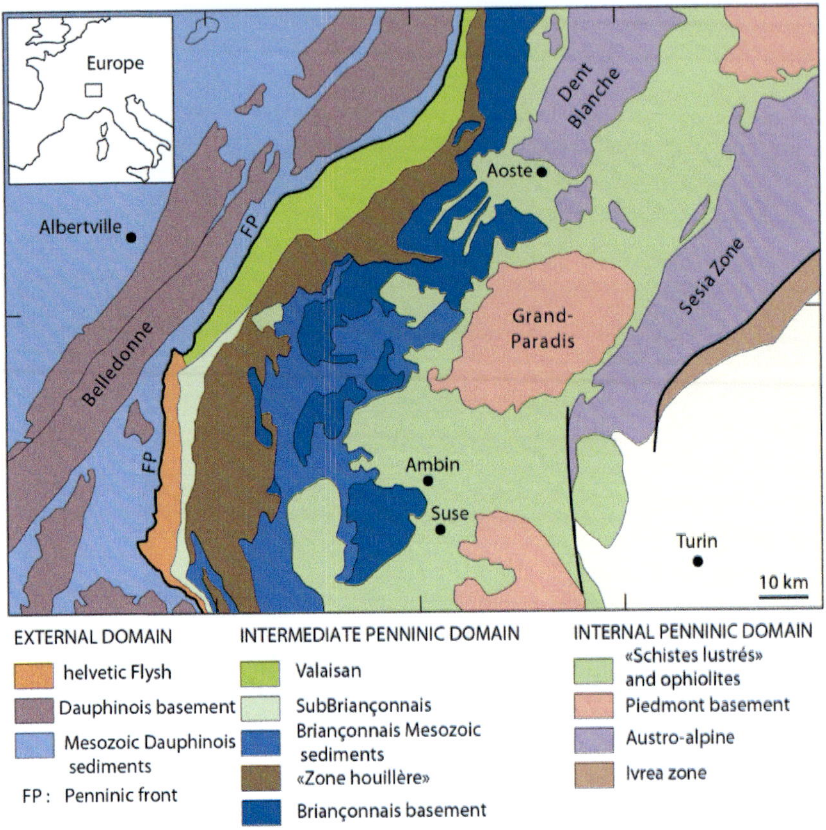

Figure 2.27. *The Briançonnais margin units of the Northwestern Alps. Map of the Briançonnais basement between Ambin and the Grand Saint-Bernard nappes (GSB) (modified from Ganne et al. 2003)*

2.3.3.3. *Subbriançonnais and Ultrabriançonnais*

On both sides of the Briançonnais domain were smaller domains that ensured the transition with the neighboring Dauphinois and Vocontian domains to the west and the Piedmont domains to the east: the Subbriançonnais and Ultrabriançonnais domains.

To the west, the Briançonnais basement is thrust over the Subbriançonnais units whose basement is unknown. The very narrow Subbriançonnais domain corresponds to a set of small units that have characteristics rather similar to those of the Dauphinois domain, at least for the base of the series (thin Triassic succession,

Figure 2.26). North of the Pelvoux, at the latitude of Chambéry, these units increase in size and then pass into the Valais zone, which will be discussed below (Barbier and Debelmas 1966).

Figure 2.28. *Panorama of the Croix-des-Têtes on the right bank of the Maurienne allowing us to observe the folding of the Subbriançonnais series (photo Y. Lagabrielle). In contrast to the Briançonnais series, the Triassic is thin, the Lias is present and there is no syn-rift emersion (tectonic contacts in red)*

To the east of the Briançonnais zone, the Acceglio zone corresponds to the Ultrabriançonnais domain. It is characterized by an extremely reduced Mesozoic series, where the Triassic was entirely removed before the deposition of the Upper Jurassic. They contain carbonaceous layers that evoke the Briançonnais Jurassic emersions and indicators of submarine erosion. They reflect the presence of steep slopes linked to tectonic activity during the Lias-Dogger. Following erosion of the Triassic dolomites, the Jurassic-Cretaceous layers were directly deposited over Permian and Werfenian volcanodetritic sediments (*Verrucano* facies) and over the Paleozoic basement (Debelmas and Lemoine 1957; Lemoine 1960, 1961; Lefèvre 1962; 1968; 1982; Michard and Henry 1988) (Figures 2.21, 2.24 and 2.27).

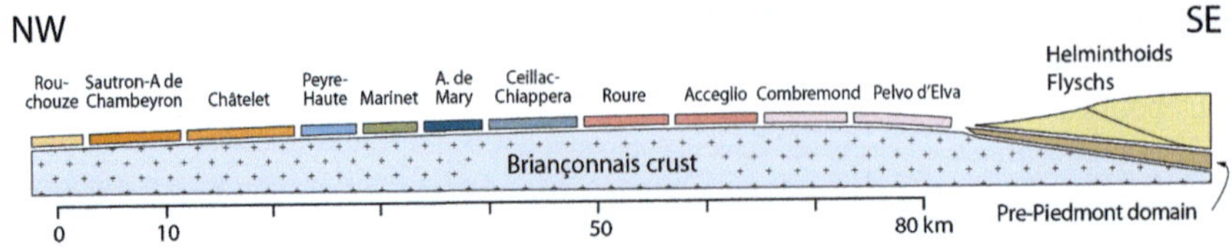

Figure 2.29. *Diversity of the Briançonnais and Ultrabriançonnais series along a NW-SE section, 45 Ma ago, south of the Western Alps. The pre-Piedmont series at the bottom of the margin is already involved in the subduction channel. Each name corresponds to a series and thus to a micro-domain of the margin, with its own stratigraphic characteristics (modified from Michard et al. 2004)*

The Acceglio-type metamorphic units are imbricated today into the heart of a plurikilometric east-diverted anticline fold (retro-thrusting stage) also affecting the Piedmont zone (Figure 2.28(B)). They also outcrop along a north-south zone, the Bande du Roure, directly against the Briançonnais zone, west of the Acceglio zone (Le Guernic 1966). In spite of their very reduced character, the Ultrabriançonnais series are varied, which has enabled the definition of internal and external units within them. Finally, the Ultrabriançonnais and Briançonnais series are thus numerous (Lefèvre 1984). This diversity is illustrated in the reconstruction of Figure 2.27 where the Roure-, Acceglio-, Combremond-, Pelvo d'Elva-type Ultrabriançonnais series are placed between the Briançonnais series and the external border of the pre-Piedmont zone (Michard et al. 2004). As also pointed by Figures 2.8 and 2.19, this figure illustrates the great geo-diversity of the domains of the pre-orogenic Alpine margin, a geo-diversity that must be seen as an expression of its very large original geographic size.

For Michard et al. (2004) and Ballèvre et al. (2020), the Variscan basement of the Acceglio zone is in continuity under the Piedmont zone (described in the following), with terrains belonging to the Briançonnais zone, well visible in the Pinerolo (*Pignerol*) region (Sanfront-Pinerolo unit, Avigad et al. 2003). These Pinerolo outcrops belong to the lowest unit known in the structure of the Western Alps. The Briançonnais basement includes Carboniferous terrains intruded by late-Hercynian granites overlain by Permian sediments. The Pinerolo unit is overlapped by more metamorphic units of the Dora Maira massif, of more internal origin, including coesite-bearing rocks equilibrated a great depth.

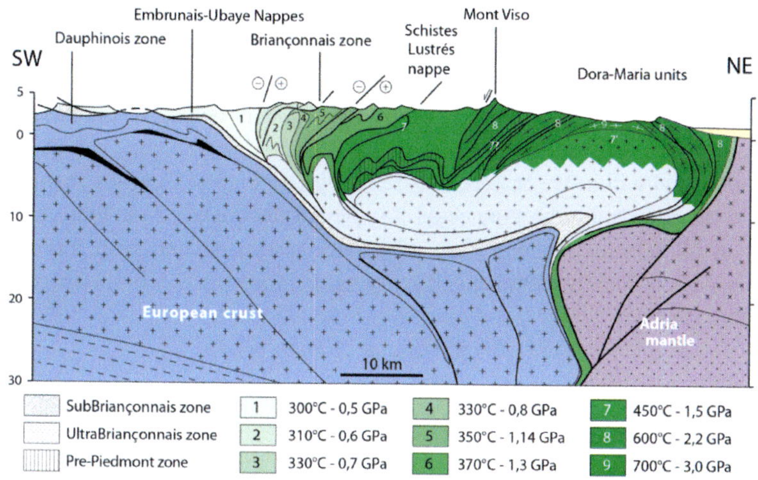

Figure 2.30. *The series composing the internal alpine margin south of the Western Alps. Geological cross-section of the Briançonnais and Piedmont zones in the South-western Alps showing the diversity of the metamorphic evolution of each unit and sub-unit of the orogenic prism. Note that the overthrusting of the internal nappes is interpreted as a consequence of the collision of the nappe stack at depth by the very strong Adria mantle (modified from Michard et al. 2004)*

Each series from the Mesozoic Briançonnais and Ultrabriançonnais domains differs from its neighbor by the presence or absence of typical terms, facies, or evidence of synsedimentary tectonic activity. In general, the specificity in the series stratigraphy coincides with the division into independent tectonic units. In the Briançon region, we can distinguish the Champcella, Roche Charnière, Peyre-Haute, Assan nappes, etc. whose stratigraphic successions differ. This also occurs in the Vanoise-Ambin. Such coincidence probably results from the structural control of the inversion by the extensional structures of the margin during the initiation of the nappe individualization (inheritance process). For Stampfli (1993), the lack of stratigraphic consistency between numerous small series is an argument for making the Briançonnais a peninsula from the south, in the form of an "*exotic terrane*" limited by multiple detachments, as also suggested by Ricou (1984).

Box 2.3. *Diversity of the Mesozoic Briançonnais and Ultrabrianconnais series*

2.3.4. *The internal crystalline massifs and their detached covers (Briançonnais domain in the broad sense)*

The internal Alpine zone also includes strongly deformed and eclogitized massifs of ortho- and paragneiss, separated from the basement of the Briançonnais zone by thin units of metasediments and magmatic rocks from the Ligure-Piemontese oceanic lithosphere forming the Schistes lustrés nappe. This is the case with the Gran Paradiso, Monte Rosa and Dora-Maira[4] massifs forming the whole of the internal crystalline massifs. In reality, these massifs belong to the basement of the most internal Briançonnais zone (in the broadest sense): they act as tectonic windows opened within the Schistes lustrés nappe during the Cenozoic exhumation. They are thus structurally connected to the Briançonnais basement under the thick layer of Schistes lustrés. They are sometimes associated with fragments of Mesozoic cover, usually detached, displaying Briançonnais, Ultrabriançonnais or pre-Piedmont affinity.

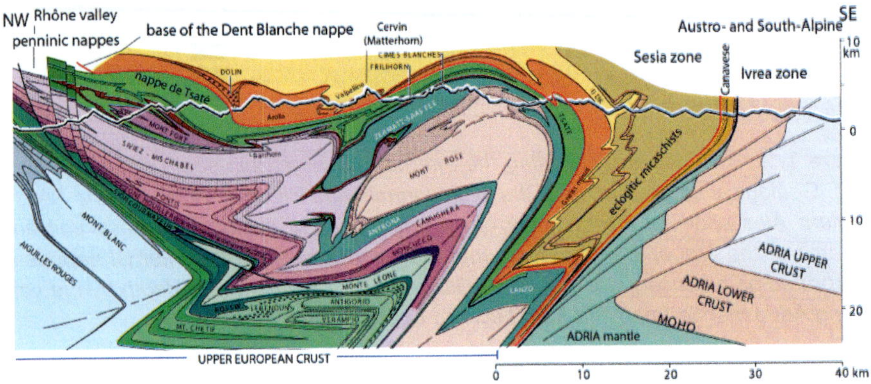

Figure 2.31. *Cross-section of the Northwestern Alps along the Prealps-Monte Rosa traverse showing the superposition of the Briançonnais basement units (in pink). Extracted from the general section of the Alps constructed by Escher et al. (1994)*

The Monte Rosa massif is a large-scale fold of Paleozoic gneisses that rest on the ophiolitic units of the Antrona zone and are covered by the ophiolites of the Zermatt zone, themselves covered by the Tsaté nappe and the Cimes Blanches nappe (Figure 2.28). Due to back-thrusting, the Monte Rosa nappe is buried beneath the Siviez-Mischabel nappe, which is itself thrust into the Zone Houillère. Figure 2.29 shows a possible reconstruction of the corresponding margin. Note that here again, the reconstructed margin displays considerable width. It is composed of a mosaic of continental blocks separated by crustal zones with thinned crust and by an oceanic realm still open in the Eocene, corresponding to the Antrona ophiolites (Steck et al.

4. The name of the Dora Maira massif does not correspond to a geographical entity. It is composed from the name of the two rivers that border it: to the north, the Dora Baltea, flowing in the Susa Valley, and to the south, the Maira, flowing in the Maira Valley.

2015). The Mesozoic cover of the basement units is found in thin, fully detached sheets such as the Furgg, Cimes Blanches, and Barrhorn units. These cover series are imbricated with continental basement sheets and some oceanic units in a very complex pattern, deformed by back-thrusting. Despite the intensity of the deformation, some successions, such as the Barrhorn series, show very marked characteristics of the Briançonnais series (Figure 2.29).

Like the Vanoise series, the Barrhorn series comprises very thin Triassic dolomites, discontinuous dark Lower-Middle Jurassic marbles with traces of emersion, pelagic light-colored marbles of Upper Jurassic age and thinly bedded Upper Cretaceous pelagic phyllitic marbles (Ellenberger 1958; Sartori 1990). The Barrhorn series is generally considered to represent the original basement cover of the Siviez-Mischabel nappe. The original cover of Monte Rosa may be preserved in part in the Furgg nappe.

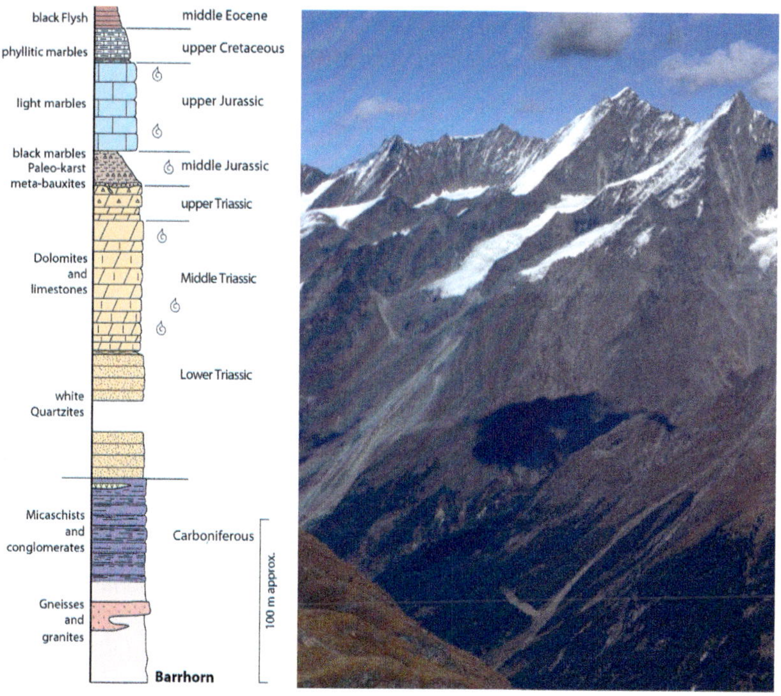

Figure 2.32. *The Barrhorn series (log modified from Steck et al. 2015), possible initial cover of the Paleozoic Siviez-Mischabel schist and gneiss unit of the Briançonnais unit (photo on right, Jürg Meyer)*

The Gran Paradiso massif is also a vast recumbent anticline of eclogitic gneisses forming a dome that rests in anomalous contact over less metamorphic Briançonnais units with Zone Houillère affinity, exposed in the small Money window (Chopin 1979; Ballèvre 1984, 1988; Brouwer et al. 2002; Le Bayon and Ballèvre 2006; Manzotti and Ballèvre 2013; Manzotti et al. 2014; Manzotti et al. 2015a, 2015b). Mesozoic Briançonnais-type cover elements are associated with the Gran Paradiso gneisses (Deville 1989).

Two stratigraphic sequences are recognized in the Money window. (1) A sequence of discontinuous, graphite-rich polygenic conglomerates interbedded with quartz-rich micaschists. This sequence is very similar to that found in the Pinerolo series at the base of the Dora Maira Massif (see below). (2) A monogenic sequence of quartz-rich conglomerates with rare interbeds of graphitic schists. Unlike the case of the Pinerolo window, the detrital zircons of the Money window do not contain Carboniferous grains and are dominated by populations older than 450-460 Ma. Thus, there were disparities in the sources of the Carboniferous basins of the future Briançonnaise Zone Houillère (Manzotti et al. 2016).

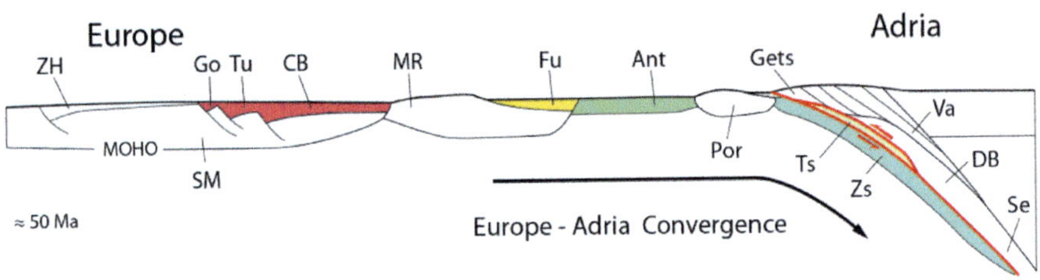

Ant: Antrona ophiolites; Be-Ru: Benisal-Ruginenta; CB: Cimes Blanches; DB: Dent Blanche; 2DK: 2nd dioritico-kinzigitic zone; Fu: Furgg; Go: Gornergrat; MR: Monte Rosa; Port: Portjengrat; Se: Sesia; SM: Siviez-Mischabel; Ts: Tsaté ophiolites; Tu: Tuftgrat; Va: Valpelline; ZH: Zone Houillère; ZS: Zermatt-Saas Fee ophiolite.

Figure 2.33. *Reconstruction of the European margin around 50 Ma, along the Monte Rosa cross-section. Situation after the burial of the Piedmont-Ligurian domain and the Sesia-Sent Blanche zone into the subduction channel (from Steck et al. 2015). This section shows the position of the future Monte Rosa units and the Grand Saint Bernard (Siviez-Mischabel) nappes*

The Dora Maira massif is less homogeneous than the Gran Paradiso massif (Vialon 1966; Michard 1967; Sandrone et al. 1993; Michard et al. 1993; Philippot 1990). It corresponds in reality to a complex of nappes made of stacking of thin basement slices separated by sheets of extremely deformed and hardly recognizable Mesozoic metasediments. A thin metamorphic eclogitic unit equilibrated at depths greater than 60–70 km (ultra-high pressure conditions) (Chopin 1984), directly overlaps a Paleozoic unit with blueschist to garnet-blueschist facies metamorphism, forming the base of the nappes stack (lower unit of Pinerolo).

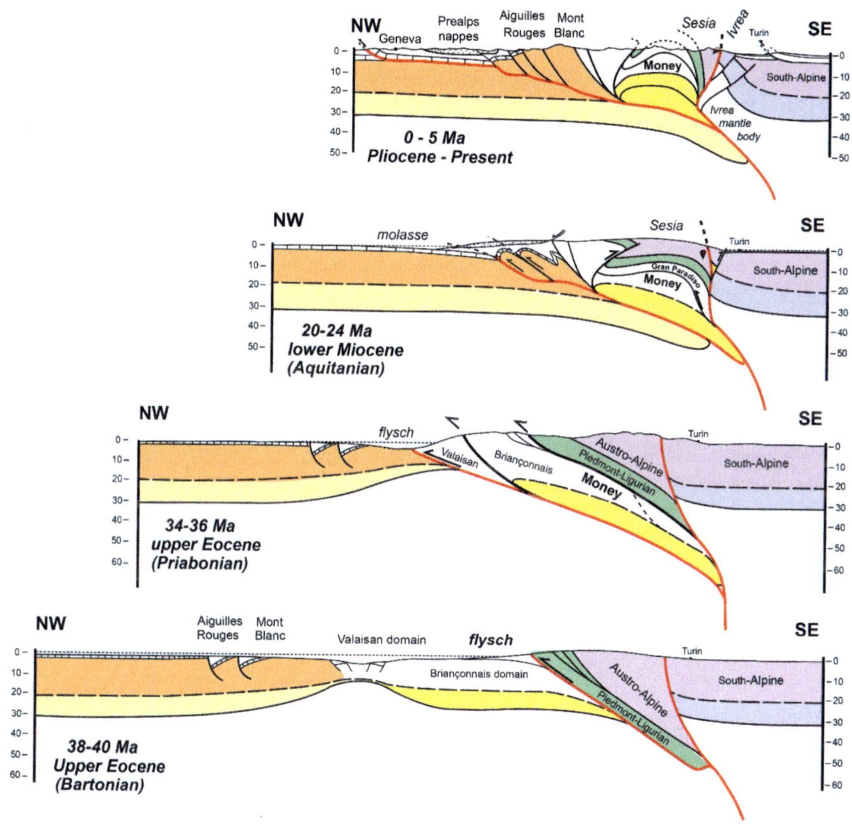

Figure 2.34. *Evolution of the Western Alps along the cross-section of the Gran Paradiso massif. This massif forms the eastern end of a vast Briançonnais continental block and is sinking into the Alpine subduction while being thrust over the Valais domain. Its early cleavage explains the Money and Gran Paradiso units' superposition. This sketch also illustrates the great width of the original Alpine margin domain (modified from Le Bayon and Ballèvre 2006)*

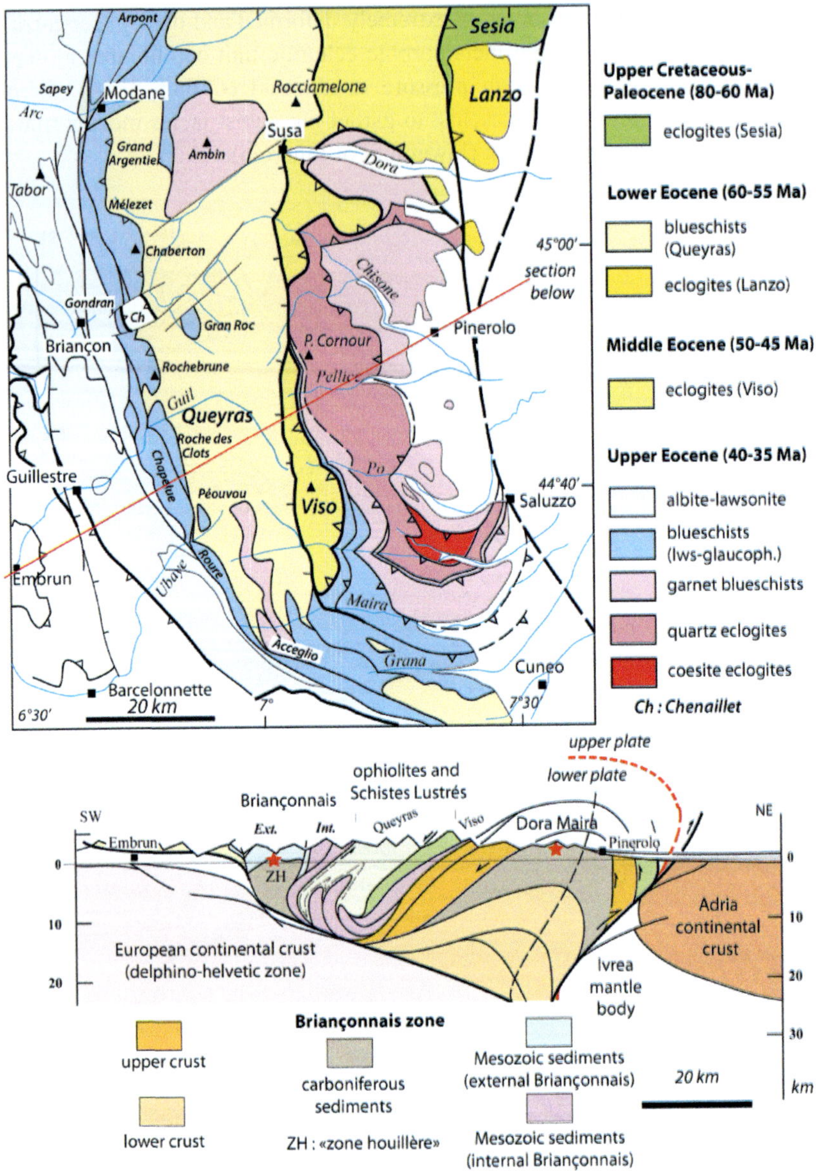

Figure 2.35. *The Briançonnais and Piedmont zones in the Southwestern Alps. Ages and conditions of alpine metamorphism and simplified cross-section (modified from Ballèvre et al. 2020). Note that the term "Internal Briançonnais" refers to the pre-Piedmont in the classical sense used in this chapter*

The metasedimentary slices involved in the Dora Maira nappe stack have affinities with the Briançonnais series, as shown by the presence of metadolomites and calcschists. They probably represent remnants of the most distal Alpine passive margin cover. The original position of the different series along a southern cross-section of the massif is hardly restored. Recent attempts at reconstruction have been proposed by Michard et al. (2004) and Ballèvre et al. (2020). These authors agree that there is continuity in the pre-Mesozoic basement of the Briançon domain with the lower Pinerolo unit (Figure 2.31). The initial continuity between the external Zone Houillère and the Pinerolo Zone is confirmed by dating the Carboniferous series of these two basement units (Ballèvre et al. 2020).

2.3.5. *The Piedmont zone: from the continental margin to the ocean*

In the Western Alps, the Piedmont (*sl*) zone corresponds to the Schistes lustrés nappe (the internal crystalline massifs treated above were historically attached to it). The Schistes lustrés nappe is made up of very abundant metasediments in which carbonate schists, called *calcschists*, associated with ophiolites dominate. Very poor in fossils and for a long time presented as monotonous, the Schistes lustrés nappe reveals, in reality, a lithostratigraphic diversity progressively highlighted during the last 50 years. Thus, before the 1970s, most authors considered that there was only one group within the Schistes lustrés. Today, two major groups can be distinguished (Lemoine and Michard 1963; Lemoine et al. 1986), from the external to the internal regions (from the continent to the ocean):

– The external Piedmont domain, or pre-Piedmont domain, corresponds to the continental bedrock series of the distal passive margin initially located more internally than the Briançonnais domain. This domain was separated from the Briançonnais domain by Ellenberger and Lemoine in 1955. Emblematic examples of this type of series are found north and south of Briançon in the Chaberton and Rochebrune massifs (Lemoine et al. 1978; Dumont 1983; Dumont et al. 1984) and south of the Dora Maira massif in the Val Grana (Ellenberger et al. 1964; Michard 1967).

– The Piedmont-Ligurian domain, or internal Piedmont domain, refers to the province of units derived from the oceanic lithosphere and is formed mainly by the original sedimentary cover of the ophiolites. This domain is the subject of Chapters 2 and 3 of the book devoted to ophiolites in the same collection (Lagabrielle and Ceuleneer, in progress).

We will concentrate hereafter on the pre-Piedmont series, but we will consider part of the ophiolite units in section 2.4. Indeed, we will see that the latter may have been emplaced at the very distal fringe of the pre-Piedmont domain, where the continental crust has been extremely thinned and where the mantle rocks have been exhumed.

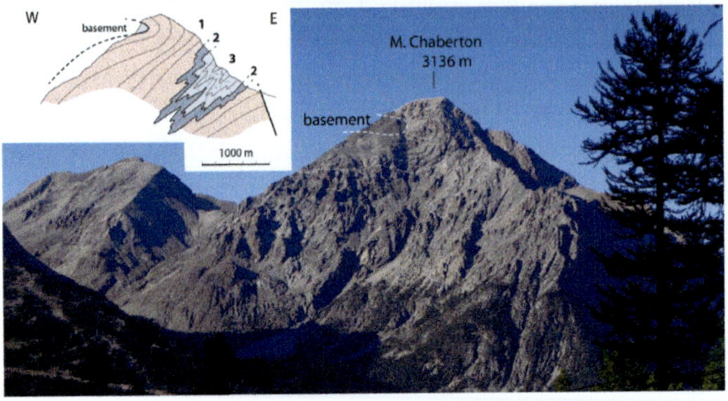

Figure 2.36. *The Chaberton seen from the Chenaillet summit (photo Corre, East is on the right). Inset, geological section according to Caron (1977) showing a residue of the Briançonnais basement pinched into a fold at the top of the pyramid, on its western flank. 1: Norian dolomites, 2: Rhaetian-Hettangian and 3: pre-Piedmont Lias*

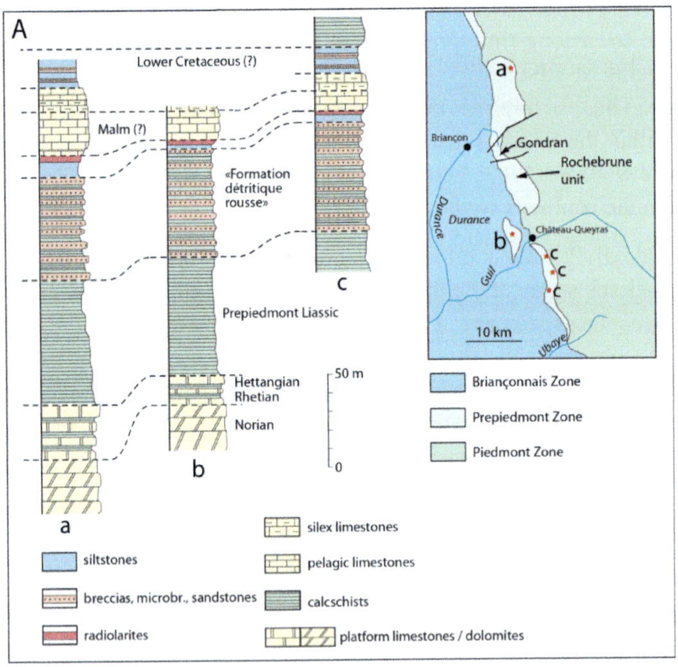

Figure 2.37. *Representative logs of three pre-Piedmont series (A–C) located in the Durance-Guil region (modified from Lemoine et al. 1978). (1–4: according to legend, 5: lower Malm jasper, 6: upper Malm gray marble, 7–9: upper Cretaceous schist, calcschist and marble). On the right, the Charvie series according to Dumont (1984)*

Figure 2.38. *The Rochebrune Range and the Fonds de Cervière, seen from the serpentinites of the Chenaillet (looking south). The pre-Piedmont units are back-thrusted on the Schistes lustrés ophiolite schists (photo Corre)*

Geological works in the pre-Piedmont complexes of the Val Grana, Chaberton, Rochebrune, Roche des Clots and Péouvou mountains, north and south of Briançon, has revealed the specificity of the pre-Piedmont series (Debelmas and Lemoine 1957; Ellenberger et al. 1967; Lemoine 1971; Caron 1977; Lemoine et al. 1978; Dumont 1983). In contrast to the Briançonnais series, the pre-Piedmont series are characterized by the lack of syn-rift emersion, and thus by a continuous Jurassic sedimentary record. However, they display Briançonnais affinities, with well-developed dolomitic and quartzite Triassic layers (Ellenberger and Lemoine 1955). The basal levels include well-developed Norian dolomites (dated by the fossil *Worthenia Contabulata*: Michard 1967) in Valgrana. Further north, these series are detached at the level of Carnian gypsum. They include limestones and dolomites of the Rhaetian-Hettangian age and a succession of Liassic limestones first finely bedded then showing in-fingering of multiple levels of breccias, microbreccias and dolomitic turbidites. This group, which locally exceeds 100 m in thickness, has been given the name "*formation détritique rousse*" (red detrital formation). It is given a Middle Jurassic age. The detrital levels with dolomite clasts of the red detrital formation are progressively enriched in micas and quartz, which indicates the progressive denudation first of the Briançonnais domain sedimentary cover and second of its basement. Meta-radiolarites (jaspers) and gray marbles (attributed to the Lower and Upper Malm) are then found, followed by black schists and calcschists correlated with equivalent Piedmont metasediments that yielded Upper Cretaceous foraminifera (Lemoine et al. 1984) (Figures 2.29, 2.39 and 2.40).

The stratigraphy and paleo-architecture of the pre-Piedmont units enable the definition of two phases in the Jurassic rifting in the distal margin. First, during the Lower Lias, normal faults cut the Triassic and Rhaetian-Hettangian carbonate platform into small tilted blocks. Deposition of breccia formations locally relates to this rifting tectonics. Second, during the Latest Lias and the Dogger, generalized subsidence of the margin occurred, triggering the deposition of the turbidites of the red detrital formation.

Even if they end like the Briançonnais series, it is necessary to underline the unique character of the pre-Piedmont series, characterized especially by the presence of the red detrital formation. The richness in breccias and calciturbidites testifies to the collapse of the internal margin during the rifting linked to the opening of the Ligurian Tethys. Finally, a part of the material that was eroded from the Brianconnais domain is now found in the detrital sediments of the pre-Piedmont domain

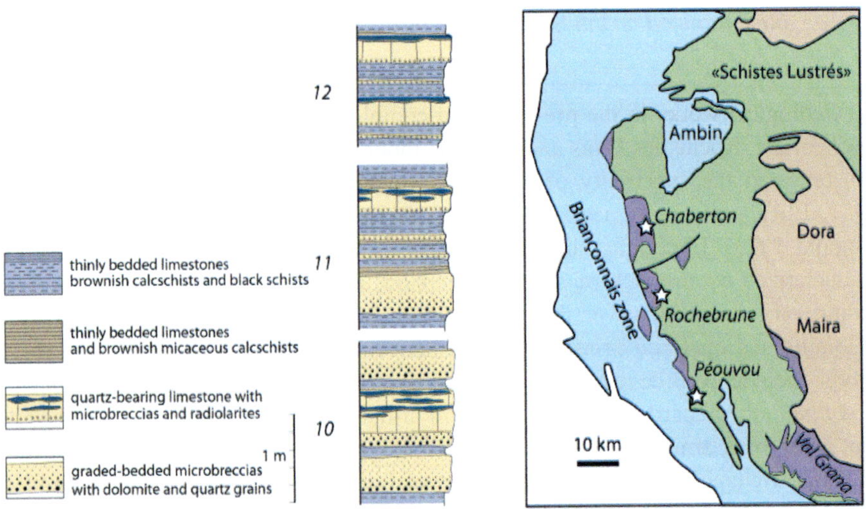

Figure 2.39. *Left: sandstone successions of the pre-Piedmont series in the red detrital formation (*formation détritique rousse*) (example of sections 10–12) (modified from Dumont 1983). Right: map showing the distribution of the pre-Piedmont series (in purple) in the Southwestern Alps (modified from Dumont 1984)*

The coherence of tectonic-sedimentary records between the Briançon and Pre-Piedmont paleo-domains allows us to propose reliable margin models in their broad acceptance. These models confirm: (1) the ubiquity of the Lias-Dogger rifting and, (2) the establishment of pelagic conditions at the end of the Jurassic in the three domains of the margin (Dauphinois, Briançonnais and Pre-Piedmont domains). In the Western Alps, these pelagic conditions persisted into the Upper Cretaceous (Dumont 1984).

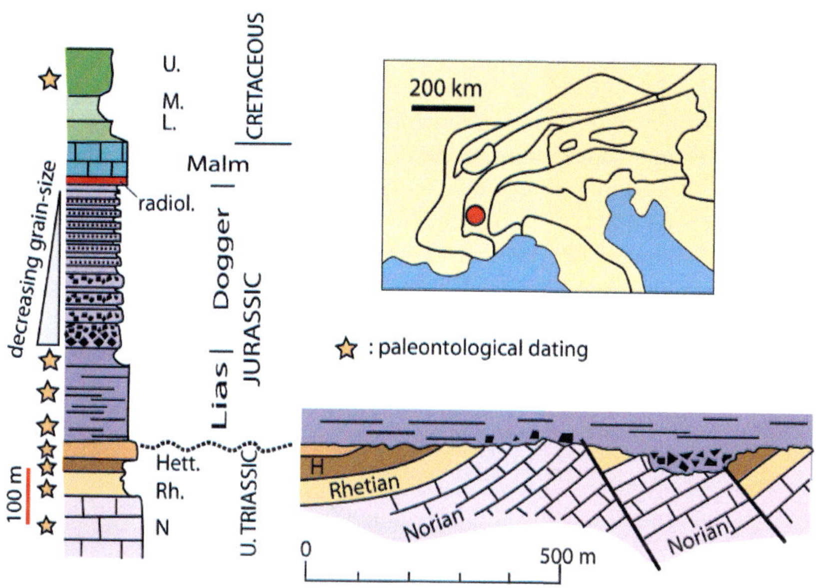

Figure 2.40. *Simplified stratigraphic log of the pre-Piedmont units in the southern part of the Western Alps and reconstruction of the corresponding margin with its two rifting phases: (1) Early Lias, rift tectonics and block tilting; (2) Late Lias, generalized subsidence and emplacement of turbidites of the "red detrital formation" in the Mid-Jurassic (modified from Lemoine et al. 2000; Lc: Limestone Lias, Ls: schisty Lias)*

2.3.6. *The Briançonnais and pre-Piedmont of the Chablais-Romande Prealps and of the Valais*

The Alpine internal zones largely overthrust the external zones and form klippes of continental and oceanic material exposed regularly in front of the Penninic thrust along the entire arc of the Western Alps. From south to north, these are: (1) the nappes of the Ligurian Maritime Alps, (2) the nappes of the Embrunais-Ubaye with the Helminthoid flysch of the Autapie and Parpaillon (and the Subbriançonnais slices involved in the tectonic sole) (Kerchove 1969), and (3) the nappe of the Chablais Prealps (in France) and their continuity in the Romande Prealps (in Switzerland). To these must be added the small Annes and Sulens klippes in the Bornes subalpine massif (Caron et al. 1967).

The Chablais and Romande Prealps are formed by the juxtaposition of units originating from the Alpine distal continental margin, the Ligurian-Piedmontese oceanic domain and probably from the opposite Adriatic margin (Badoux and Mercanton 1962; Baud 1972). They are tectonically emplaced over the Ultrahelvetic terrains to the east and over the Oligocene "red molasse" to the west. At the base, they display terrains of Briançonnais and Subbriançonnais affinity forming the Rigid Median Prealps and the Plastic Median Prealps respectively. These terrains are tectonically overlain by sediments of more internal origin, forming the pre-Piedmont Brèche nappe (Caron and Weidmann 1967). The klippe ends with the upper nappe of the Prealps, formed by four subunits: the Sarine, the Dranses, the Simme, and the Gets nappes (Caron 1972). The first two units are comparable to the Helminthoid flysch of the Southwestern Alps and Ligurian Apennines (Kerckhove 1969). The Gets nappe contains oceanic material from the Ligurian Tethys. The Simme nappe presents flysch-type facies reworking sediments having an affinity with the South Alpine margin. This makes it the westernmost sedimentary evidence of Adriatic origin at the external front of the Alpine belt.

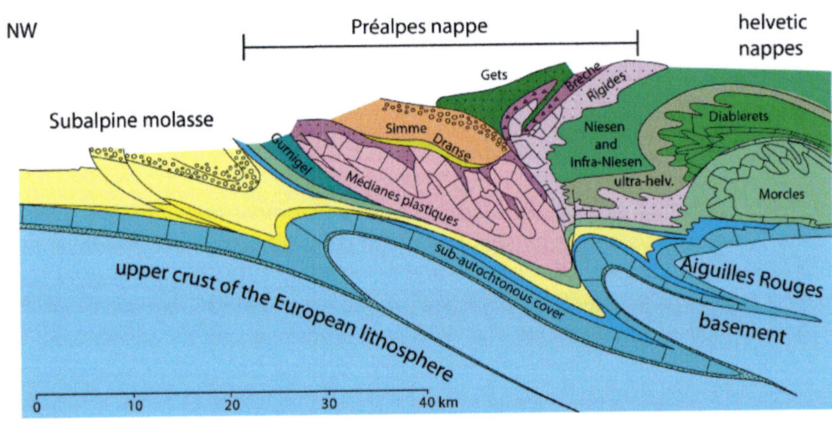

Figure 2.41. *The Prealps nappes in their tectonic framework. Cross-section taken from the general geological section of the Alps constructed by Escher et al. (1994)*

The Dogger sediments of the Rigid and Plastic Median Prealps include very shallow marine facies (Dogger with Mytilus) and conglomerates with coal seams. It is locally rich in iron oxides (siderolitic facies), red and ochre alteration clays, bauxite and dolomitic breccias that infiltrate the Triassic dolomites. As in the Briançonnais of the Western Alps, these emersion facies demonstrate that Alpine rifting is associated with a surrection of the distal margin.

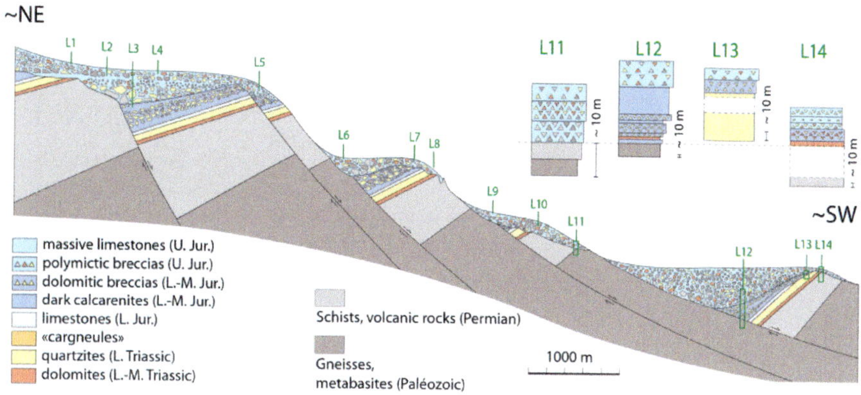

Figure 2.42. *Reconstruction of the European paleomargin in the original location of the Mont-Fort nappe (Pantet et al. 2020). This restoration, based on lithostratigraphic sections, shows the abundance of breccia facies of the Lower-Middle Jurassic age that recorded the dislocation of the margin during Alpine rifting. Affinities are strong with the Prealps Brèche nappe and the Roche des Clots-Grande Hoche pre-Piedmont units*

In the Brèche nappe, the Dogger layers are turbidites deposited in a deep-sea fan environment. The source of the detritus was a ridge separating a so-called "Helvetic" basin from the Piedmont-Ligure Ocean. This bathymetric high, in a Briançonnais position, corresponds to the uplifted region of the Median Prealps. It persisted following the collapse of the margin during the Jurassic rifting (Ringgenberg et al. 2001). The Brèche nappe thus exhibits strong affinities with the pre-Piedmont terrains of the Roche des Clots-Grande Hoche unit characterized by the turbidites of the "red detrital formation". However, the coarse facies of the Brèche nappe indicate that the corresponding domain was closer to the Brianconnaise slope than were the pre-Piedmont formations of the Cottian Alps.

The Mont-Fort nappe (Figure 2.25) is located in the Valais region, east of the Prealps, and is part of a larger group of nappes known as the Grand Saint-Bernard nappe. The Mont-Fort nappe displays strong affinities with the Brèche nappe (Figure 2.34). According to some authors, although it is now located more internally, this nappe originated from a more external margin domain than the Brèche nappe and was separated by an ancient ridge from the latter (Stampfli et al. 2001). This paleogeography is reversed for Pantet et al. (2020) who propose that the Brèche nappe was located closer to the domain of the present-day Prealps. These authors identify the specific character of the Brèche nappe whose origin is in a domain called "Prepiedmont", corresponding to the pre-Piedmont of the South-western Alps.

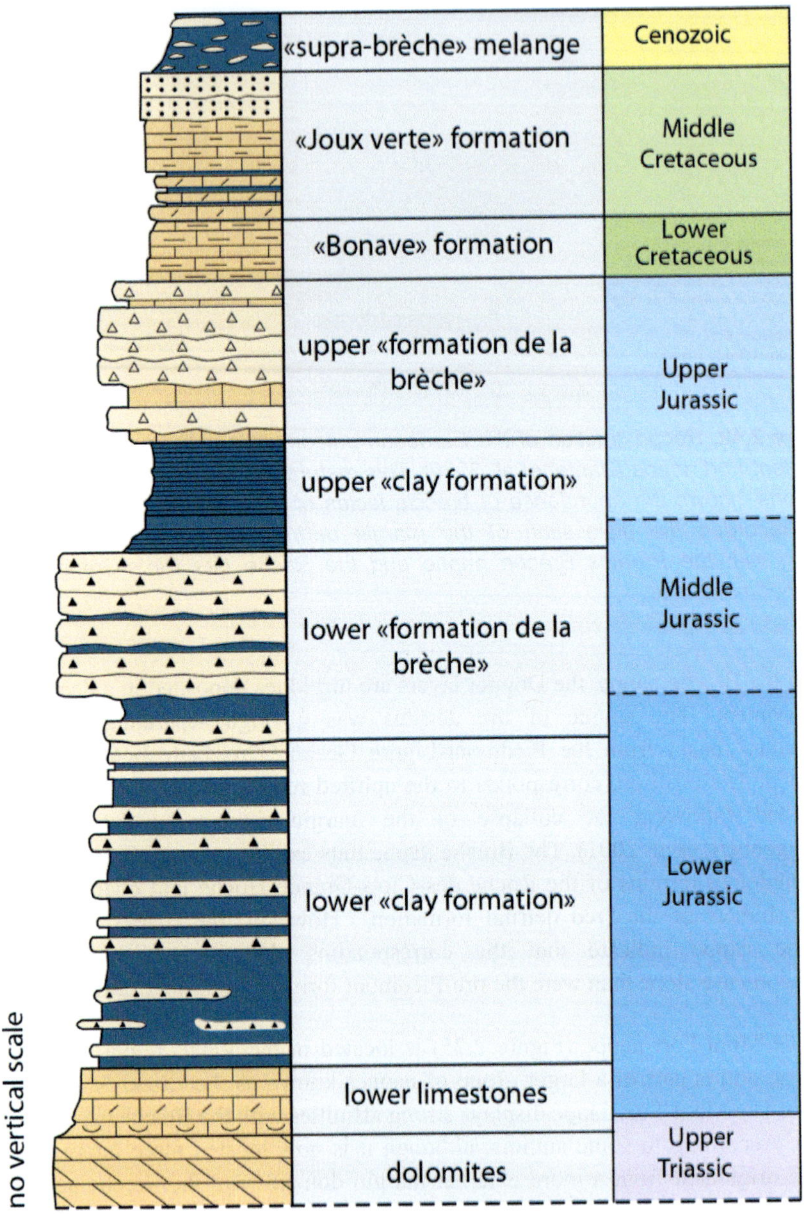

Figure 2.43. *Simplified log of the pre-Piedmont series of the Brèche nappe (Prealps) (from Pantet et al. 2020)*

The pre-Piedmont and Briançonnais domains were distinguished by Ellenberger and Lemoine as early as 1955 following their work in the Vanoise. Later, during his research around Briançon, M. Lemoine confirmed that the pre-Piedmont domain should be distinguished from the Briançonnais domain because of its pre-Malm content. The presence of a thick, detrital Liassic formation indicates that it belongs to the pre-Piedmont domain. Swiss geologists define the Briançonnais domain as the part of the European margin more or less separated from it by the opening of the Valaisan basin. This definition does not take into account the details of the Mesozoic cover stratigraphy; it is based only on the continuity of the Paleozoic basements in the position of a "Briançonnais microblock". This is a definition of a palinspatic order (which refers only to the initial position of the block in question). For this reason, the Mesozoic covers of these basements have been qualified as Briançonnais, except for the extreme distal margin sometimes called the "pre-Piedmont ramp" (Stampfli 2001). To the north of the Vanoise, on the transversal of the Prealps, the units of the Brèche and Mont-Fort nappes have often been included in the internal Briançonnais, although they have strong affinities with the pre-Piedmont domain in the sense of Lemoine. To get around these terminology problems, we can follow Pantet et al. (2020) who use the term Briançonnais sensu lato (or broad sense) for the whole margin, including the Subbriançonnais and the "pre-Piedmont" (Figure 2.36). Ballèvre et al. (2020) use the term "internal Briançonnais" to designate the pre-Piedmont domain in the Southwestern Alps. This definition of the internal Briançonnais is not that of Lemoine's, as we have seen, nor does it correspond to those of specialists of the Ambin-Vanoise-Mont-Pourri units (Ellenberger 1958; Jaillard 1988, 1989). According to these authors, the so-called internal Briançonnais units have marked affinities with the Ultrabriançonnais domain, and not with the pre-Piedmont domain.

Box 2.4. *Pre-Piedmont and internal Briançonnais: different views*

In spite of inconsistencies due to questions of terminology between geologists of different schools and experiences, the reconstructions of the European margin on the transversal of the Prealps and the Valais show a succession of classical domains, also well-identified in the Southwestern Alps. The Subbriançonnais and Briançonnais domains form paleogeographical relays toward the Piemont-Liguria oceanic domain. However, one important question does not stand up to the analysis of the literature, namely the definition of the pre-Piedmont in the Penninic zones located in Switzerland. This question has been discussed in detail in the box above. A second important point concerning the differences between the north and southwestern internal zones is worth emphasizing here. In the Southwestern Alps, the Dauphinois domain lies directly west of the Briançonnais. By contrast, in the Swiss Alps, a new domain (of oceanic or proto-oceanic origin) relaying the Subbriançonnais domain to the north, is now present east of the Helvetic domain. This domain is the Valais domain, presented in the following section.

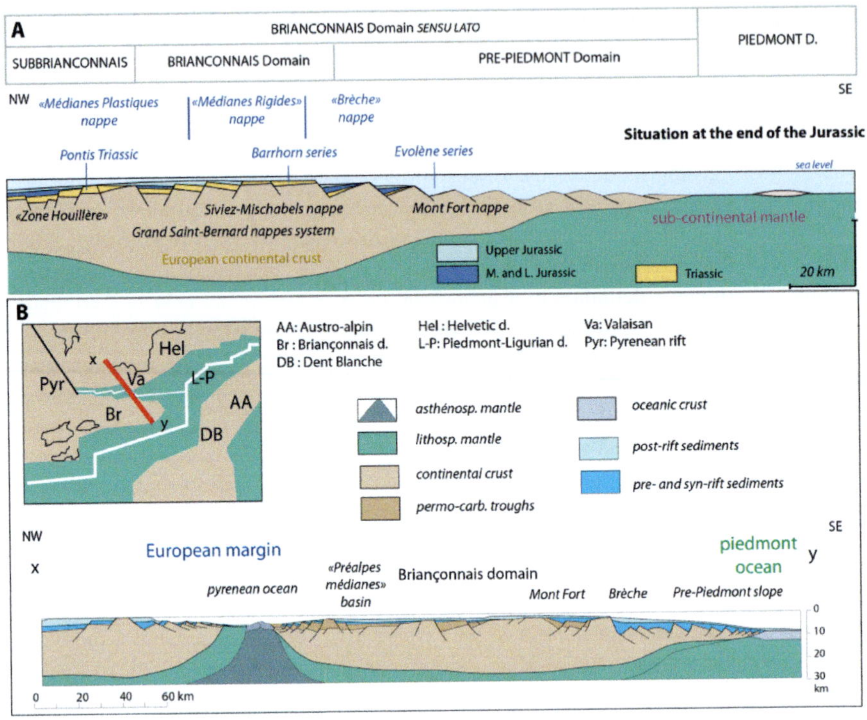

Figure 2.44. *Position of the various units of the Prealps on the European paleomargin according to Pantet et al. (2020) (section in A) and according to Stampfli et al. (2001) (map and section in B). The xy-section shows an Albian situation for the domains now included in the Alpine and Pyrenean orogens*

2.3.7. *The Valais domain*

The relicts of the Valais basin form the Valais Zone developed from the northern end of the Subbriançonnais Zone at the Petit-Saint-Bernard Pass (Versoyen Complex on the French-Italian border) to the Engadine Window at the eastern end of the Central Alps (Tasna Complex in Graubünden, Switzerland) (see full references in Fudral 1973; Debelmas 2001; Loprieno et al. 2010; Beltrando et al. 2012). The Valais zone is bounded to the northwest by the frontal Penninic thrust and to the southeast by the Briançonnais thrust. It is generally accepted today that the Valaisan basin opened up within the European margin by isolating a strip of the continent often called the "Briançonnais microblock" (Trümpy 1980). In general, opening models show a V-shaped basin that closes to the south.

The Valais domain has been defined as a single subsidence basin, located north of the Briançonnais and hosting marine series made up of breccias, conglomerates and fine sandstone-carbonate flysch (calcschists). This basin corresponds to the Valaisan geosyncline of Haug (1909), also highlighted by Trümpy (1955). The coarser sediments, recording strong geodynamic activity, contrast with the relative "calm" of the neighboring Subbriançonnais series, which justified the identification of a specific palaeogeographic zone.

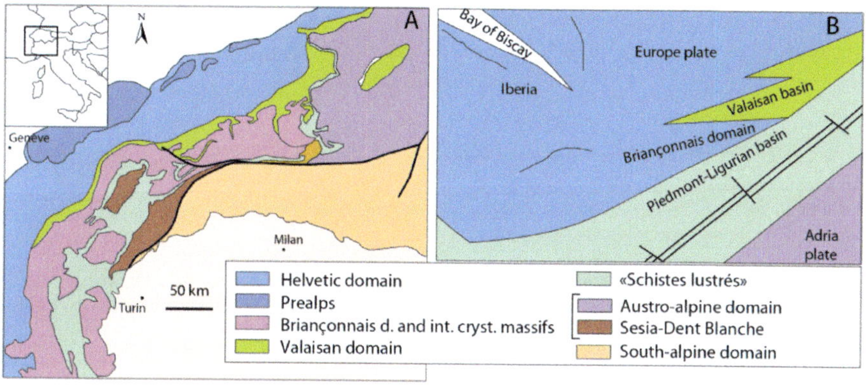

Figure 2.45. *Location of the Valais zone in the Western Alpine belt and Cretaceous paleogeographic reconstruction showing the junction with the Piemont-Liguria Ocean. Note that, in this representation, the Valaisan oceanic domain closes to the southwest without connecting to the Bay of Biscay (modified from Beltrando et al. 2012). Compare with Figures 2.40 and 2.43 in which this connection is represented*

The calcschists of the Valais zone outcropping in Switzerland in the Grisons region, are called the "Bünderschieffer". In the Tarentaise valley (France), the Valaisan metasediments form the "Tarentaise breccias" or Roignais-Versoyen unit, subdivided into several subunits forming an internal Valaisan and an external Valaisan. The latter includes the Moutiers, Roc de l'Enfer and Petit Saint Bernard units (Fudral 1973, Figure 2.46). The basement of the Valaisan flysch is composed of Carboniferous, Permian, thin carbonate Triassic and a very small Jurassic series with a Subbriançonnais character. The internal Valaisan corresponds to the Versoyen unit, the basement of which contains greenschist facies mafic and ultramafic material including pillow lavas associated with breccias of various magmatic rocks. Because of the affinity of some of these outcrops with an ophiolitic complex, it has long been thought that the Valais Basin may locally have an oceanic basement (Loubat 1968, 1975, 1984; Loubat and Antoine 1965; Loubat and Delaloye 1984).

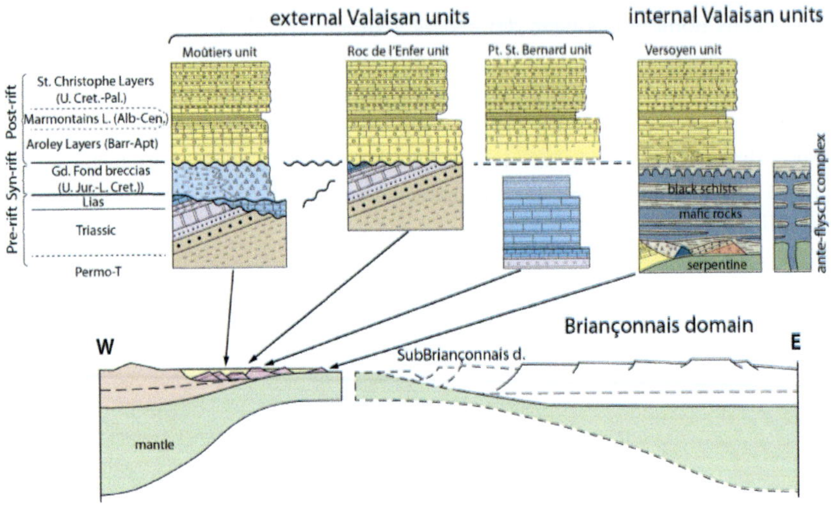

Figure 2.46. *Stratigraphy and paleoarchitecture of the Valais Basin: mantle exhumation and individualization of extensional allochthons according to Loprieno et al. (2011)*

The Valaisan flysch belongs to a trilogy that applies to all the series of the internal and external Valaisan. This "Valaisan trilogy" includes, from bottom to top, the Aroley, Marmontains and Saint-Christophe layers. Similar formations lying on top of a Triassic-Liassic series and also associated with serpentinites and black schists are found further north, at the front of the Briançonnais basement units, in the Visp and Brig regions (Burri 1979).

Reliable ages of the Valaisan trilogy have been obtained through paleontological dating. The minimum age of its basal part is Turonian (*Globotruncana lapparenti,* Antoine 1965 and *Globotruncana lineiana*, Fudral 1973, 1980, 1998) in the Aroley Formation (Gidon 1996). However, for some authors, the Valaisan trilogy started earlier, in the Barremian-Aptian (Loprieno et al. 2011). It continued until the Eocene.

In the internal Valaisan (Versoyen unit), the flysch basement consists of serpentinites associated with breccias at Collet des Rousses, basement fragments (granitoids) at Punta Rossa, and with gabbros at Pointe du Clapet. On these magmatic rocks rests the ante-flysch complex made up of basic sills intrusive in black schists (Cannic et al. 1995; Cannic 1996). This complex is in direct contact with mantle rocks at Mount Miravidi. The ages obtained by the U/Pb method on zircons from magmatic rocks range from 337 ± 4.1 Ma (Masson et al. 2006) and 309 ± 6 Ma (Schärer et al. 2000) to 272 ± 2 Ma and 267 ± 1 Ma (Beltrando et al. 2007).

This implies that the intrusions of the ante-flysch complex cannot be considered the result of Cretaceous magmatism (Mugnier et al. 2008). For Beltrando et al. (2007; 2012), the dated magmatic rocks are part of the classic Permian bimodal complex. It should be noted that serpentinites and sills of the ante-flysch complex remain undated. In addition, the presence of belemnites in the matrix of the Collet des Rousses breccia indicates a Mesozoic age for the flysch basement (Antoine 1971). However, Jurassic ages are reported from the Misox zone (161.0 ± 5.9 Ma; Liati et al. 2005) and Cretaceous ages are known in the Central Alps (Chiavenna ophiolites at ca. 93 Ma) and in the Balma unit in the Western Alps (93.4 ± 1.7 Ma) (Liati and Froitzheim 2006).

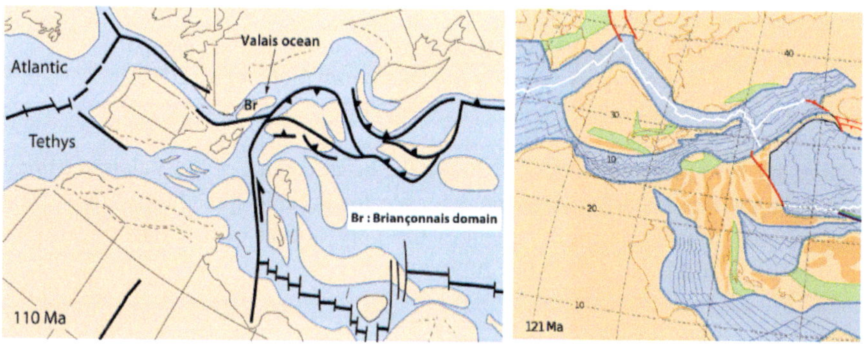

Figure 2.47. *How the Valais Ocean isolated the Briançon peninsula. For Ricou et al. (1996) (left) and Stampfli and Hochard (2009) (right), the Valais Ocean connected to the northern Pyrenean zone making the Briançonnais domain a peninsula barely attached by the future land of Corsica to the northern Iberian plate. In these configurations, around 120–110 Ma (Early Aptian-Albian), the Alpine realm displayed the maximum extension of its pre-orogenic oceanic domains*

Many questions concerning the Valais zone have remained unanswered for a long time, especially those concerning the basement/cover relationships, which has led to contradictory geodynamic interpretations. Since the age of the Valaisan trilogy is well constrained, interpretations are mainly dependent on the age and significance of its basement.

According to the authors, the opening of the Valais Basin occurred either in the Upper Jurassic, like the Piemont-Liguria Ocean or in the Lower-Middle Cretaceous (Aptian-Albian), when the Iberia plate began its eastward drift leading to the opening of the Pyrenean Basin and the Bay of Biscay (Frisch 1979; Stampfli 1993). In this case, the Valais Ocean was not closed to the southwest and ocean spreading was relayed by a transform fault or an active ridge to the northern Pyrenean zone and then to the North Atlantic Ocean (e.g. Stampfli et al. 2002) (Figure 2.43).

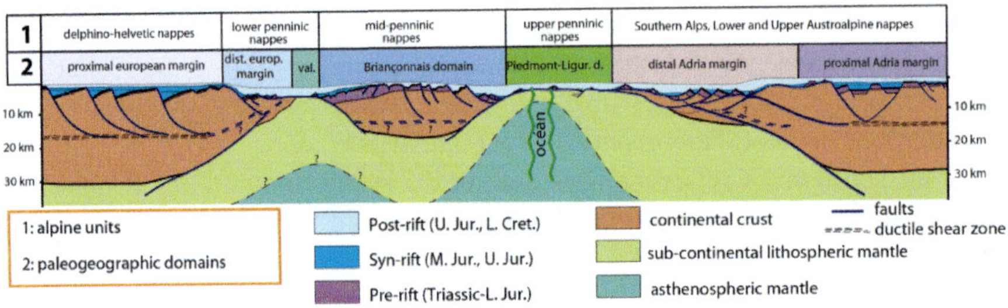

Figure 2.48. *Reconstructed section of the Alpine domain along the Valais Ocean cross-section (modified from Mohn et al. 2010)*

Note that for Masson et al. (2002, 2008), the basal contact of the Valaisan trilogy is not stratigraphic. He argues that the evidence for oceanic spreading of the Valais domain is not reliable and he challenges the very notion of a Valais zone. This hypothesis was still being debated in 2020.

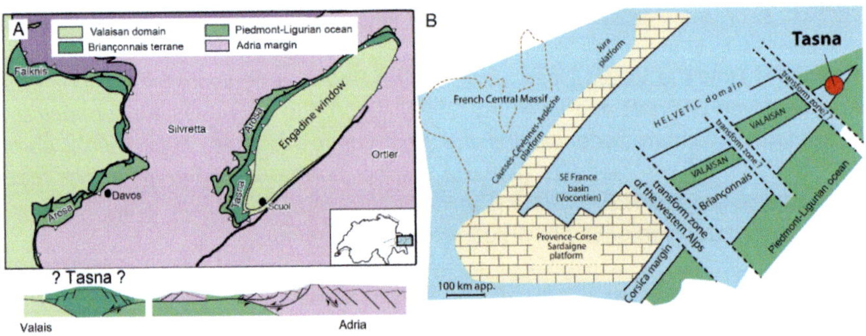

Figure 2.49. *A) Simplified map of the Engadine window and cross-section showing the position of the distal Tasna margin on the Briançonnais block at the eastern edge of the Valais domain (modified from Florineth and Froitzheim 1994). B) Initial position of the Tasna complex according to Lemoine et al. (2000)*

Interpretations by Mohn et al. (2010), Loprieno et al. (2011) and Beltrando et al. (2012) make the External Valaisan the distal continental margin of Europe, in a paleogeographic position similar to the location of the pre-Piedmont domain at the margin of the Piemont-Liguria Ocean. In their models (Figures 2.42, 2.44 and 2.75), the Internal Valaisan contains the mantle rocks exhumed in the very distal part of the margin. Serpentinites with possible traces of magmatism (undated) and continental tectonic breccias may have formed along with a crust-mantle detachment at the foot of the passive margin.

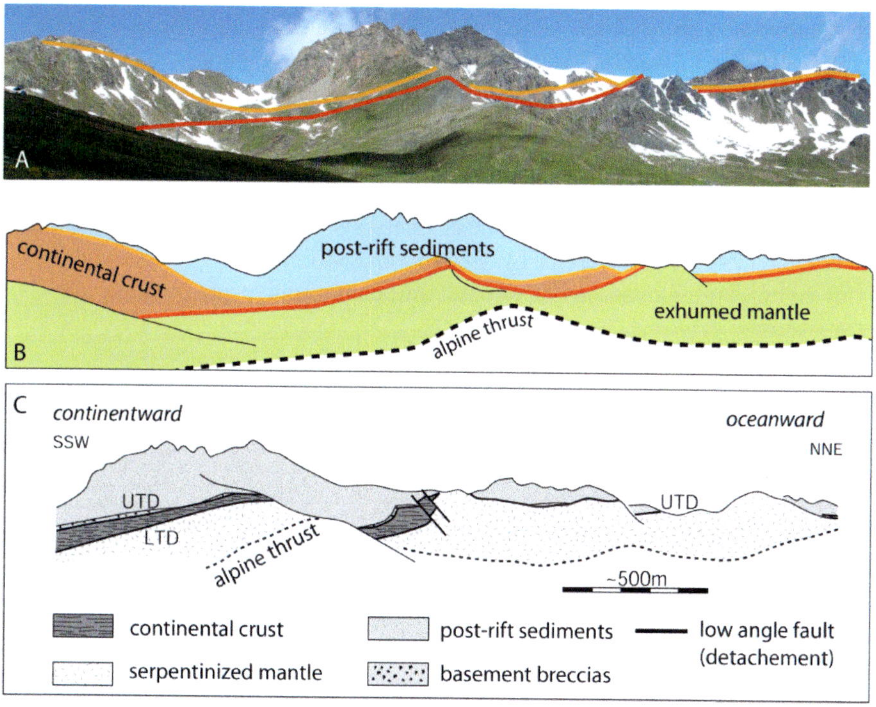

Figure 2.50. *The Tasna detachment system from Beltrando et al. (2014) (A and B) and its original interpretation from Florineth and Froitzheim (1994) (C). UTD: Upper Tasna detachment, LTD: Lower Tasna detachment*

For the above authors, the Valais trilogy represents a post-rift formation. Such interpretations imply that the Valaisan rifting occurred during the Cretaceous, in a Subbrianconais domain already affected by Jurassic rifting. To support their interpretation, the authors first rely on information obtained in the Tasna complex (see below) showing the exhumation of the mantle north of the Valais domain. They then refer to models of the opening of the future North Pyrenean zone, during the same period, also involving the exhumation of the sub-continental mantle (Lagabrielle and Bodinier 2008; Lagabrielle et al. 2020).

The Lower Engadine window is located at the boundary of the Eastern Alps. It is open immediately east of the Arosa-Platta zone and gives access to the Tasna nappe, which represents a remnant of the very distal Briançonnais margin on which the Pennic and Austroalpine units of the Grisons lie tectonically. The western edge of the window provides an exceptional SSW-NNE section showing a 5 km long transition between the hyper-thinned continental crust and the exhumed sub-continental mantle (Ribes et al. 2019). The continental crust forms a wedge of highly deformed material including amphibolites, gabbros, tonalites, and migmatites, thinning to the north (Florineth and Froitzheim 1994). This wedge is bounded at its base by an extensional detachment separating it from the sub-continental mantle represented by highly serpentinized peridotites (Froitzheim and Rubatto 1998). Another detachment caps both the continental basement and the mantle rocks. This second detachment intersects the first one (Florineth and Froitzheim 1994). It is overlain by tectonic and sedimentary breccias and clays, followed by Lower Cretaceous limestones, thus confirming its pre-orogenic age.

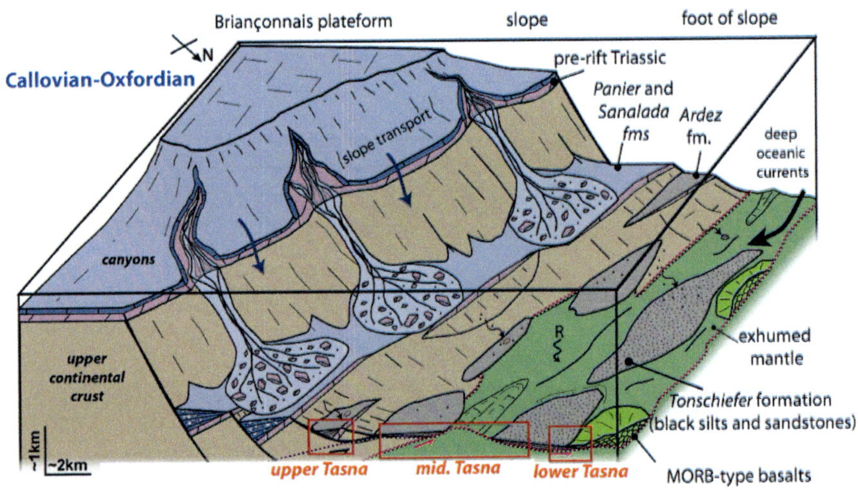

Figure 2.51. *Reconstruction of the bathymetry of the Tasna margin according to Ribes et al. (2019) (looking SW)*

Paleogeographical reconstructions by Ribes et al. (2019) for the Tasna margin show that the oldest sediments deposited on the continental crust and exhumed mantle are of Upper Jurassic age (Tonschiefer Formation). In Figure 2.51, clastic sediments from the Briançonnais platform are transported perpendicular to the axis of the basin, especially along canyons (Paniers and Saldana formations). Silico-clastic sediments derived from the erosion of the pre-rift basement are deposited in the axis of the basin (Figure 2.51).

The Tasna complex escaped subduction and Alpine metamorphism. Therefore, it has preserved in its structure good evidence of mantle exhumation on the floor of the Valais basin thanks to the activity of detachment faults (Florineth and Froitzheim 1994; Manatschal et al. 2007; Ribes et al. 2019). In this sense, this complex constitutes a global reference for the evolution of non-magmatic continental margins. We revisit these concepts in section 2.4.

2.3.8. *The Maritime and Ligurian Alps*

The Briançonnais and Piedmont zones are also present in the south of the Western Alps, at the junction with the Apennines, in the Maritime and Ligurian Alps. These zones are structurally associated with the Helminthoid Flysch nappe constituted by two external and internal subunits, a basal complex and sandstone series. The Ligurian Briançonnais has remained behind the Helminthoid Flysch while the Subbriançonnais units lie under this nappe (Figure 2.47).

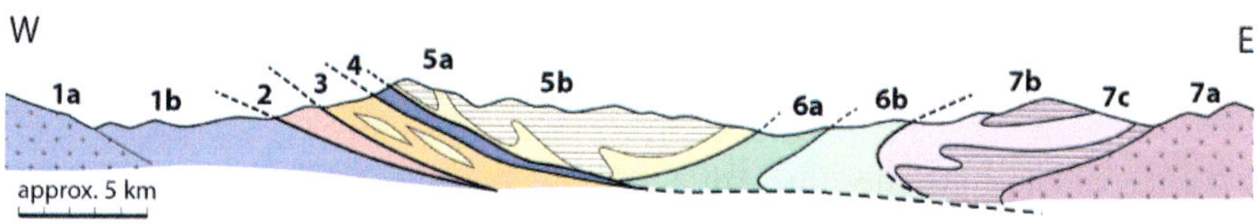

1) autochthonous external zone: Argentera massif (1a) and its cover (1b);

2) Para-autochthonous unit of the Col de Tende;

3) Baiardo flysch;

4) Subbriançonnais units;

5) Helminthoid flysch nappe (5a, basal complex; 5b, sandstone and carbonate series);

6) Helminthoid flysch nappe, internal units (6a, Moglio-Tetico and 6b, Alassio);

7) Briançonnais zone (7a, Permo-Carbonif.; 7b, Mesozoic and 7c, Cretaceous-Eocene).

Figure 2.52. *Section of the Ligurian Maritime Alps between the Argentera massif and the Briançonnais basement (modified from Debelmas 1974)*

2.4. The Alpine passive paleo-margin of the Adria plate

2.4.1. *Introduction*

The conjugate margin that moved away from the European margin during the opening of the Piemont-Liguria Ocean corresponds to the northern and western edge of the Adria plate. Its evolution in the Alpine orogenic prism was totally different from that of Europe. During the Alpine convergence, the Adria plate behaved as the upper plate, whereas Europe, the lower plate, was largely subducted, as shown by the orogenic and metamorphic evolution of the Briançonnais domain. The Adria lithosphere was thus characterized by the absence of Alpine metamorphism. A discussion still remains about the case of the Sesia domain which could represent the western subducted edge of Adria. A recent hypothesis integrating older interpretations proposes a solution in which Sesia represents continental blocks tectonically detached from the Adria margin along detachment faults in the Ivrea Zone (Debelmas 1974; Babist 2006; Manzotti et al. 2014) (see section 2.4).

North of the Alpine belt, the Adria plate terrains represent most of the Eastern Alps. These include units of Austroalpine basement, and its pre- and syn-rift cover. The latter forms a large portion of the Northern Calcareous Alps (NCA). It is detached on Permo-Triassic salt and thrusted far to the north, at the front of the chain, over the peri-Alpine molasse basin (Figures 2.48 and 2.52).

To the south of the Alps, the Adria margin constitutes the basement of all the Apennine units, as well as the basement of the Padan basin (named after the Po river). The Apennine nappes and the nappes of the Southern Alps are thrusted in opposite directions over the Padan basin. In this respect, it should be noted that some authors suggest that right under the Southern Alps, subduction of the Adria plate beneath the European plate already initiated (Schmid et al. 2008) (Figure 2.48).

The deep crust of the Adria plate outcrops in the Ivrea Zone. The Ivrea zone is separated from the eclogitic units of Sesia by the Canavese zone. The latter represents the distal, hyper-extended Adriatic margin. For some authors, it represents the suture of an oceanic basin located between Sesia and Ivrea. Finally, the Austroalpine units derived from the Adria continental crust are also involved in the upper nappes of the Central Alps.

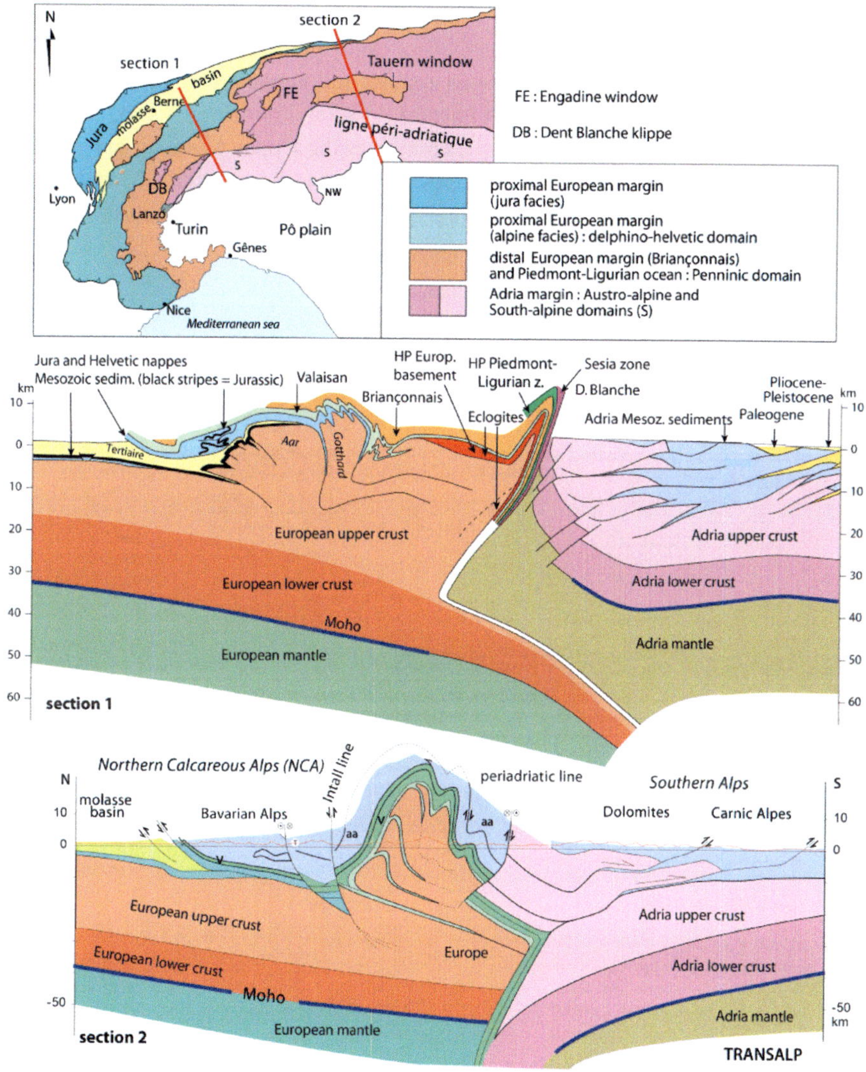

Figure 2.53. *Cross-sections of the Central, Northern, and Southern Alps showing the architecture of continental and oceanic paleo-domains, the subduction of Europe beneath Adria (cross-section 1), and the onset of subduction of Adria beneath Europe (cross-section 2). Simplified sections from Schmid et al. (2004, 2008)*

2.4.2. *The Central Alps*

In eastern Switzerland, in the Grisons region, the nappe structure was formed during the Upper Cretaceous, as in the Eastern Alps (and not in the Cenozoic, as in the Western Alps). From top to bottom, it comprises the Upper Austroalpine (Ötztal, Silvretta-Sesvenna and Campo-Ortler nappes), the Lower Austroalpine (Bernina and Err nappes), and the South Penninic nappes (Arosa zone, Platta-Totalp nappe and Forno-Malenco complex).

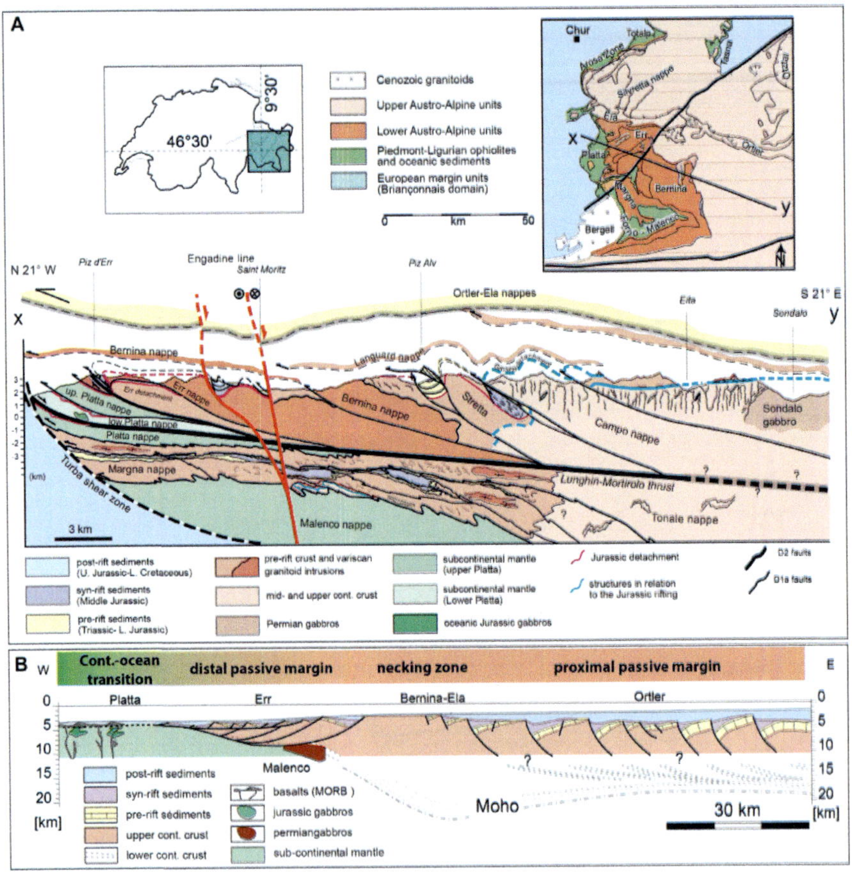

Figure 2.54. *Cross-section of the nappe system in the eastern Central Alps (A) and position of the units within an idealized continental margin (B). The Ortler-Ela nappe is the highest. It overlies the Bernina nappe, which in turn overlies the Err nappe, which is itself thrust over the Platta nappe (modified from Manatschal and Bernoulli 1999 and Mohn 2010, 2011)*

The highest Austroalpine nappes are derived from the proximal Adriatic margin and the lowest ones from the distal margin. The Penninic nappes, on the other hand, are derived from the Piemont-Liguria Ocean (Totalp) or from the ocean-continent transition (Platta) (Dietrich 1969, 1970; Eberli 1988; Froitzheim 1988; Conti et al. 1994; Manatschal and Nievergelt 1997; Hermann et al. 1997; Desmurs et al. 2001; Mohn 2010; Mohn et al. 2010, 2011, 2012; Incerpi et al. 2019) (Figure 2.49).

The Ortler nappe is composed of Triassic to Cretaceous sediments with small units of Paleozoic basement overlain by Permo-Triassic sediments. Extensional pre-orogenic deformation related to the Liassic rifting is well recorded by significant thickness variations of the syn-rift Liassic sediments on either side of steep Liassic normal faults. The basins bounded by these faults are filled with breccias and calciturbidites interbedded with marls and limestones belonging to the Allgaü Formation (Eberli 1988) (Figure 2.50). This formation mainly results from the erosion of a Triassic carbonate platform and thus shows clear affinities with the pre-Piedmont detrital formation of the Roche des Clots-Grande Hoche series of the Western Alps.

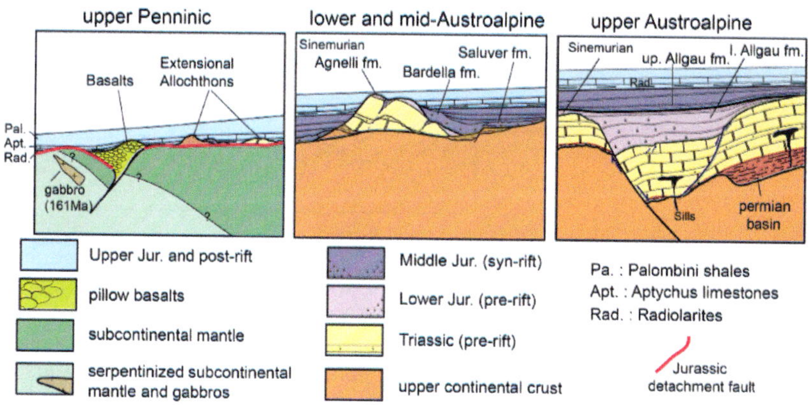

Figure 2.55. *Diagrams showing the sedimentary architecture of key sectors of the reconstructed Adriatic paleomargin in the Central Alps (Mohn et al. 2010)*

The Allgaü Formation contains a Toarcian manganese-rich median layer, separating two parts. The lower part fills grabens opened thanks to steep faults cutting through the Triassic pre-rift carbonate platform. The upper part of the formation seals these faults. This horizon allows correlations with similar sequences in the Northern Calcareous Alps and the Southern Alps (Mohn et al. 2010; Ribes et al. 2019).

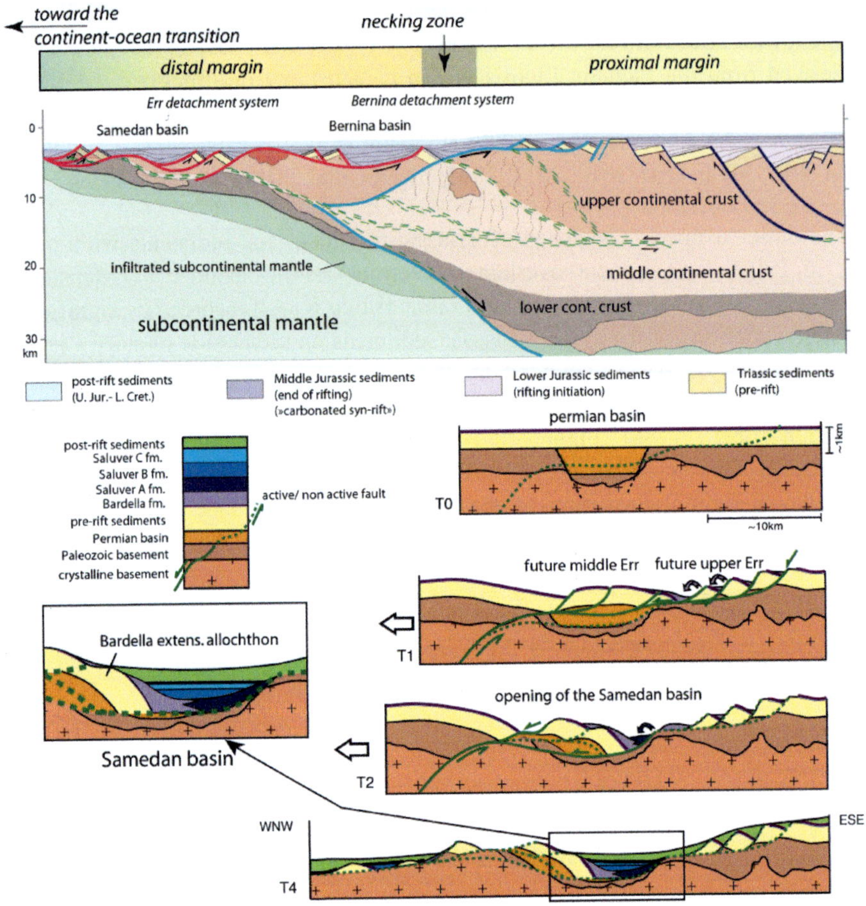

Figure 2.56. *Architecture of the distal Adriatic paleomargins of the Central Alps and position of the Samedan Basin in relation to the activity of the Err detachment. The lower diagrams provide details of the evolution of the Samedan Basin during the Jurassic. This basin is controlled by the structural inheritance of a Permian basin (from Masini et al. 2011)*

After the deposition of the Allgaü (syn-rift) Formation, radiolarites were deposited that marked the establishment and generalization of the post-rift period. These radiolarites are Bathonian to Oxfordian in age (Bill et al. 2001) and constitute a major landmark in the Alpine Tethys. Indeed, they occur both within the series of the Adriatic margin (upper and lower Austroalpine) and over the ophiolites of the Piedmont-Ligurian domain. They can be correlated to the nodular limestones of the

"Guillestre marbles" type in the Briançonnais domain. They represent the post-rift level since they were deposited after oceanic spreading when the margin was no longer tectonically active.

In the lower Austroalpine terranes of the Err nappe, the basal part of the Allgaü Formation is lacking. The Triassic sediments occurred as extensional allochthons abandoned on the continental crust at the continent-ocean transition of the Adriatic margin by an active detachment fault in the Jurassic, during the deposition of the syn-rift Agnelli Formation of the Pliensbachian (Froitzheim and Eberli 1990; Froitzheim and Manatschal 1996; Manatschal and Nievergelt 1997). The most spectacular extensional allochthons are observed in the Bardella block and the Piz Laviner. The detachment cuts through Triassic and Liassic to Pliensbachian terranes. The same detachment is overlain by radiolarites and basalts, while MOR-type gabbros intruding the mantle have been dated at 161 ±1 Ma (Schaltegger 2002). In the Samedan paleo-basin, corresponding to the middle and upper parts of the Err nappe, the geometry of various stratigraphic markers such as the top of the Agnelli Formation and the base of the radiolarites (TAF and RC in Figure 2.51) indicates that the detachment operated for 20 Ma. In this syn-rift basin, the clastic sedimentary sequence between TAF and RC is made up of two members (Bardella and Saluver) whose grain size changes from coarse to very fine upward. In fact, the Bardella Formation includes breccias reworking the Triassic and the Lower Jurassic carbonates including the syn-rift Agnelli Formation. Above the Bardella Formation lie conformably the Saluver A and B Formations, rich in fine sandstones, then the Saluver C Formation, rich in black schists and distal turbidites with rare intercalations of conglomerates (Masini et al. 2011). The angular relationships between these different formations indicate greater tilting than generated by conventional tilted blocks bounded by steep faults. Such a higher extensional rate is the consequence of detachment faulting.

The Platta nappe was located in a more western position than the Err nappe, i.e. toward the oceanic domain. It contains ophiolites and represents the region floored with exhumed ultrabasic subcontinental rocks lying at the foot of the very distal margin, at the transition with the Piemont-Ligurian Ocean. Extensional continental basement klippes tectonically overlie the exhumed mantle. This peculiar setting was acquired prior to the deposition of the first basalts and oceanic sediments. The peridotites involved in the Platta nappe show characters of the sub-continental mantle with a low rate of partial melting, that has been impregnated (re-fertilized) by MORB-type magmas during exhumation. In section 2.5 of this chapter, we review these concepts and discuss the advances in our understanding of the structure of non-magmatic passive margins.

2.4.3. *The Eastern Alps*

In the Eastern Alps, the bulk of the outcrops corresponds to the Austroalpine nappes, all of which are of Adriatic origin and grouped within the Lower, Middle and Upper Austroalpine. The stacking of the nappes occurred during the Late Cretaceous following east-west kinematics. However, units of European and Piedmont-Ligurian origin are visible in tectonic windows opened by erosion and through strong anticlinal folds affecting nappe stacking. The folding of the nappe stack occurred during the Cenozoic following south-north kinematics. Tectonic windows are from west to east: the Lower Engadine (described above), Hohe Tauern and Rechnitz windows on the Austro-Hungarian border. In the Hohe Tauern window, the Penninic domain is represented by ophiolites from the Piemont-Liguria Ocean[5]. From north to south, the Eastern Alps include: (1) the Germanic molasse deposited in the peri-Alpine trench, (2) the Helvetic nappes, (3) the flysch zone (Cretaceous-Eocene), (4) the North Calcareous Alps (NCA): a group of three units is generally defined here: the Bavarian, Tyrolean and Juvavian nappes, (5) the grauwackes zone and the lower Austroalpine units.

At the edge of the Hohe Tauern window, the lower Austroalpine unit is reduced to Triassic chips of more or less mylonitized quartzite or dolomite and a tectonic slice of Paleozoic quartzo-phyllades ("Innsbruck Phyllades"). The middle Austroalpine unit developed immediately to the west in the Mesozoic massifs of the Stubai Alps (secondary cover), and in the Pötztal Alps (ancient crystalline basement). The Mesozoic series stratigraphically overlying the gneisses and micaschists of the basement consists of a thick formation of Triassic dolomites up to 1000 m thick, overlain by limestones and calcschists of Liassic to the Jurassic age. The upper Austroalpine unit tectonically overlies the middle Austroalpine unit. It consists of Paleozoic (Steinach nappe) and Mesozoic terranes, mainly Triassic in age (Blaser Dolomite) (Geyssant 1973). In order to fully understand the architecture of the Alpine paleomargins in the Eastern Alps region, it is important to take into account the allochthony of the NCAs. It has long been known that the latter represents the cover of a partially unknown domain. The Tyrolean nappes may represent the cover of the large detached Grauwackes nappe. The Bavarian nappes represent the cover of the Silvretta basement (north of the Engadine window). The original location of the Juvavian nappes remains undetermined.

5. Note that for some authors such as R. Trümpy, whose interpretations are rather "autochthonistic", the ophiolites are derived from several separate oceanic domains, arranged in echelon. Thus, for this author, an isolated Tauern Ocean opened within the future Eastern Alps that led to the Tauern ophiolites (Lemoine and Trümpy 1987).

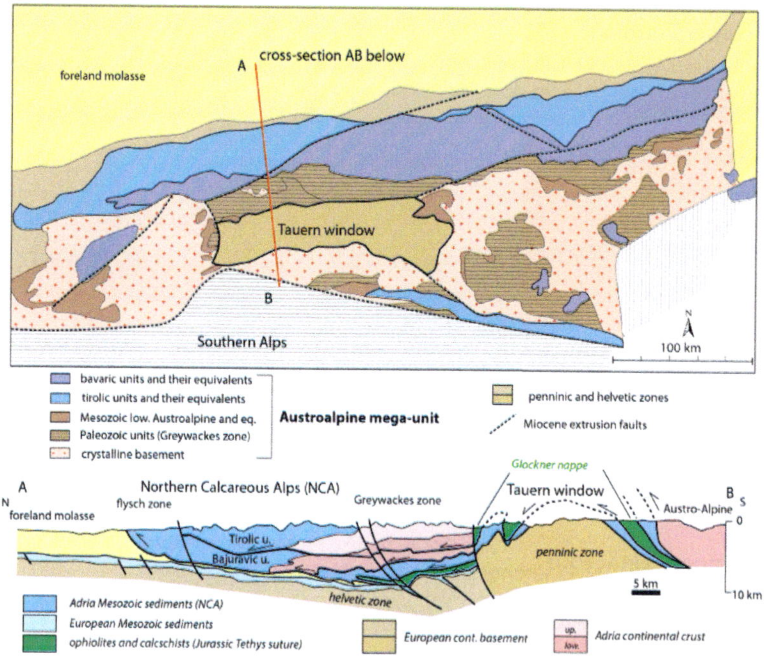

Figure 2.57. *Geological map of the Eastern Alps (modified from Missoni and Gawlick 2011) and highly simplified N-S section according to the Salzburg geological map at 1/200,000*

Figure 2.58. *Simplified stratigraphic column of the western NCAs from Tanner et al. (2003). Panorama of the Dachstein limestone in the Hoher Dachstein (photo Loitzel)*

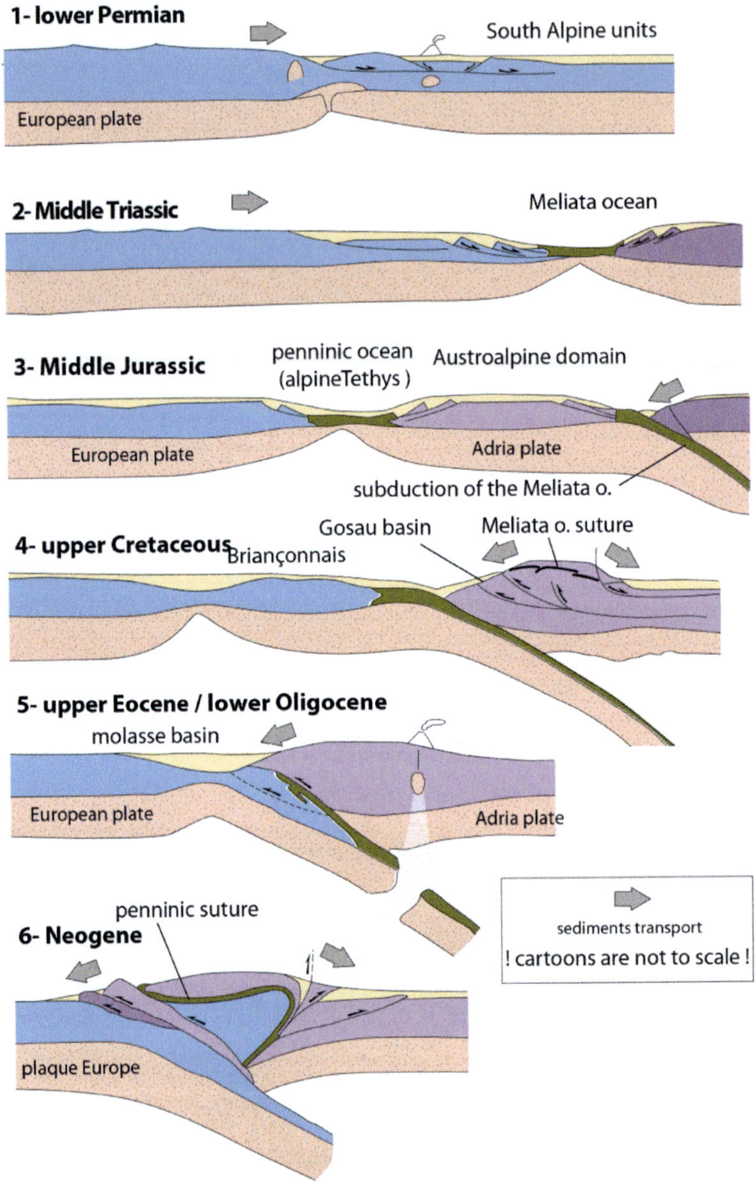

Figure 2.59. *Permian to Neogene evolution of the Eastern Alps (modified from Neubauer et al. 2007). In this model, the Gosau formation results from the erosion of the reliefs related to the Meliata Ocean suture*

The restoration of the structures of the Eastern Alps requires placing the NCA domain behind the domain corresponding to the Middle and Lower Austroalpine nappes. The Jurassic margin of the Piemont-Liguria Ocean at the edge of the Austroalpine domain has disappeared in the Eastern Alps (only its distal part remains visible in the Platta ophiolites of the Central Alps). Therefore, the original location of the NCA has to be placed rather close to the Southern Alps as suggested on the paleogeographic diagram in Figures 2.54 and 2.56.

The Permo-Mesozoic series of the NCA is carbonate-dominated, reaching a thickness of 3–6 km. In the western NCAs, the dolomitic layers of the *Hauptdolomit* lie over Permian sediments with a locally evaporite-rich layer and a detrital Lower Triassic. The Jurassic and Lower Cretaceous formations are relatively thin. They include radiolarites, Aptychus limestones and marls, indicative of a locally deep environment, following irregular subsidence and dismembering of the Triassic platform during the Lias-Dogger, in an evolution similar to that of the Briançonnais domain. Shallow areas and basins were individualized during the Malm. The Lower Cretaceous sediments often contain sandstones. The series ends with continental and detrital layers of the Gosau Formation deposited in Turonian times over an erosional surface intersecting compressional structures (Stern et al. 2011).

At its typical locality in the Gosau Basin (Salzkammergut), the Gosau Formation comprises conglomerates and polygenic sandstones up to 2500 m thick. In some areas, the unconformity is sealed by rudist limestones built on bauxite paleokarsts. Coal and terrestrial fossil layers are numerous (lower Gosau Fm.).

Higher in the series, in the Campanian, sedimentation was dominated by nearshore fine sandstones marking numerous marine incursions (upper Gosau Fm.) (Hofer et al. 2011). Note that the Lower Cretaceous includes olistoliths (pre-Gosau phase) announcing the Upper Cretaceous tectonic phase and its associated main Gosau unconformity (Janoschek and Matura 1980).

Heavy minerals isolated from the Gosau Formation include spinels and blue amphiboles associated with serpentinite clasts (Winkler 1996). These minerals result from the erosion of a mountain range involving ophiolites and exhumed high-pressure units. The Gosau Formation is thus a major marker of alpine geodynamics as it reflects the burial and subsequent exhumation of an eastern Alpine oceanic domain as early as the beginning of the Early Cretaceous.

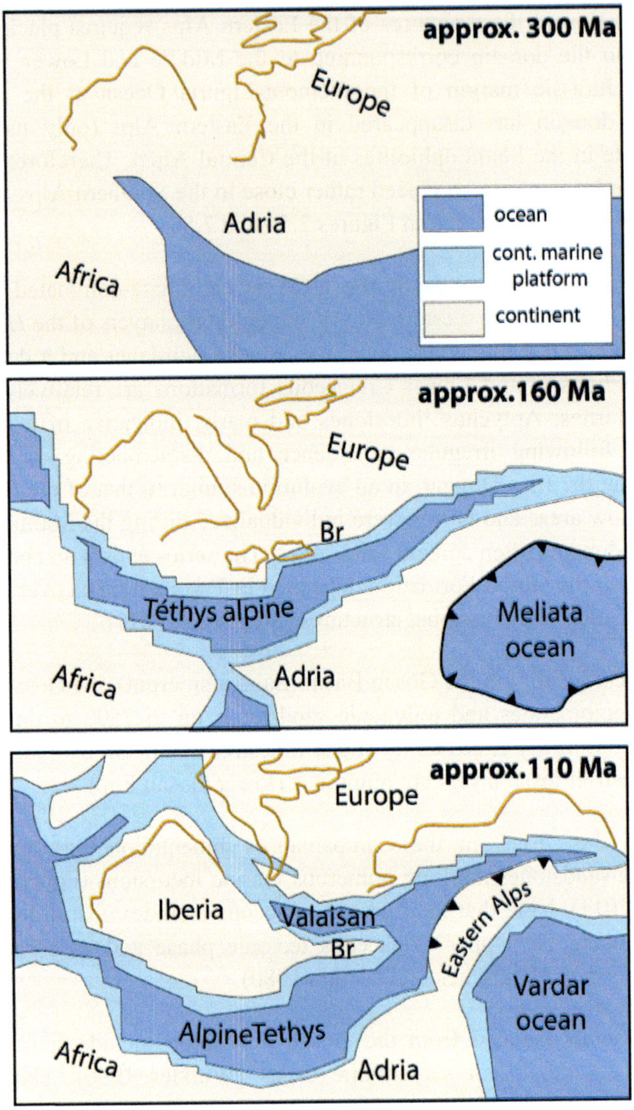

Figure 2.60. *One among several possible models for the evolution of the Eastern Alps (model by Stampfli et al. 2002, modified from Beltrando et al. 2007). Note that the subduction beneath the Adriatic margin of the Eastern Alps occurred as early as 110 Ma. Br: Briançonnais*

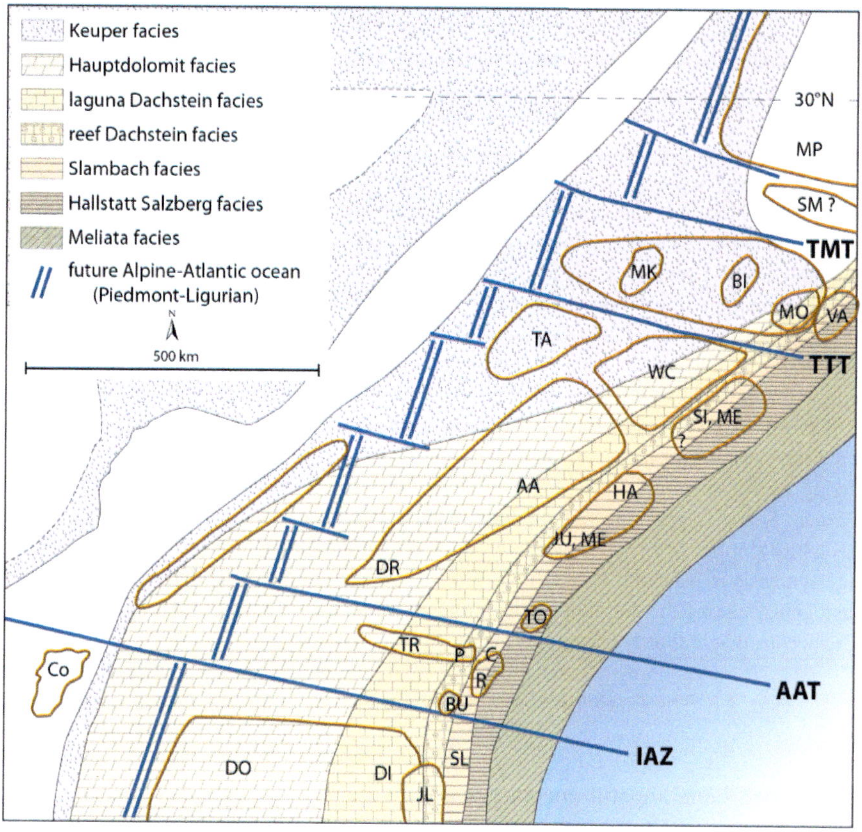

IAZ: future Iberia-Adria zone; AAT: future Austroalpine-Adria transform; TTT: future Tisza-Tatra transform; TMT: future Tisza-Modesia transform; AA: Austroalpine; BI: Biharia; BR: Briançonnais; BU: Bükk; C: Csovar; Co: Corsica; DI: Dinarides; DO: Dolomites; DR: Drau range; HA: Hallstatt; JU: Juvavicum; JL: Julian Alps; Me: Meliaticum; MK: Mecsec; MO: Moma unit; MP: Moesian platform; P: Pilis-Buda; R: Rudabanyaicum; SI: Silicicum; SL: Slovenian trench; SM: Serbian-Macedonian unit; TA: Tatricum; TO: Tornaicum; TR: Transdanubian chain; VA: Vascau unit; WC: Central Western Carpathians.

Figure 2.61. *Paleogeography of the Eastern and Southern Alps within the framework of the NeoTethys during the Upper Triassic. Distribution of Hauptdolomit, Dachstein, and Hallstatt facies discussed in the text. Note the position of the future Piedmont-Ligurian opening (Alpine Tethys) and the Dolomites of the Southern Alps (modified from Missoni and Gawlick 2010)*

The tectonic phase during which the stacking of the Austroalpine nappes took place is termed *Eoalpine*. It occurred prior to and during the sedimentation of the Gosau layers. Numerous radiochronological dating of the metamorphic units of the Eastern Alps indicates burial at 110–90 Ma (see age synthesis in Manzotti et al. 2014). We recall that in this time frame, the Briançonnais domain is characterized by pelagic sedimentation (*calcaires en plaquettes*) and remains far from any orogenic influence. The same applies to the Delphino-Helvetic domain. The Piedmont-Ligurian domain is also under deep marine conditions, but the abundance of ophiolitic debris in the post-ophiolitic series of probable Upper Cretaceous age is interpreted by some authors as the mark of initiation of nappe formation in relation to an early shortening of the oceanic lithosphere domain (Lagabrielle and Polino 1988; Deville et al. 1992). This reflects the onset of intra-oceanic subduction in the Alpine Tethys. This hypothesis is consistent with an interpretation of ophiolitic detritus well represented in the Apennine Ligurian nappes in the Upper Cretaceous (Cenomano-Turonian Casanova complex, Granjacquet and Haccard 1977; Labesse 1981; Lagabrielle 1987). It should also be recalled here that radiochronological dating of high-pressure metamorphism shows that the Sesia continental block and the Zermatt Saas-Fee ophiolites were buried in subduction beneath the Adriatic margin (in a geometry similar to that of the Eastern Alps), during the Late Cretaceous (90–70 Ma, see age synthesis in Manzotti et al. 2014). This is evidence of a progression of the Eoalpine tectogenesis along the Adriatic border, which became a convergent margin in the Lower Cretaceous that is shortly after its formation as a passive margin.

Box 2.5. *The Eoalpine phase and the behavior of the Alpine margins*

However, there are still uncertainties and various interpretations concerning the pre-Gosau and syn-Gosau tectogenesis of the Eastern Alps are sometimes opposed. Indeed, the Gosau layers could have been fed either by: (1) an Eoalpine chain linked to the suture of the Meliata Ocean (Neotethys) under the Adria margin, in the continuity of the subduction of the Meliata Ocean (Missoni and Gawlick 2010), or (2) by an Oman-type chain linked to the subduction of the Ligurian Tethys as early as the Lower Cretaceous as proposed by Winkler (1996). Here we present the evolutionary model of Neubauer et al. (2007) (Figure 2.55), which favors a feeding of the Gosau basins from the Meliata Ocean suture. The model shows the development of the Meliata Ocean margins after the Permian, followed by the individualization of the Austroalpine domain induced by the opening of the Piedmont-Ligurian (so-called Pennic) Ocean during the Jurassic. The subduction of this ocean occurred very rapidly under the Austropalpine margin from the base of the Upper Cretaceous.

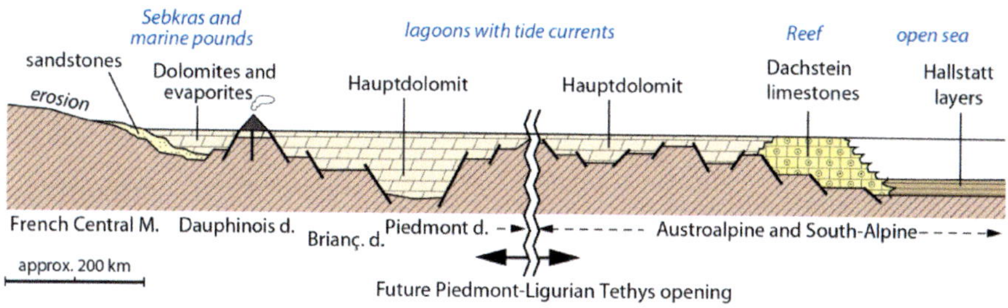

Figure 2.62. *The Alpine Triassic margin along a theoretical W-E transect from the French Massif Central to the Eastern Alps, according to Lemoine et al. (2000). This sketch shows how the Upper Triassic lithologies reflect the architecture of the margin, largely opened at the time eastward, toward the Neotethys (Meliata Ocean)*

In the central and eastern NCAs, the Norian dolomites of the *Hauptdolomit* pass laterally to the Dachstein limestones. These 1000 m thick Norian limestones are not of lagoonal origin like the Norian dolomites but are reef-dominated deposits. They pass laterally to the Hallstatt layers rich in pelagic clays and chert, strongly disturbed by underlying Permian salt diapirs, and then to the pelagic layers of the Triassic Meliata ocean. This represents a full "Neo-Tethyan" logic (Figure 2.57). This logic is reversed during the development of the passive margin linked to the opening of the Jurassic Piedmont-Liguria Ocean. At this time, the deepest pelagic zones are now located to the northwest and no longer to the southeast (Figure 2.56). Central Europe was now connected via the Ligurian Tethys to the central Atlantic Ocean. Such a geodynamical shift triggered a profound change in the depositional environments and the ecosystems.

Box 2.6. *The Mesozoic of the Eastern Alps: the transition from a Neotethyan margin open to the east, to a margin open to the north toward the Atlantic*

2.4.4. *The Southern Alps*

The Southern Alps expose the basement of the Austroalpine margin, whose western end forms the Ivrea zone, and its Mesozoic sedimentary cover. This margin is also known as the South Alpine margin. The Mesozoic sedimentary cover was transported along with large south-verging thrusts while being little deformed internally. It, therefore, shows large, well-preserved extensional structures, where the Lias-Dogger tectonics related to rifting is particularly well observed.

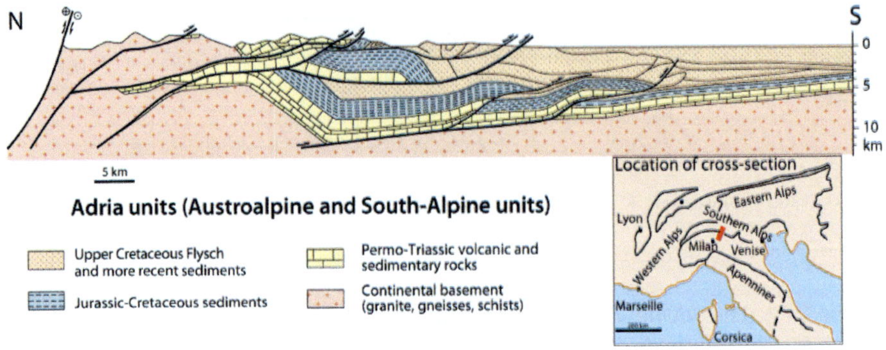

Figure 2.63. *Cross-section of the Southern Alps (modified from Manatschal and Bernoulli 1999). Note the south-verging displacements of the nappes*

The southern Alpine part of the Adriatic paleo-margin is limited to the east (the external side) by the carbonate platform of Friuli, which remained a marine shallow zone during the Jurassic and the Cretaceous. This would be the eastern equivalent of the Causses of the Massif Central (Lemoine et al. 2000). To the west (inner side), toward the ancient Piemont-Ligurian Ocean, the southern Alpine paleomargin is bounded by the Canavese domain, which corresponds to the distal margin (Figure 2.60). In this major suture zone, poor quality outcrops and intense Alpine deformation prevent a reliable understanding of ancient structures. Between Friuli and Canavese, the margin presents a succession of highs and basins opened during the rifting of the Alpine Tethys. North of Venice is the Belluno Basin, opened between the Trentino platform (west) and the Friuli platform (east), with an almost N-S direction. North of Milan, in the Lombardy Alps, rifting has led to the individualization of basins and ridges of smaller dimensions but displaying similar orientations (Bernoulli et al. 1990).

The Dolomite region of northern Italy is emblematic of the pre-rift sedimentary cover of the southern Alpine margin (Figures 2.61 and 2.62). This cover includes at the base a thick dolomitic level of Upper Triassic age, the *Hauptdolomit* or Main Dolomite, deposited during the Norian, 230–210 Ma ago. Deposition occurred in a tropical climate in a lagoon of a shallow coastal sea, undergoing tidal movements, and regular subsidence leading to sediment accumulation up to 1000 m thick in places.

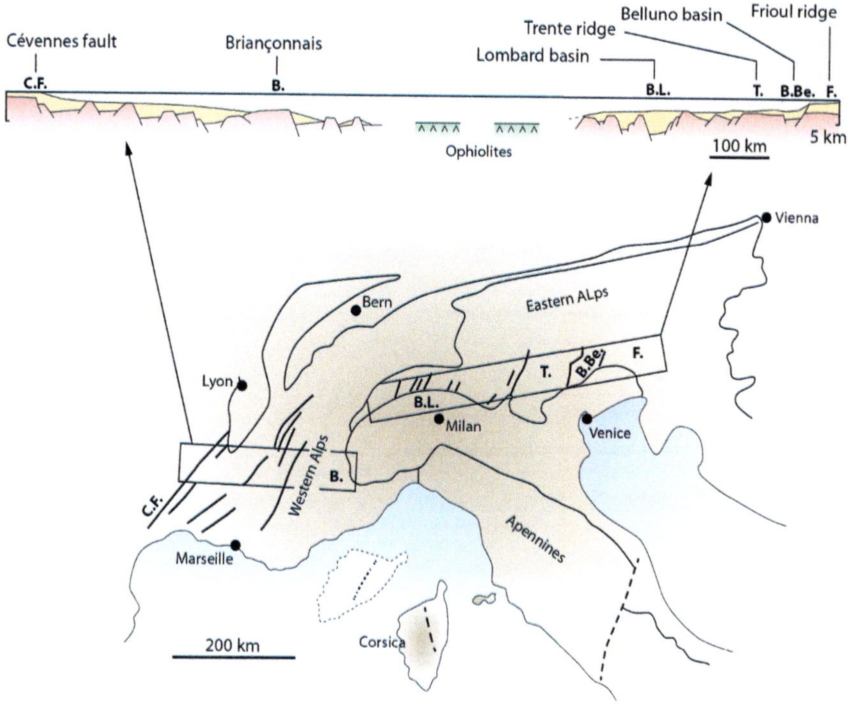

Figure 2.64. *Jurassic margins of the Western and Southern Alps: a simple diagram from Lemoine et al. (2000). The map, which avoids the structures related to the Adria-Africa-Europe convergence, helps consider the margins of Europe and Adria both in their present-day configuration and in their original architecture as a conjugate system, separated by the Piemont-Ligurian Ocean*

The rifting period of the southern Alpine margin extended from the Upper Triassic to the Dogger but distinct phases cannot be defined at the regional scale. Significant faults that played a role during the Alpine rifting and inversion have been recognized at the boundary between the Trentino horst and the Lombard basin (Lake Garda fault) or the western edge of the Lombard basin (Lake Maggiore and Lugano faults) (Figures 2.63 and 2.64). Each fault controlled the sedimentation of a basin in which accumulated several thousand meters of detrital and pelagic marine sediments with breccia intercalations (Bertotti et al. 2011).

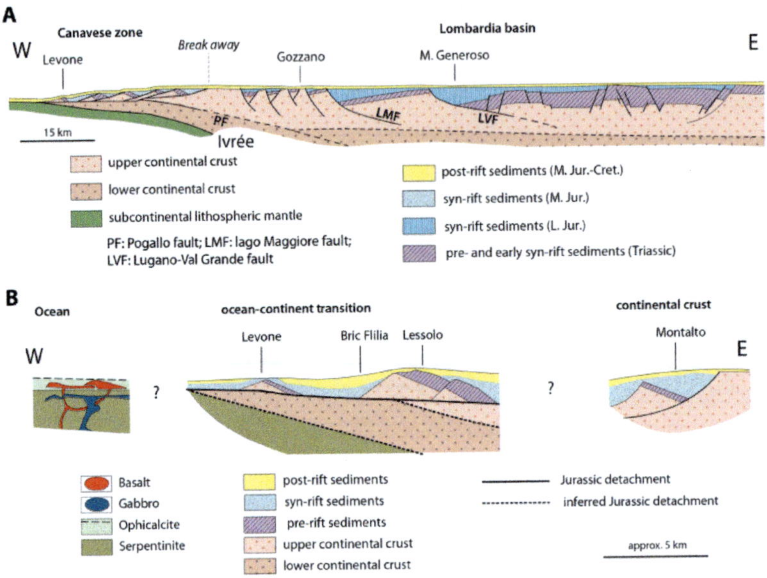

Figure 2.65. *Reconstruction of the most distal Adriatic margin: the Canavese zone (modified from Ferrando et al. 2004)*

Figure 2.66. *The Brenta Dolomites are located north of Lake Garda, near Trento. They reflect the slow thermal subsidence of the southern Alpine Triassic margin (photo Vellere)*

In Monte Generoso (southern Switzerland, Ticino transect) and in the Ortler nappe in the Eastern Alps (Grischun transect), the Jurassic margin basins are 10–30 km wide and filled with detrital sediments ranging in thickness from hundreds of meters to several kilometers (Figures 2.63 and 2.64). The nature of the infill is variable, but the dominant sediments are turbidites and debris flows interbedded with hemipelagic sediments (Eberli 1988).

Figure 2.67. *The Hohe Tauern and the Dolomites. On the left, the picture is taken from the west to the east (photo: Rosenberg). The large valley is the Val Pusteria which corresponds to a segment of the Periadriatic Line. The snow-covered reliefs are the Hohe Tauern, the Dolomites form the beige peaks in the background on the right. On the right, a close aerial view of the Dolomites (photo Dumont)*

In the Lombard Basin of Monte Generoso, controlled by the Lugano fault, extensional tectonics started at the end of the Triassic and ceased in the Toarcian, when sedimentation became pelagic and very reduced, characterized by Ammonitico Rosso facies (Bernoulli 1964). In the Belluno Basin and the Trentino ridge, the post-Triassic sedimentation is composed of more or less dolomitized limestones.

During the Middle Jurassic (170 Ma) the Trentino platform subsided, triggering the deposition of red nodular limestones (lower Ammonitico Rosso). At the same time, in the Belluno Basin, silex-bearing clay formations were deposited. The Friuli platform then became a source of abundant limestone debris feeding carbonate turbidites progressively filling the Belluno Basin and overlapping the edge of the Trentino platform (Vajont calcarenites). The accumulation of the turbidite flows ceased in the Upper Jurassic with the sedimentation of fine silex-bearing limestones followed by a new red nodular level, the Upper Ammonitico Rosso. The pelagic sedimentation continued in the Upper Cretaceous with fine carbonate muds (now the white Biancone limestones) and then by red clays (Scaglia Rossa) recording the arrival of terrigenous material due to the surrection of emerged zones in relation to the first Alpine folds of the Eoalpine phase.

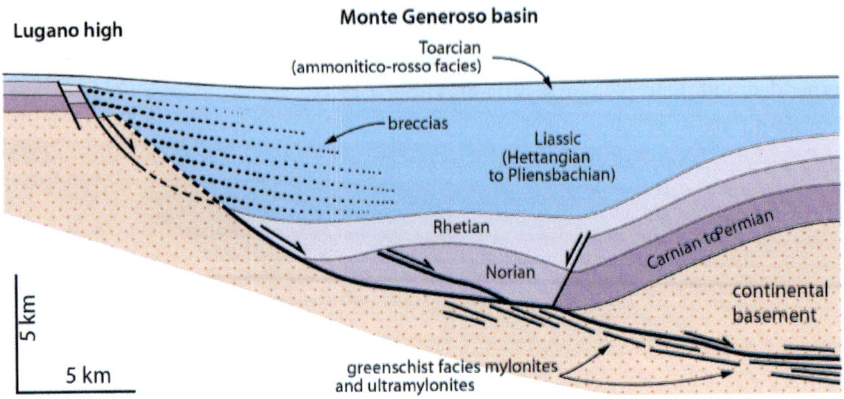

Figure 2.68. *The Lugano fault and the Monte Generoso basin. The thickness of the Liassic sediments (Rhetian to Pliensbachian) reaches several kilometers. The Lugano fault was a listric fault that progressively exhumed the Ante-Permian basement. Faulting started during the Norian and ceased at the time of deposition of the Toarcian ammonitico rosso facies (modified from Lemoine et al. 2010).*

There are similarities in the rifting phase record between the European and Adriatic conjugate margins. In both cases, collapse due to syn-rift tectonics began in the Liassic times and post-rift pelagic conditions occurred during the Malm and persisted into the Cretaceous. But there is a major difference between the two margins. The Briançonnais domain underwent uplift and emersion during the Dogger with karstification of the Triassic and Lower Jurassic sediments, whereas at the same time, conditions remained deep, below the CCD (calcite compensation depth), on the South Alpine margin, as evidenced by the deposition of radiolarites (Lemoine et al. 1986). Such contrasted evolution allowed Lemoine et al. (1987) to propose the first asymmetric rifting model favoring displacement along with a large extensional detachment beneath the European margin (hanging wall or upper plate) and leading to the uplift of the Adriatic margin (footwall or lower plate). This fault accommodated the exhumation of the sub-European mantle that formed the ocean floor of the Ligurian Tethys (see section 2.4).

Box 2.7. *Conclusion to the study of the European and Adriatic conjugate margin series: similarities and differences*

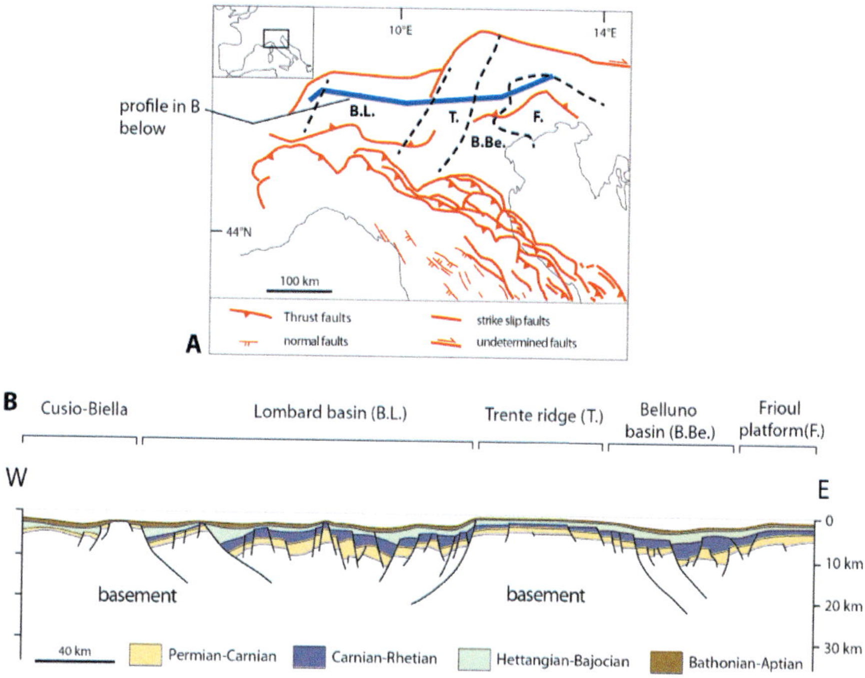

Figure 2.69. *The Adriatic margin of the Southern Alps (northern Italy). (A) location map of section B showing the large paleomargin structures (basins and ridges), preserved in the Alpine upper plate. (B) W-E section of the reconstructed structures before Alpine inversion (modified from Bera and Carminati 2009)*

2.5. Architecture of the Alpine paleomargins: from the model of tilted blocks to the concept of crustal hyper-thinning

The first major stratigraphic and paleogeographic syntheses of the Alpine paleomargins at the scale of the Mediterranean regions were carried out immediately after the advent of plate tectonics when the concepts of the continental margin and oceanic domain became evident. The new marine geophysical data and the first deep drillings of the DSDP program have provided new points of comparison that were quickly exploited at the scale of the Neotethys.

In a key work, Bernoulli and Jenkyns (1974) draw up a synthetic table of the principal sedimentary facies of the Mesozoic carbonate series of the Alps and the peri-Alpine domains, from the Maghreb to the Dinarides (Figure 2.66). They thus defined, at a European scale, the age of the transition from the Triassic carbonate platform to the Jurassic collapse and the generalization of pelagic facies. They showed that most of the extensional tectonics took place with block tilting during the Lias. The subsidence that continued in the Middle Jurassic was heterogeneous and some blocks remained in shallow positions. At the top of these blocks, condensed series were deposited (such as red beds and Ammonitico Rosso-type sediments), on the other hand, sedimentation of turbidites was dominant in the basins. The authors point to the great similarities between this evolution and that of the western Central Atlantic. They conclude with the existence of a vast marine trench cutting the Pangea and running from Indonesia to the Caribbean during the Jurassic. The rough outlines of this trench fit well with the trace of the Haug geosynclines.

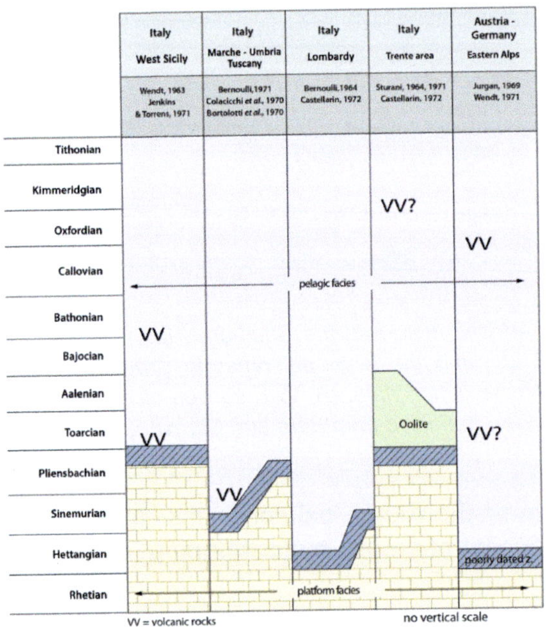

Figure 2.70. *Early major margin-to-margin correlations in the peri-Mediterranean domain. Five lithostratigraphic columns taken from the original table of Bernoulli and Jenkyns (1974)*

In the area of the future Western Alps, this evolutionary pattern perfectly applies to the European margin. Furthermore, rifting is shown to have started in proximal areas of the margin and to have progressively affected the distal margin regions (Lemoine et al. 1986; Eberli 1988; Manatschal and Bernoulli 1998). Rifting structures were first recognized in the Dauphinois domain, particularly in the

Bourg-d'Oisans region. There, ancient steep normal faults could be mapped in detail and their syn-sedimentary evolution was demonstrated by facies analysis at fault contact as discussed in section 2.3.2. These faults were initially drawn with a planar attitude with a high dip, and their deeper portions were not represented. Then, in the light of discoveries made along present-day passive margins, the attitude of the faults changed in the alpine margin reconstructions. They became listric in recent models and are shown to connect to subhorizontal ductile horizons in the middle crust (Figure 2.67). The representations provided by the structures of the South Alpine margin, such as those of the Monte Generoso basins, have been important scientific benchmarks together with geophysical images from the present-day oceans.

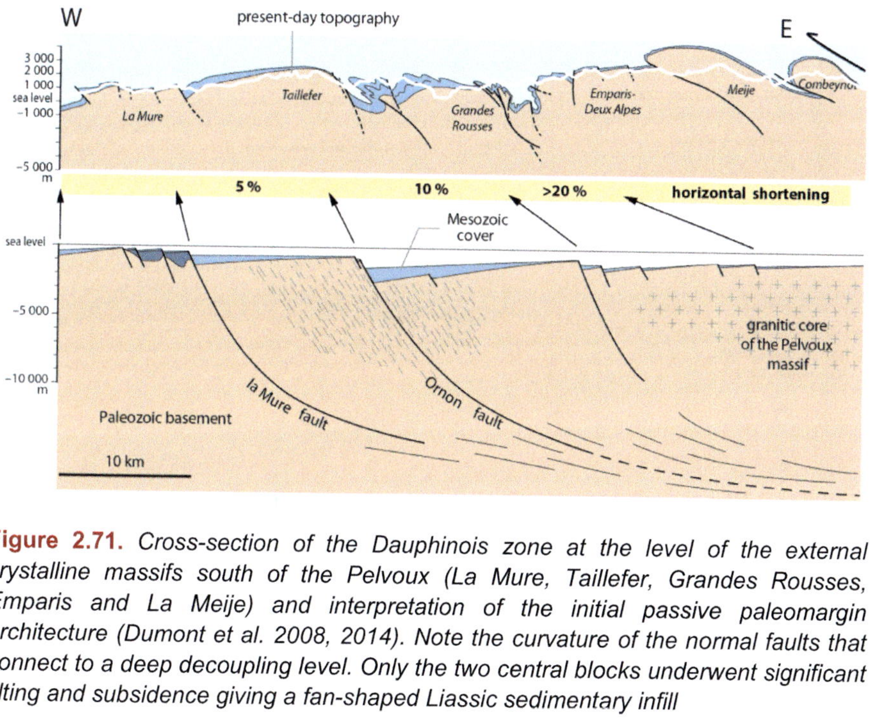

Figure 2.71. *Cross-section of the Dauphinois zone at the level of the external crystalline massifs south of the Pelvoux (La Mure, Taillefer, Grandes Rousses, Emparis and La Meije) and interpretation of the initial passive paleomargin architecture (Dumont et al. 2008, 2014). Note the curvature of the normal faults that connect to a deep decoupling level. Only the two central blocks underwent significant tilting and subsidence giving a fan-shaped Liassic sedimentary infill*

The first thinning models considered for the continental crust used simple tilted blocks giving a homogeneous structure along the entire margin (Figure 2.14). They were built on the basis of seismic profiles off the coasts of Armorica and then the coasts of Iberia, especially Galicia. Here, mantle rocks exhumed at the foot of a margin were discovered for the first time in 1978, during dredging by the M/V Le Suroit off the Banc de Galice at a depth of 5100 m (Figure 2.68). The dredge brought a compact orange clay, identified as a highly altered peridotite, onto the bridge. Knowledge of this distal margin domain has been completed by numerous geophysical surveys and several deep-sea drillings carried out within the frame of the ODP and iODP programs (legs 103, 149 and 173). The recovered data collectively show that the continental crust of the margin is separated from the mantle by a major shear zone, a detachment, that allowed the exhumation of peridotites, as well as tectonic lenses of lower crustal material (Boillot et al. 1987).

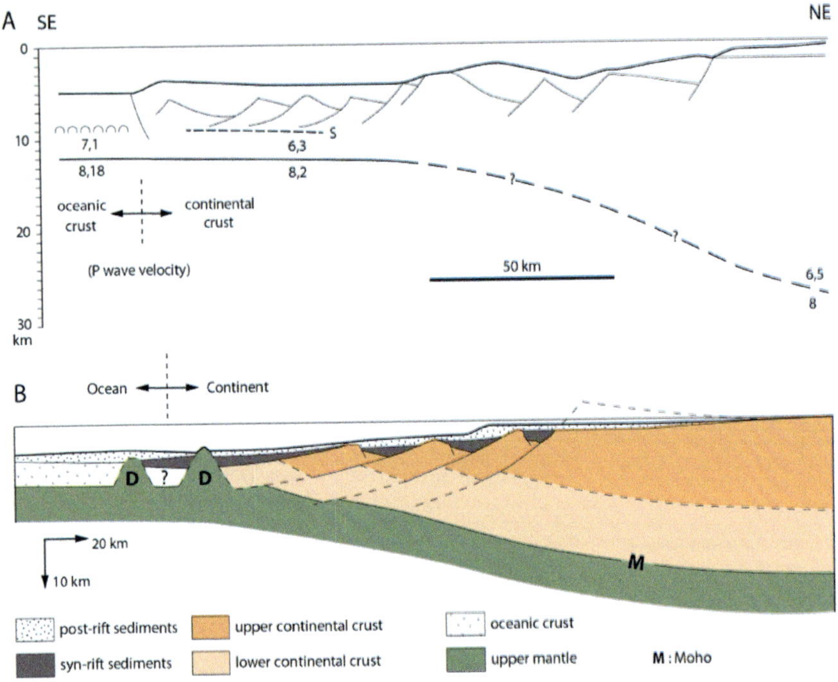

Figure 2.72. *Early interpretations of passive margin seismic profiles. Top, a sketch of the first tilted blocks, from de Charpal et al (1978). Bottom, one of the first theoretical profiles through a passive margin, integrating the early constraints delivered by the study of the Iberian margin. The dredged mantle rocks at the foot of the distal margin (hill 5100) were then interpreted as a serpentinite diapir (after Boillot et al. 1980)*

The discovery of peridotites at the foot of the Iberia distal margin was a major event in the history of Alpine geology. This provided consistency to observations that would have been meaningless without modern knowledge of the architecture of passive margins. The first model of mantle denudation along a shallow dipping major normal fault was published by Lemoine, Boillot and Tricart in 1987. This model is the result of the collaboration between two alpine geologists, promoters of the concept of mantle denudation in the Piemont-Liguria Ocean (Lemoine and Tricart) and the marine geologist who discovered mantle exposures at the foot of the passive margin (Boillot). The model caused a sensation and was well accepted. Indeed, it provided a consistent interpretation of a complex geological puzzle that fed interrogations for the community of geologists favoring at that time the Penrose model for the origin of ophiolites (see the volume on ophiolites that will be part of this collection, Lagabrielle and Ceuleneer, in progress).

This new hypothesis was proposed on the basis of observations obtained on a present-day margin, and also took into account the probable narrowness of the Piedmont-Ligurian basin[6].

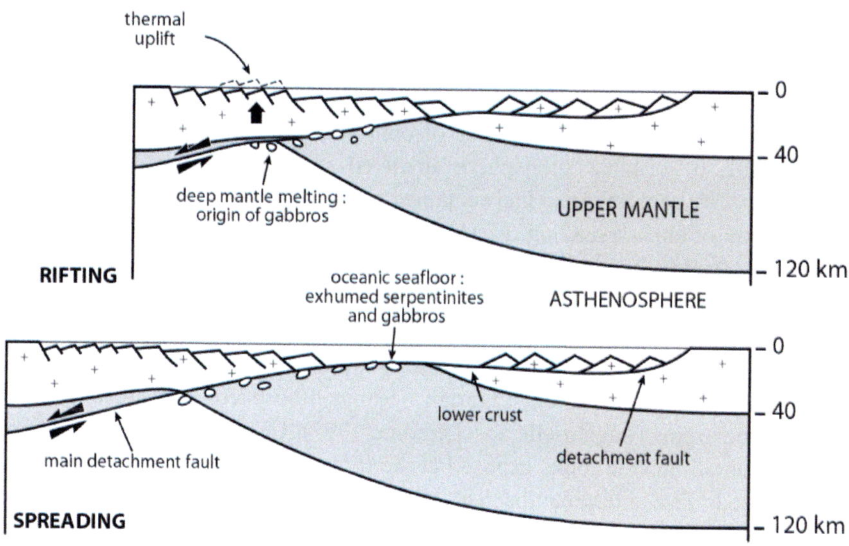

Figure 2.73. *The early model of mantle exhumation applied to the Western Alps (Lemoine et al. 1987). This totally asymmetric model accounts for displacement along a single normal fault. It is based on the contrasting behavior of the European (left) and Adriatic (right) margins. The European margin is characterized by the uplift and the erosion of the Briançonnais during the Middle Jurassic, in relation to the thermal evolution of the footwall of the fault. The whole system then underwent generalized subsidence when the mantle reached the ocean floor*

6. The absence of magmatic markers of Alpine subduction and the narrowness of the space available in the kinematic framework between Africa and Eurasia imposed by oceanic magnetic anomalies have always led to the idea that the Piemont-Liguria Ocean was at most a few hundred kilometers wide.

The model by Lemoine et al. (1987) (Figure 2.69) perfectly took into account the following facts:

– the ophiolites of the Alps show a serpentinite and gabbroic bedrock similar to the floor of the non-volcanic distal passive margins with an exhumed mantle;

– in the ophiolites, the gabbros are intrusive in the mantle and are locally exhumed and emplaced on the seafloor prior to the deposition of basalts;

– the lower plate of the system must show the deep part of the crust: this is the case in the Ivrea Zone;

– the stratigraphy of the Briançonnais domain is consistent with the thermal uplift expected to occur in the hanging wall during the activity of the detachment fault.

This model applies to the initiation of the oceanic opening; however, it cannot account for the diversity and abundance of ophiolites which represent a significant initial volume of oceanic lithosphere involved in the collision. It was then abandoned by M. Lemoine who had remained attentive to new constraints obtained from the study of slow-spreading oceanic ridges. Indeed, in the 1990s, these ridges provided multiple examples of geological relationships consistent with reconstructions of the Tethys ocean floor based on alpine ophiolites (Lagabrielle and Cannat 1990). Taking up and completing his observations on ophiolites with Lagabrielle, Lemoine then considered that an important part of the ophiolites of the Alps derived from the Piedmont-Ligurian oceanic lithosphere and not only from passive paleomargins (Lagabrielle and Lemoine 1997). Certainly, for these authors, the hypothesis of "mantle rocks exposed at the foot of distal passive margin" could not be rejected. This interpretation had to be considered for specific cases where continental basement material is present together with the mantle rocks. Such continental material may occur as breccias or olistholiths and extensional allochthons. Breccias and olistholiths are well known in the Ligurian ophiolites and extensional allochthons are described in the Platta and Tasna nappes. Here, the mantle and the continental material are still observed with pre-orogenic relationships.

A second revival for comparisons between the present and ancient Alpine margins came from the Central Alps with the discovery of oceanic paleo-detachments well exposed in the field. The theoretical model of mantle exhumation was given substance by the direct observation of structures and rocks formed along shallow dipping extensional detachment faults. Florineth and Froitzheim (1994) (Figure 2.70) published the first field observations showing that the continent–ocean transition of the Briançon margin is preserved in the Tasna nappe. The authors emphasize two fundamental observations:

– the presence of cataclasites at the top of a thin granite and gneiss sheet represents the detachment capping a thinned crustal tectonic lens exposed at the tip of the margin;

– the presence of ophicalcites at the top of the mantle rocks, stratigraphically overlain by Cretaceous marine sediments. This feature implies that the subcontinental mantle has been exposed to the seafloor during the final stages of the Valais Ocean spreading.

It seems clear that without the understanding of current distal margin systems deduced from marine geoscience campaigns, these observations could not have been understood and brought into consistency. The study of passive margins is thus a good example to show the continent versus ocean interfertilization in geological sciences.

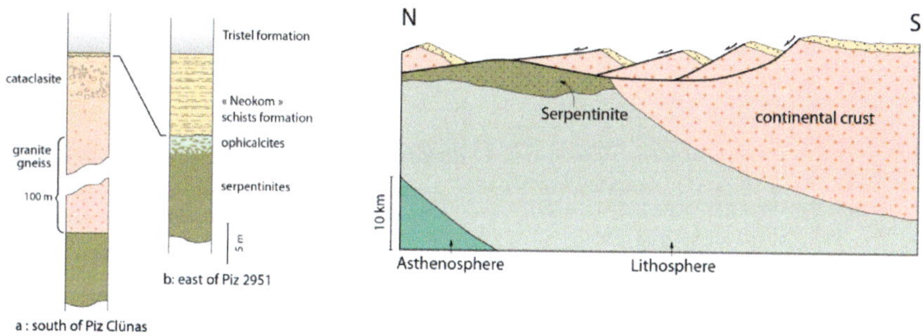

Figure 2.74. *Two diagrams from the pioneering publication of Florineth and Froitzhein (1994) reporting the discovery of cataclasites and ophicalcites in the Tasna system (Engadine window, Central Alps). These observations have provided references for later interpretations of the structure of the Alpine passive margins. Left, two logs showing the relationship between basement and sediments in the most distal margin. Right, an early model of the architecture of an Alpine passive paleo-margin characterized by the exhumed mantle*

Searches for equivalents of the Tasna complex in the Grisons region (Graubünden) have shown that structures inherited from the Adriatic distal margin formation are also preserved in the Err nappe and in the lower nappes containing ophiolitic material (Platta and Totalp units) (Figure 2.71). The presence of sedimentary breccias reworking cataclasites (tectonic-sedimentary breccias) and crustal extensional klippen resting on the mantle rocks allowed Manatschal and Nievergelt (1997) and Desmurs et al. (2001) to propose that the Grisons ophiolites represent equivalents of an oceanic floor emplaced through mantle denudation processes at the foot of the Iberian margin. The authors describe a system of detachment faults that originate in the upper crust and progressively reach the subcontinental mantle triggering its exposure up to the seafloor. This interpretation is consistent with geochemical data obtained from mantle rocks (Müntener et al. 2004). Note that the lower crust of the Adria plate is not exposed in the Err and Platta systems. It only appears in the Lower Margna and Malenco zones and the Ivrea Zone.

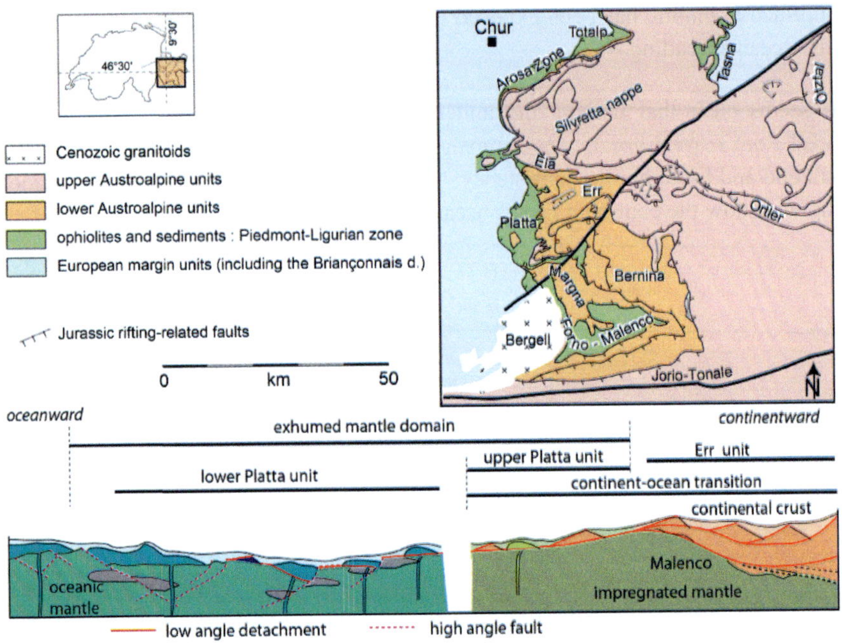

Figure 2.75. *Mantle exhumation at the foot of the Alpine passive paleomargins. Examples from the Adriatic margin in the Central Alps. Significance of the ophiolites of Platta and Malenco. Explanations in the text (modified from Müntener et al. 2004)*

Since the 1980s, passive margin formation models have therefore evolved considerably (Péron-Pinvidic and Manatschal 2009; Sutra et al. 2013) (Figure 2.72). From a simple model involving a few tilted blocks, we moved to a more complex mode of mixing: (a) large, shallowly dipping faults (simple shear deformation), (b) ductile thinning distributed across large volumes of rock (pure shear deformation), and (c) mantle exhumation at the foot of the margin, associated with the presence of klippen abandoned by detachment faults cross-cutting the distal margin (extensional allochthons) (Manatschal 2004; Manatschal et al. 2007). These advances in our knowledge of passive margin evolution are illustrated in Figure 2.72, which parallels an interpreted section of the Galician margin with the reconstructed section of the preserved Adriatic margin in the Grisons region. It is clear that the general concepts and rheological constraints used to produce these figures are similar.

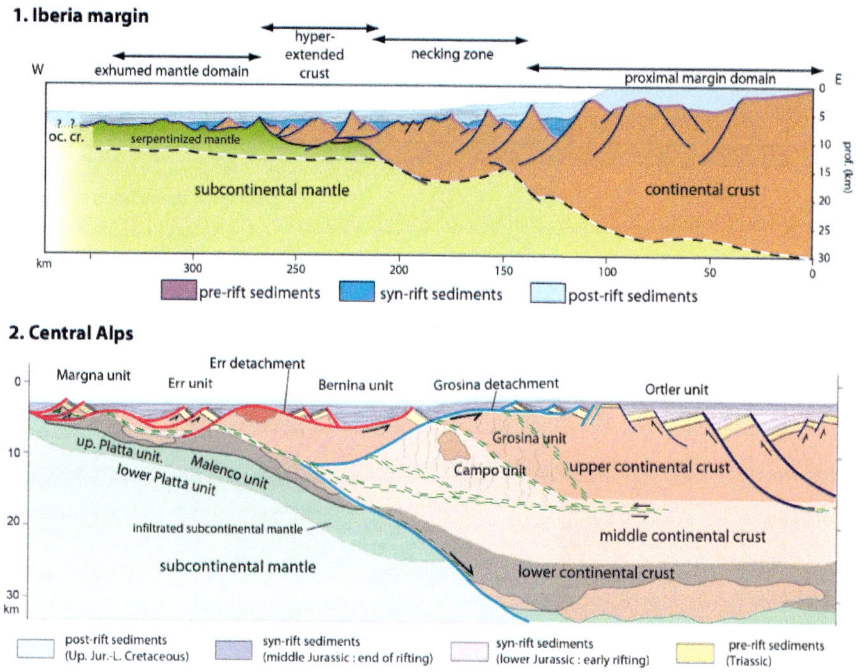

Figure 2.76. *Comparison between the simplified profile of the Iberian Atlantic margin inferred from the interpretation of seismic and ODP drilling data (1) (modified from Sutra et al. 2013) and the reconstructed profile for the Adriatic margin involved in the Grisons nappe system (Platta, Err, Bernina and Campo nappes) (2) (from Manatschal and Mohn, pers. comm.)*

Mechanically, geologists pay attention to the lithological and therefore to the rheological stratigraphy applying to models of passive margin or Alpine analogs. In general, in the Alps, the modeled crust includes a brittle surface layer (upper quartz-feldspar crust), a middle layer that behaves ductilely because of moderate temperatures allowing the plasticity of quartz (>350°C, middle quartz-feldspar crust), and a lower layer, more resistant due to its anhydrous and mafic character (mafic lower crust). This lower layer results essentially from the underplating of Permian magmas leading to multiple gabbroic intrusions. This mafic lower crust seems to be restricted to the future Alps (it is unknown in the Pyrenean realm). The three crustal layers rest on the resistant sub-continental mantle lying under the Moho and showing increasing plasticity with depth. Mantle and crust thus constitute a four-layer rheological model.

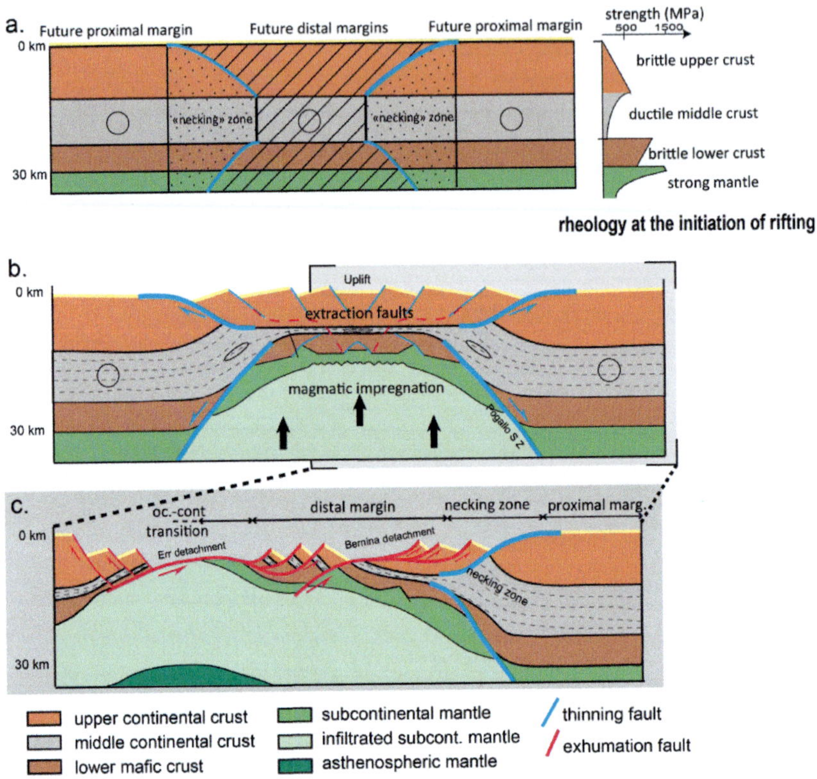

Figure 2.77. *A possible model for the evolution of continental crust during the formation of the Alpine margins of the Piemont-Liguria Ocean (Mohn et al. 2012)*

Because of this four-layer rheology, the application of tensional stress to the model in order to simulate rifting leads to the individualization of four domains in the passive margin (Mohn et al. 2012, Figure 2.73): (1) an upper (proximal) margin made of thick crust with some tilted blocks, (2) a middle margin corresponding to so-called necking zone and concentrating a large portion of the overall thinning by faulting and homogeneous stretching (pure and simple shearing), (3) a deep (distal) margin where the crust is considerably thinned and (4) the ocean–continent transition zone where the sub-continental mantle is possibly exhumed.

Since the 2000s, numerous interpretations of the Alpine internal zones have been proposed involving models of mantle exhumation in the distal paleomargin. Onland analogs provide easy access to fossil margin systems whose present-day equivalents are lying at great depth and under kilometer thick accumulations of marine sediments. This led to significant advances in our understanding of continental rifting and of magma-poor passive margins evolution. Some illustrated examples are proposed in the following.

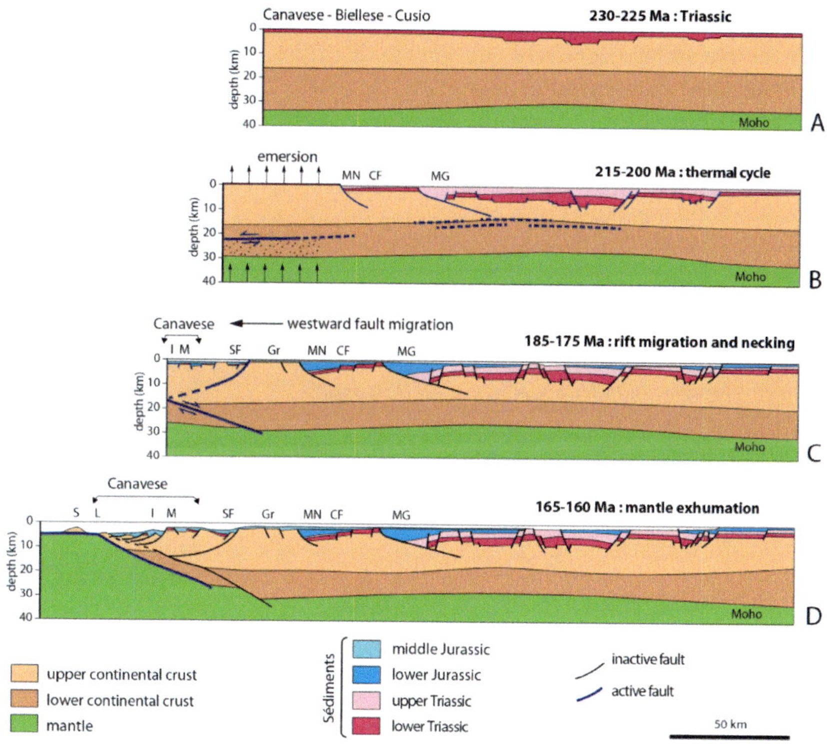

Figure 2.78. *Crustal-scale evolution of the southern Alpine margin with the integration of stratigraphic and structural data from Bertotti et al. (2003) into a general framework (Bertotti et al. 2011; Beltrando et al. 2015)*

1) The large view of the present margins has allowed Alpine geologists to rethink the paleo margins on a crustal or even lithospheric scale. This is the case for the Adriatic (South Alpine) margin, whose evolution has been reconstructed on thermomechanical grounds by integrating older stratigraphic and structural data together with the constraints of petrological studies of the pre-rift magmatism

(Figure 2.74) (Beltrando et al. 2015). The Canavese zone, the site of maximum thinning of the Adriatic margin thus appears limited by faults whose location is controlled by Permian magmatism.

2) The proliferation of examples of extensional allochthons on present-day magma-poor passive margins brought about a sort of revolution in the study of the internal zones. These examples enabled the interpretation of thin sheets made up of highly sheared continental material, or larger enigmatic units, in terms of paleo-extensional allochthons of any size. Three cases can be exemplified:

a) The micaschists, gneisses and granitoids of the Versoyen complex (Punta Rossa Formation), interpreted as a strongly dissected Variscan basement, may in fact represent allochthons tectonically lying on the exhumed mantle on the floor of the Valais Ocean (Beltrando et al. 2010). In such a case, major shearing of these units would have occurred along with the Cretaceous extensional detachment and would not be related to the Alpine compression (Figure 2.75).

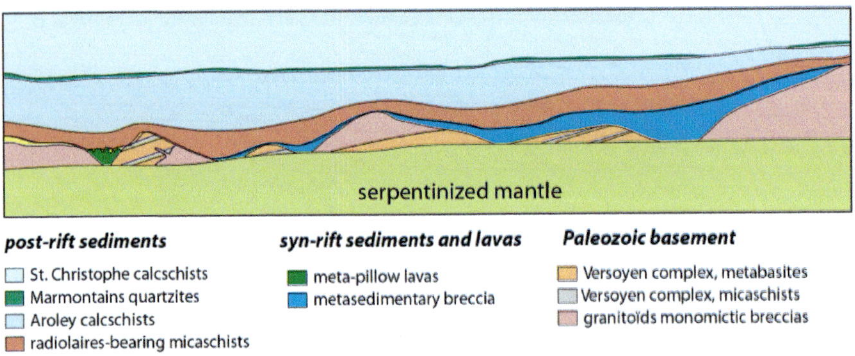

Figure 2.79. *Reconstructed cross-section of the Valais Ocean basement with extensional allochthons of continental material abandoned on the serpentinized mantle due to displacement along a detachment fault (Beltrando et al. 2010)*

b) Let us also mention the case of the Dora Maira Massif, whose various superimposed units could be derived from vast margin foot klippen composed of a continental basement and its Mesozoic cover (Ballèvre et al. 2020). This type of interpretation accounts well for the specific characteristics of each unit. The stacking of these different units within the Alpine tectonic wedge ensures its remarkable variability and avoids any regional and local-scale cylindrism (Figure 2.76).

c) In Alpine Corsica, the reinterpretation of the content of the Schistes lustrés series and their sub-autochthonous basement has led several authors to propose a new evolutionary model of the distal paleo-margin (Meresse et al. 2013). Indeed, the presence of detrital material of continental origin in the sedimentary series associated with the distal margin is regarded as indicative of the presence of extensional allochthons emplaced during the extreme thinning of the margin (Figure 2.77).

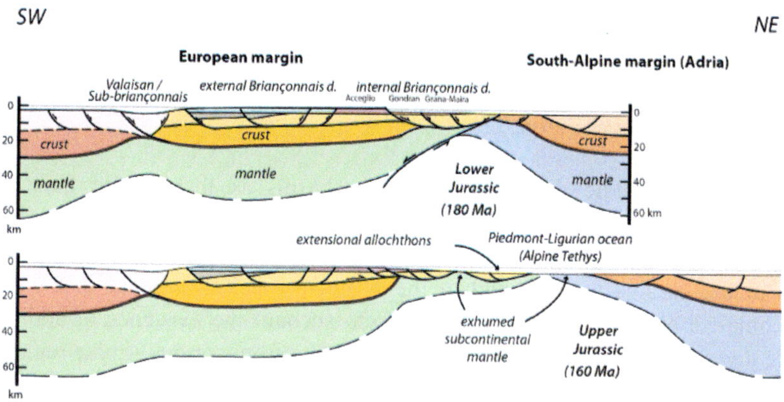

Figure 2.80. *The Piemont-Liguria Ocean margins in the future Southwestern Alps. Note the architecture of the distal regions where "extensional allochthons" of continental material appear between the zones of the exhumed mantle (modified from Ballèvre et al. 2020)*

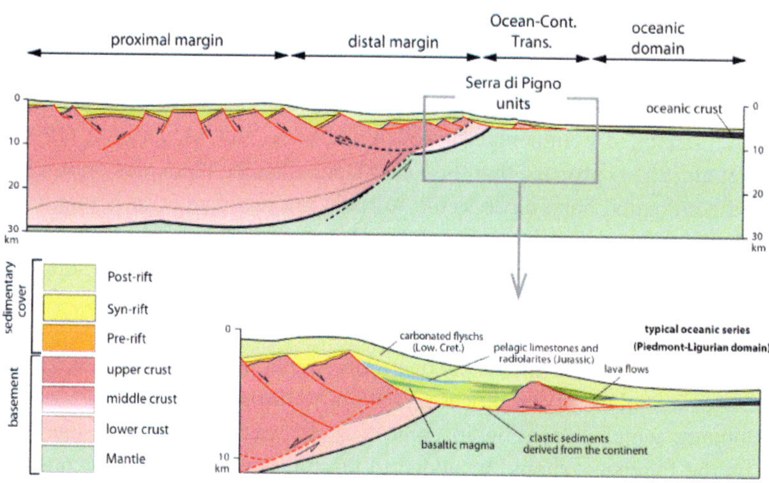

Figure 2.81. *A reconstruction of the Tethyan passive margin of Corsica showing "extensional allochthon" of continental material lying on the exhumed mantle (modified from Meresse et al. 2013)*

2.6. By way of conclusion, the contours of the Piedmont-Ligurian oceanic paleo-domain: how many passive continental margins have been inverted in the Alpine belt?

As discussed above, before the Cenozoic orogeny, the Alpine system is characterized by numerous small continental blocks, separated by basins floored by more or less typical oceanic lithosphere. But how many oceanic basins and how many conjugated continental margins bordering these basins really existed between Africa, Iberia, Adria and Europe?

Figure 2.60 (middle diagram) shows the basins as they would have been arranged at the end of the syn-rift period, during the Oxfordian, when oceanization in the Piedmont-Ligurian domain had already begun. If we want to construct a sketch of the passive margins and of the continent–ocean boundary that was the most complete at that time, we must take into account the existence of additional oceanic diverticula, as they are imposed by field data from various Alpine regions. It is very possible that a branch of the Piemont-Liguria Ocean opened in the future Betic Cordilleras realm, isolating a microcontinent off the coast of Africa, as suggested by Puga et al. (2017) and Molli and Malavieille (2011) (Figures 2.82(C) and (D)). This microcontinent is also considered in reconstructions from Corsica, where it is named the Nebbio microcontinent (Molli 2008; Figure 2.82(D')).

Further north, many authors have suggested that the Sesia-Dent Blanche block separated from the Adriatic margin due to the opening of a narrow N-S trending oceanic basin. This basin is considered the origin of the small Canavese ophiolitic remnants. It is also interpreted as the source of the Lanzo ophiolites (Debelmas 1975, Figures 2.82(A) and (B)). More recent interpretations show affinities with this concept of "separated continental stripe". Froitzheim and Manatschal (1996) and Manzotti et al. (2014) propose that the Sesia-Dent Blanche block does not represent an isolated continental horst close to the Adriatic margin. Rather, it is regarded as a very large "extensional allochthon" abandoned over the exhumed mantle zone at the foot of the hyper-extended Adria passive margin (Figure 2.78(E)) (see supplements in section 2.3).

The basins listed above are basins or exhumed mantle zones of the Jurassic age. Since the continental extension processes continued during the Early Cretaceous, the Valais domain, which according to some authors was connected to the Atlantic domain by the Pyrenees, should be added to these basins in the final reconstruction of the pre-orogenic basins (see Figures 2.40 and 2.52) (Ricou et al. 1996; Stampfli and Hochard 2009; Handy et al. 2010).

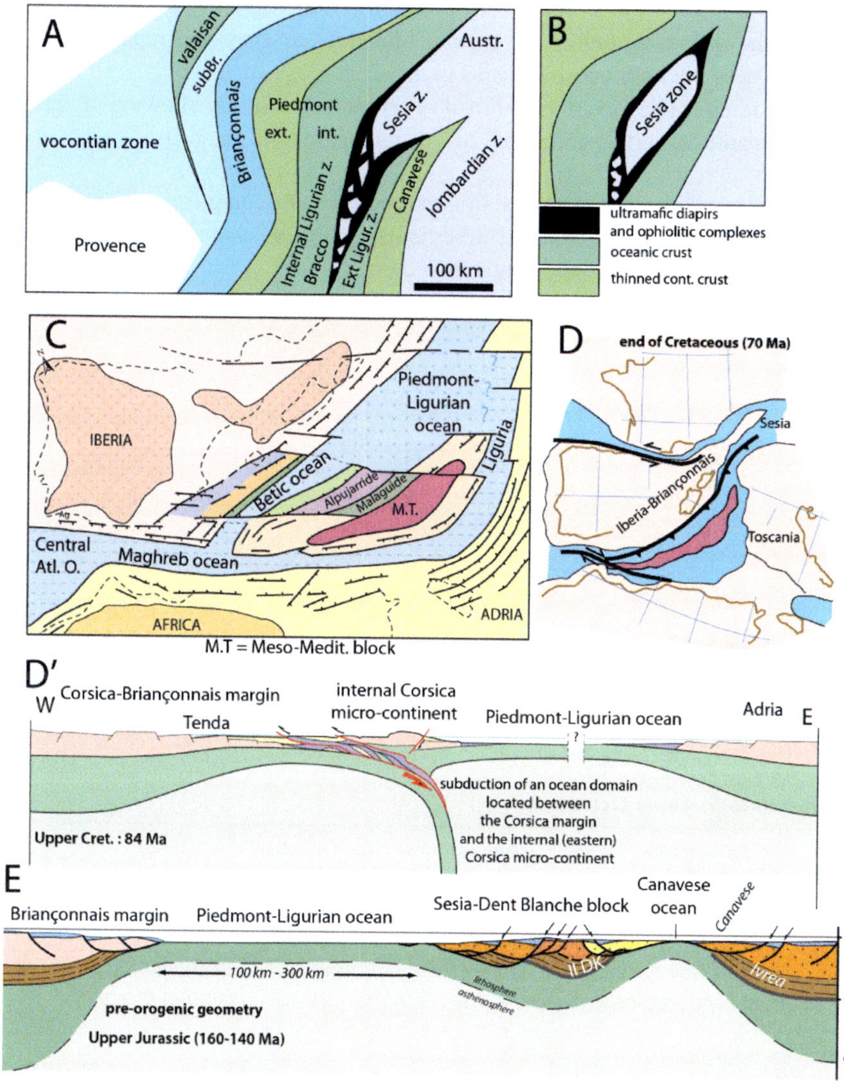

Figure 2.82. *How many oceanic basins before Alpine compression? (A and B) Two hypotheses for the position of the Sesia basement and its relationship with the Adria (modified from Debelmas 1975). (C) A small Betic ocean off Iberia isolating a Meso-Mediterranean (M.T.) micro-continent (modified from Puga et al. 2017). (D and D') A continental strip east of the Briançonnais (in red) (modified from Molli 2008; Molli and Malavieille 2011). (E) Location of the Sesia-Dent Blanche block relative to the Canavese margin in relation to the presence of a Canavese paleo-ocean (modified from Manzotti et al. 2014)*

In Figure 2.83, we propose an original paleogeographical reconstruction including the greatest number of oceanic domains as they are reported in the literature. These are both oceanic basins in the strict sense (that is emplaced from an ocean spreading ridge, even an ultra-slow ridge) or regions showing a mantle basement (mantle exhumed at the foot of a distal continental margin).

In our reconstruction, the system of Cretaceous basins comprises a main oceanic axis connected to triangular-shaped diverticula, acting as short spreading systems that propagate into pre-fractured continental domains.

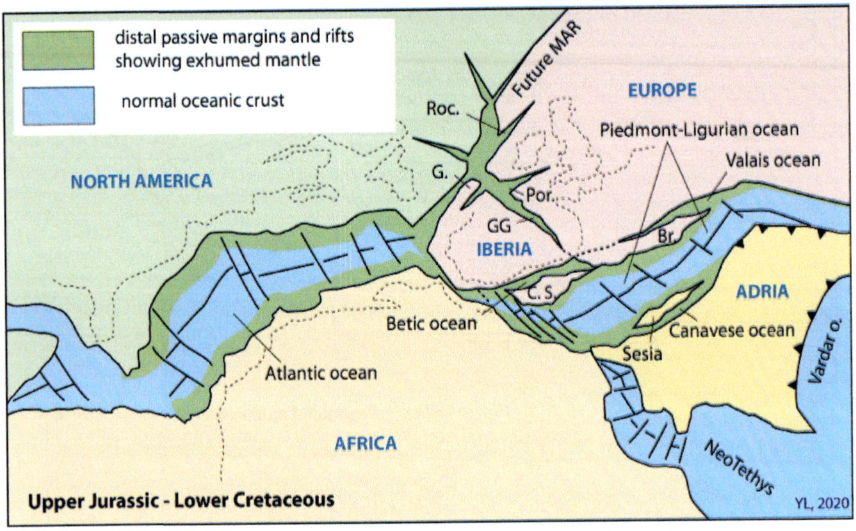

Br. : Briançonnais; Future MAR : Future North Mid-Atlantic Ridge; GG : Bay of Biscay; C.S. : Corsica-Sardinia block; G. Galicia Bank; Por.: Porcupine basin; Roc. : Rockall basin

Figure 2.83. *A novel paleogeography of the Atlantic and Alpine domain in the Early Cretaceous period, taking into account all the oceanic or mantle-floored basins cited in the literature (see discussion in the text)*

A good analog of the pre-Alpine situation reconstructed here is probably provided by the North Atlantic Ocean at the onset of the Eocene. Indeed, the central ocean spreading axis seems to be well established, but numerous attempts at oceanization were made laterally by propagating rifts. These attempts resulted in the denudation of the subcontinental mantle in the Bay of Biscay, off Newfoundland, in the Rockall Basin, the Porcupine Basin and the Labrador Sea (future abandoned ridge). The future North Mid-Atlantic Ridge finally propagated between Scotland and Greenland.

All these attempts at ocean spreading have thus multiplied the areas of hyper-extended crust and exhumed mantle. They created basins exhibiting various shapes with relatively contrasted and rapid evolution. The margins of these basins receive significant amounts of sediments (see the Lombardy basins for example) able to build voluminous mountains during subsequent tectonic contraction. These basins that characterize the North Atlantic region are excellent examples for better understanding the variety of ophiolites and paleogeographic domains in the Western, Central, Eastern, and Southern Alps (Figure 2.81).

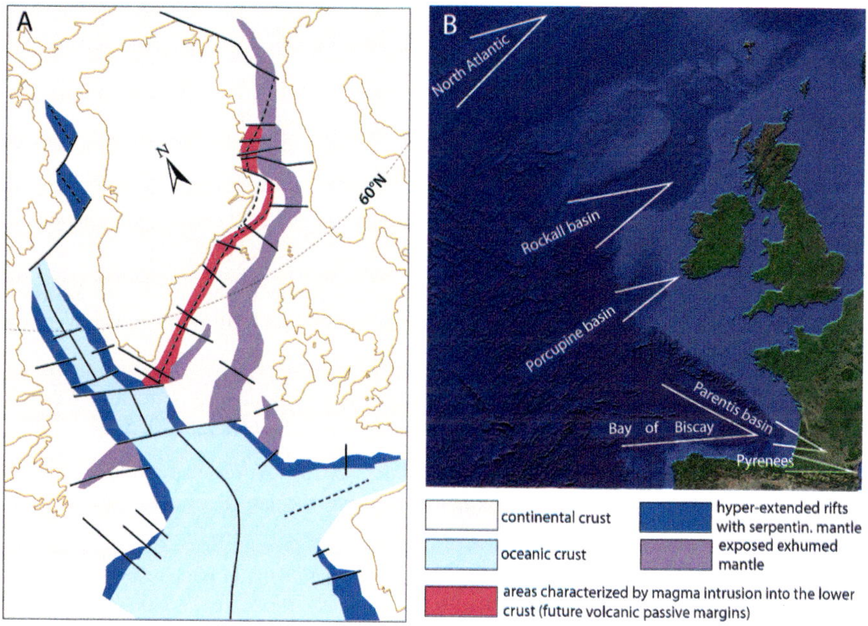

Figure 2.84. *(A) Eocene North Atlantic (modified from Lundin and Doré 2011), a pattern of exhumed ultrabasic bottom zones, oceanic spreading ridges, and propagators opening V-shaped notches within the continental lithosphere. (B) The succession of Cretaceous-Eocene propagators of the European platform from the South Armorican margin (basins of the North Pyrenean zone, Parentis Basin, Bay of Biscay) to the Ireland–Scotland margin (Porcupine basin and Rockall basin) (bathymetry from Google Earth)*

2.7. Acknowledgments

This chapter has been written both on the basis of the knowledge provided by my teachers, Professors and colleagues and on the basis of my own Alpine experience. I would like to thank Marc Tardy, René Blanchet, Jean Aubouin, Marcel Lemoine,

Pierre Tricart, Riccardo Polino, Serge Fudral, Renaud Caby, Eric Deville, Michel Ballèvre, Catherine Mével, Othmar Müntener and many others. Congratulations go to Geoffroy Mohn and Thierry Dumont who provided very constructive and helpful comments to the first version of this chapter. Their experience in the Alpine margins contributed to greatly improving the text, figures, and references of an early draft. Special thanks go to Bernadette Coléno who realized most of the figures. I wish to thank G. Mohn, M. Ballèvre, S. Schmid, and J.L. Epard who provided some of their original figures that I could adapt in this chapter.

2.8. References

Adline, G. (1982). Les spilites potassiques triasiques de la bordure occidentale du massif des Ecrins Pelvoux (Alpes françaises) : aspect volcanologique et problèmes de carbonatation. Thesis, Université Scientifique et Médicale de Grenoble.

Agard, P. and Lemoine, M. (2003). *Visages des Alpes : structure et évolution géodynamique.* Editions de la Carte Géologique du Monde (CCGM), UNESCO, Paris.

Agard, P., Jolivet, L., Goffé, B. (2001). Tectonometamorphic evolution of the Schistes Lustrés Complex: Implications for the exhumation of HP and UHP rocks in the Western Alps. *Bull. Soc. Géol. Fr.*, 5, 617–636.

Andrieux, J., Frizon De Lamotte, D., Braud, J. (1989). A structural scheme for the Western Mediterranean area in Jurassic and Early Cretaceous times. *Geodin. Acta,* 3, 5–15.

Angrand, P., Mouthereau, F., Masini, E., Asti, R. (2020). A reconstruction of Iberia accounting for W-Tethys/N-Atlantic kinematics since the late Permian-Triassic. *Solid Earth*, 11, 1313–1332.

Antoine, P. (1965). Sur l'existence du Crétace supérieur daté dans la nappe des Brèches de Tarentaise au Nord des Chapieux (Savoie). *C. R. Acad. Sc. Paris*, 261, 3640–3642.

Antoine, P. (1971). La zone des Brèches de Tarentaise entre Bourg-Saint-Maurice (vallée de l'Isère) et la frontière italo-suisse. *Travaux du Laboratoire de Géologie de l'Université de Grenoble*, 9, 367.

Antoine, P. (1972). Le domaine pennique externe entre Bourg-Saint-Maurice (Savoie) et la frontière italo-suisse. *Géologie Alpine*, 48, 5–40.

Antoine, P., Barféty, J.C., Vivier, G., Debelmas, J., Desmons, J., Fabre, J., Loubat, H., Vautrelle, C. (1992). Carte géologique de France à 1/50 000. Bourg Saint-Maurice [Online]. Available at: http://ficheinfoterre.brgm.fr/Notices/0727N.pdf.

Arnaud H. (1966a). Contribution à l'étude du Diois nord-oriental. *Géologie Alpine*, 42, 7–31.

Arnaud H. (1966b). Contribution à l'étude géologique des plateaux du Vercors méridional. *Géologie Alpine*, 32, 34–51.

Arnaud H. (1981). De la plate-forme urgonienne au bassin vocontien : le Barrèmo-Bédoulien des Alpes occidentales entre l'Isère et le Buëch (Vercors méridional Diois oriental et Dévoluy). *Géologie Alpine*, 12, 804.

Arnaud-Vanneau, A., Arnaud, H., Meunier, A.-R., Séguin J.-C. (1987). Caractères des transgressions du Crétacé inférieur sur les marges de l'océan ligure (SE de la France et Italie centrale). *Mém. Géol. Univ. Dijon*, 11, 167–182.

Avigad, D., Chopin, C., Le Bayon, R. (2003). Thrusting and extension in the southern Dora-Maira ultra-high pressure massif (Western Alps): View from below the coesite-bearing unit. *J. Geol.*, 111, 1, 57–70.

Ayrton, S. (1980). La géologie de la zone Martigny – Chamonix (versant suisse) et l'origine de la nappe de Morcles (un exemple de subduction continentale). *Eclog. Géol. Helv.*, 3, 1, 137–172.

Azambre, B., Rossy, M., Albarède, F. (1992). Petrology of the alkaline magmatism from the Cretaceous North-Pyrenean rift zone (France and Spain). *Eur. J. Mineral.* 4(4), 813–834.

Babist, J., Handy, M.R., Konrad-Schmolke, M., Hammerschmidt, K. (2006). Pre-collisional, multistage exhumation of the subducted continental crust: The Sesia Zone, western Alps. *Tectonics*, 25, TC6008.

Badoux, H. and Mercanton, C.H. (1962). Essai sur l'évolution tectonique des préalpes médianes du Chablais. *Eclog. Géol. Helv.*, 55(1), 135–188.

Balestro, G., Festa, A., Dilek, Y. (2019). Structural architecture of the Western Alpine ophiolites, and the Jurassic seafloor spreading tectonics of the Alpine Tethys. *Journal of the Geological Society*, 176, 913–930.

Ballèvre, M. (1984). Cinématique éoalpine et alpine des unités majeures sur la transversale du Grand Paradis (Alpes occidentales). *C. R. Acad. Sc. Paris*, II(299), 981–984.

Ballèvre, M. (1988). Collision continentale et chemins P-T. L'unité pennique du Grand Paradis (Alpes occidentales). *Mém. Doc. Centre Armoricain d'étude structurale des socles*, 19, 340.

Ballèvre, M., Manzotti, P., Dal Piaz, G.V. (2018). Pre-Alpine (Variscan) inheritance: A key for the location of the future Valaisan Basin (Western Alps). *Tectonics*, 37, 786–817.

Ballèvre, M., Camonin, A., Manzotti, P., Poujol, M. (2020). A step towards unraveling the paleogeographic attribution of pre-Mesozoic basement complexes in the Western Alps based on U–Pb geochronology of Permian magmatism. *Swiss J. Geosci.*, 113, 12.

Barbier, R. and Debelmas, J. (1966). Réflexions et vues nouvelles sur la zone subbriançonnaise au N du Pelvoux. *Trav. Lab. Géol. Univ. Grenoble*, 42, 97–107.

Barféty, J.C. and Gidon, M. (1983). La stratigraphie et la structure de la couverture dauphinoise au Sud de Bourg d'Oisans. Leurs relations avec les déformations synsédimentaires jurassiques. *Géologie alpine*, 59, 5–32.

Barféty, J.C. and Gidon M. (1984). Un exemple de sédimentation sur un abrupt de faille fossile : le Lias du versant est du massif du Taillefer (Zone dauphinoise, Alpes occidentales). *Revue de géologie dynamique et de Géographie physique*, 25(4), 267–276.

Barféty, J.C., Gidon, M., Lemoine, M., Mouterde, R. (1979). Tectonique synsédimentaire liasique dans les massifs cristallins de la zone externe des Alpes occidentales françaises : la faille du col d'Ornon. *C. R. Acad. Sc., Paris*, 289, 1207–1210.

Bassoulet, J.-P., Boulin, J., Colchen, M., Marcoux, J., Mascle, G., Montenat, C. (1980). L'évolution des domaines téthysiens au pourtour du bouclier indien du Carbonifère au Crétacé. In *Géologie des chaînes alpines issues de la Téthys*, Aubouin, J. Debelmas, J., Latreille, M., Congrès géologique international (eds). BRGM, Orléans.

Baud, A. (1972). Observations et hypothèses sur la géologie de la partie radicale des Préalpes médianes. *Eclog. Géol. Helv.*, 65(1), 43–55.

Baumgartner, P.O. (1987). Age and genesis of Tethyan Jurassic Radiolarites. *Eclog. Geol. Helv.*, 80, 831–879.

Beccaluva, L., Dal Piaz, G.V., Macciotta, G. (1984). Transitional to normal MORB affinities in ophiolitic metabasites from the Zermatt-Saas, Combin and Antrona units, Western Alps: Implications for the paleogeographic evolution of the Western Tethyan Basin. *Geol. Mijnb*, 63, 65–177.

Bell, C. and Butler, R.W.H. (2017). Platform-basin transitions and their role in Alpine-style collision systems: A comparative approach. *Swiss J. Geosci.*, 110, 535–546.

Beltrando, M., Rubatto, D., Compagnoni, R., Lister, G. (2007). Was the Valaisan basin floored by oceanic crust? Evidence of Permian magmatism in the Versoyen Unit (Valaisan domain, NW Alps). *Ofioliti*, 32, 85–99.

Beltrando, M., Frasca, G., Compagnoni, R., Vitale Brovarone, A. (2012), The Valaisan controversy revisited: Multi-stage folding of a Mesozoic hyper-extended margin in the Petit St. Bernard pass area (Western Alps). *Tectonophysics.*, 579, 17–36.

Beltrando, M., Manatschal, G., Mohn, G., Dal Piaz, G.V., Vitale Brovarone, A., Masini, E. (2014). Recognizing remnants of magma-poor rifted margins in high-pressure orogenic belts: The Alpine case study. *Earth Sci. Rev.*, 131, 88–115.

Beltrando, M., Stockli, D.F., Decarlis, A., Manatschal, G. (2015). A crustal-scale view at rift localization along the fossil Adriatic margin of the Alpine Tethys preserved in NW Italy. *Tectonics*, 34, 1927–1951.

Bernoulli, D. (1964). *Zur Geologie des Monte Generoso (Lombardische Alpen)*. Kommerly & Frey AG, Switzerland.

Bernoulli, D., Bertotti, G., Froitzheim, N. (1990). Mesozoic faults and associated sediments in the Austro-Alpine-South Alpine passive continental margin. *Mem. Soc. Geol. It.*, 45, 25–38.

Bertotti, G. (2001). Subsidence, deformation, thermal and mechanical evolution of the Mesozoic South Alpine rifted margin. *Geol. Soc. London Sp. Pub.*, 187, 125–141.

Berra, F. and Carminati, E. (2010). Subsidence history from a backstripping analysis of the Permo-Mesozoic succession of the Central Southern Alps (Northern Italy). *Basin Res.*, 22, 952–975.

Bill, M., Bussy, F., Cosca, M., Masson, H., Hunziker, J.C. (1997). High-precision U-Pb and $^{40}Ar/^{39}Ar$ dating of an Alpine ophiolite (Gets nappe, French Alps). *Eclogae Geol. Helv.*, 90, 43–54.

Bocquet-Desmons, J. (1974). Le socle briançonnais de Vanoise (Savoie) : arguments en faveur de son âge anté-alpin et de son polymétamorphisme. *C. R. Acad. Sc. Paris*, II(278), D, 2601–2604.

Boillot, G., Grimaud, S., Mauffret, A., Mougenot, D., Kornprobst, J., Mergoil-Daniel, J., Torrent, G. (1980). Ocean-continent boundary off the Iberian margin: A serpentinite diapir west of the Galicia Bank. *Earth Planet. Sci. Lett.*, 48, 23–34.

Boillot, G., Recq, M., Winterer, E.L., Meyer, A.W., Applegate, J., Baltuck, M., Bergen, J.A., Comas, M.C., Davies, T.A., Dunham, K. et al. (1987). Tectonic denudation of the upper mantle along passive margins: A model based on drilling results (ODP Leg 103, western Galicia margin, Spain). *Tectonophysics*, 132, 335–342.

Bourbon, M. (1980). Evolution d'un secteur de la marge nord-téthysienne en milieu pélagique : la zone briançonnaise près de Briançon, entre le début du Malm et l'Eocène inférieur. PhD Thesis, Strasbourg.

Bourbon, M., Chenet, P., Faure, P., Gidon, M., Réau, J. (1976). Importance et signification des érosions mésozoïques à la marge externe de la zone briançonnaise (Alpes du Sud). *C. R. Acad. Sc. Paris*, 283, 1269–1272.

Bousquet, R., Oberhansli, R., Goffe, B., Wiederkehr, M., Koller, F., Schmid, S.M., Schuster, R., Engi, M., Berger, A., Martinotti, G. (2008). Metamorphism of metasediments at the scale of an orogen: A key to the Tertiary geodynamic evolution of the Alps. In *Tectonic Aspects of the Alpine-Dinaride-Carpathian System*, Siegesmund, S., Fügenschuh, B., Froitzheim, N. (eds). The Geological Society, London.

Boutoux, A., Bellahsen, N., Lacombe, O., Verlaguet, A., Mouthereau, F. (2014). Inversion of pre-orogenic extensional basins in the external Western Alps: Structure, microstructures and restoration. *J. Struct. Geol.*, 60, 13–29

Brouwer, F.M., Vissers, R.L.M., Lamb, W.M. (2002). Structure and metamorphism of the Gran Paradiso massif, western Alps, Italy. *Contrib. Mineral. Petrol.*, 143, 450–470.

Bucher, S., Schmid, S.M., Bousquet, R., Fügenschuh, B. (2003). Late-stage deformation in a collisional orogen (Western Alps): Nappe refolding, back-thrusting or normal faulting. *Terra Nova*, 15, 109–117.

Burri, M. (1979). Les formations valaisannes dans la région de Visp. *Eclog. Geol. Helv.*, 72, 789–802.

Caby, R. (1968). Contribution à l'étude structurale des Alpes occidentales : subdivisions stratigraphiques et structure de la zone du Grand-Saint-Bernard dans la partie sud du Val d'Aoste (Italie). *C. R. Acad. Sci., Paris*, D, 273, 999–1002.

Cannic, S. (1996). L'évolution magmatique et tectono-métamorphique du substratum du domaine Valaisan (complexe du Versoyen, Alpes Occidentales) : implications dans l'histoire alpine. Unpublished PhD Thesis, Université de Grenoble.

Cannic, S., Lardeaux, J.M., Mugnier, J.L., Hernandez, J. (1995). Tectono-metamorphic evolution of the Roignais-Versoyen Unit (Valaisan domain, France). *Eclog. Geol. Helv.*, 89, 321–343.

Caron, C. (1972). La nappe supérieure des Préalpes : subdivisions et principaux caractères du sommet de l'édifice préalpin. *Eclog. Geol. Helv.*, 65(1), 57–73.

Caron, J.M. (1977). Lithostratigraphie et tectonique des schistes lustrés dans les Alpes cottiennes septentrionales et en Corse. *Sci. Géol. Mém.*, Strasbourg, 48.

Caron, J.M. and Gay, M. (1977). La couverture mésozoïque du massif d'Ambin, transition entre le domaine briançonnais et le domaine piémontais ? *Eclog. Geol. Helv.*, 70, 643–666.

Caron, C. and Weidmann, M. (1967). Sur les flyschs de la région des Gêts (Haute-Savoie). *Eclog. Geol. Helv.*, 60(2), 357–398.

Caron, C., Charollais, J., Rosset, J. (1967). Eléments autochtones et éléments allochtones du soubassement des klippes des Annes et de Sulens. *Géologie alpine*, 43, 47–61.

Cavazza, W., Roure, F., Spakman, W., Stampfli, G.M., Ziegler, P. (eds) (2004). *The TRANSMED Atlas, the Mediterranean Region from Crust to Mantle*. Springer, Berlin.

Chopin, C. (1979). De la Vanoise au massif du Grand Paradis. Thesis, Université Pierre et Marie Curie – Paris VI.

Chopin, C. (1984). Coesite and pure pyrope in high-grade blueschists of the Western Alps: A first record and some consequences. *Contrib. Mineral. Petrol.*, 86, 107–118.

Conti, P., Manatschal, G., Pfister, M. (1994). Synrift sedimentation, Jurassic and Alpine tectonics in the central Ortler Nappe (Eastern Alps, Italy). *Eclog. Geol. Helv.*, 87, 63–90.

De Charpal, O., Guennoc, P., Montader, L., Roberts, D.G. (1978). Rifting, crustal attenuation and subsidence in the Bay of Biscay. *Nature*, 275(5682), 706–711.

De Graciansky, P.C. and Lemoine, M. (1988). Early Cretaceous extensional tectonics in the southwestern French Alps: A consequence of North-Atlantic rifting during Tethyan spreading. *Bull. Soc. Géol. Fr.*, IV(5), 733–737.

Debelmas, J. (1955). Contribution à la connaissance du massif de Peyre-Haute au sud de Briançon (Hautes-Alpes) : les montagnes de l'Alpavin et de l'Agnelil. *Bull. Serv. Carte Géol. France*, 246B, 53, 1–27.

Debelmas, J. (1974). Les Alpes franco-italiennes. In *Géologie de la France*, Debelmas, J. (ed.). Doin, Paris.

Debelmas, J. (1975). Réflexions et hypothèses sur la paléogéographie crétacée des confins alpino-apenniniques. *Bull. Soc. Géol. Fr.*, 18(6), 1002–1013.

Debelmas, J. (1977). Deux coupes transversales des Alpes franco-italiennes. *Schweiz. Mineral. Petrogr. Mitt.*, 56, 561–665.

Debelmas, J. (2001). La zone subbriançonnaise et la zone valaisanne savoyarde dans le cadre de la tectonique des plaques. *Géologie Alpine*, 77, 3–8.

Debelmas, J. and Kerkhove, C. (1980). Les Alpes franco-italiennes. *Géol. Alpine*, 56, 21–58.

Debelmas, J. and Lemoine, M. (1957). Calschistes piémontais et terrains à faciès briançonnais dans les hautes vallées de la Maïra et la Varaita (Alpes cottiennes, Italie). *C. R. Acad. Sci. Paris*, 38–40.

Debelmas, J. and Lemoine, M. (1970). The Western Alps: Paleogeography and structure. *Earth Sci. Reviews*, 6, 221–256.

Debelmas, J., Debeerst, P., Arnaud, H., Desmons, J., Fudral, S., Gidon, M., Grandchamp, P., Kerckhove, C. (2011). *L'exploration géologique des Alpes franco-italiennes*. Presses des Mines, Paris.

Dercourt, J., Zonenshain, L.P., Ricou, L.E., Kazmin, V.G., Le Pichon, C., Knipper, A.L., Grandjacquet, C., Sbortshikov, I.M., Geyssant, J., Lepvrier, C. et al. (1986). Geological evolution of the Tethys from the Atlantic to the Pamirs since the Lias. *Tectonophysics*, 123, 241–315.

Dercourt, J., Ricou, L.E., Vrielinck, B. (1993). *Atlas Tethys, Paleoenvironmental Maps*. Gauthier-Villars, Paris.

Dercourt, J., Gaetani, M., Vrielynk, B., Barrier, E., Biju-Duval, B., Brunet, M.-F., Cadet, J.P., Crasquin, S., Sandulescu, M. et al. (2000). *Atlas Peri-Tethys, Paleogeographical Maps*. CCGM/CGMW, Paris.

Deville, E. (1987). Etude géologique en vanoise orientale (Alpes occidentales françaises, Savoie). De la naissance à la structuration d'un secteur de la paléo-marge européenne et de l'océan tethysien : aspects stratigraphiques, pétrographiques et tectoniques. Unpublished PhD Thesis, Université Savoie Mont Blanc, Chambéry.

Deville, E. (1989). La couverture occidentale du massif du Grand Paradis (Alpes occidentales françaises) : données nouvelles et conséquences paléogéographiques. *C. R. Acad. Sci. Paris*, 309, 603–610.

Deville, E., Fudral, S., Lagabrielle, Y., Marthaler, M., Sartori, M. (1992). From oceanic closure to continental collision: A synthesis of the "Schistes lustrés" metamorphic complex of the Western Alps. *Geol. Soc. Am. Bull.*, 104, 127–139.

Dumont, T. (1983). Le chainon de Rochebrune au sud-est de Briançon : évolution paléogéographique et structurale d'un secteur de la zone piémontaise des Alpes occidentales. PhD Thesis, Université Scientifique et Médicale de Grenoble.

Dumont, T. (1984). Le Rhétien et le Lias inférieur prépiémontais : enregistrement sédimentaire du passage des carbonates de plateforme triasiques au Jurassique hémipélagique lors du début du rifting téthysien. *Géologie Alpine*, 60, 13–25.

Dumont, T. (1988). Late Triassic-Early Jurassic evolution of the western Alps and of their European foreland; initiation of the Tethyan rifting. *Bull. Soc. Géol. Fr.*, 8(4), 601–611.

Dumont, T. (1998). Sea-level changes and early rifting of a European Tethyan margin in the Western Alps and SE France. In *Mesozoic and Cenozoic Sequence Stratigraphy of European Basins*, de Graciansky, P.-C., Hardenbol, J., Jacquin, T., Vail, P.R. (eds). SEPM Society for Sedimentary Geology, Tulsa.

Dumont, T., Lemoine, M., Tricart, P. (1984). Perénnité de la sédimentation pélagique du Jurassique supérieur jusque dans le Crétacé supérieur au dessus de la croûte océanique téthysienne ligure : la série supra-ophiolitique du Lac des Cordes (Zone piémontaise des Alpes Occidentales au SE de Briançon). *C. R. Acad. Sci. Paris*, 299, 1069–1072.

Dumont, T., Champagnac, J.D., Crouzet, C., Rochat, P. (2008). Multistage shortening in the Dauphiné zone (French Alps): The record of Alpine collision and implications for pre-Alpine restoration. *Swiss Journal of Geosciences*, 101, S89–S110.

Dumont, T., Schwartz, S., Guillot, S., Simon-Labric, T., Tricart, P., Jourdan, S. (2012). Structural and sedimentary records of the Oligocene revolution in the Western Alpine Arc. *Journal of Geodynamics*, 56–57, 18–38.

Eberli, G.P. (1988). The evolution of the southern continental margin of the Jurassic Tethys Ocean as recorded in the Allgau Formation of the Austroalpine nappes of Graubunden (Switzerland). *Eclog. Geol. Helv.*, 81, 175–214.

Ellenberger, F. (1953). La série du Barrhorn et les rétrocharriages penniques. *C. R. Acad. Sci., Paris*, 236, 218–220.

Ellenberger, F. (1958). Etude géologique du pays de Vanoise. *Mémoire explicatif de la Carte géologique de France*, 561.

Ellenberger, F. (1963). Trias à faciès briançonnais de la Vanoise et des Alpes occidentales. *Colloque sur le Trias, Montpellier 1961, Mémoire B.R.G.M.*, 15, 215–231.

Ellenberger, F. and Lemoine, M. (1955). Les faciès prépiémontais et le problème du passage de la zone du Briançonnais aux Schistes lustrés piémontais. *C. R. Somm. Soc. Géol Fr.*, 146–148.

Ellenberger, F., Michard, A., Sturani, C. (1964). Découvertes d'ammonites et observations stratigraphiques dans les "Schistes lustrés" du Val Grana (Alpes cottiennes). *C. R. Acad. Sci., Paris*, 3(259), 3047–3050.

Escher, A., Masson, H., Steck, A., Epard, J.L., Marchant, R., Marthaler, M., Sartori, M., Venturini, G. (1994). Coupe générale des Alpes occidentales du graben de la Bresse à la zone d'Ivrée. *Section des Sciences de la Terre, BFSH*, Lausanne.

Ferrando, S., Bernoulli, D., Compagnoni, R. (2004). The Canavese zone (internal Western Alps): A distal margin of Adria. *Schweiz. Mineral. Petrog. Mitt.*, 84, 237–256.

Florineth, D. and Froitzheim, N. (1994). Transition from continental to oceanic basement in the Tasna nappe (Engadine window, Graubünden, Switzerland): Evidence for Early Cretaceous opening of the Valais ocean. *Schweiz. Mineral. Petrog. Mitt.*, 74, 437–448.

Frisch, W. (1979). Tectonic progradation and plate tectonic evolution of the Alps. *Tectonophysics*, 60, 121–139.

Froitzheim, N. (1988). Synsedimentary and synorogenic normal faults within a thrust sheet of the Eastern Alps (Ortler zone, Graubünden, Switzerland). *Eclog. Geol. Helv.*, 81, 593–610.

Froitzheim, N. and Eberli, G.P. (1990). Extensional detachment faulting in the evolution of a Tethys passive continental margin, Eastern Alps, Switzerland. *Geol. Soc. Am. Bull.*, 102, 1297–1308.

Froitzheim, N. and Manatschal, G. (1996). Kinematics of Jurassic rifting, mantle exhumation, and passive-margin formation in the Austroalpine and Penninic nappes (eastern Switzerland). *Geol. Soc. Am. Bull.*, 108, 1120–1133.

Froitzheim, N. and Rubatto, D. (1998). Continental breakup by detachment faulting: Field evidence and geochronological constraints (Tasna nappe, Switzerland). *Terra Nova*, 10, 171–176.

Fudral, S. (1973). Contribution à l'étude de l'unité de Moutiers (zone des brèches de Tarentaise) entre le vallon du torrent du Cormet d'Arêches et le hameau des Chapieux (Savoie) – Alpes françaises. Thesis, Université Scientifique et Médicale de Grenoble.

Fudral, S. (1980). Une nouvelle interprétation de l'unité de Salins (zone des Brèches de Tarentaise) au nord-ouest de Bourg-Saint-Maurice (Savoie). Conséquences structurales. *C.R. Acad. Sci. Paris*, 290, 1333–1336.

Fudral S. (1998). Etude géologique de la suture téthysienne dans les Alpes franco-italiennes nord occidentales de la Doire Ripaire (Italie) à la région de Bourg-Saint-Maurice (France). *Géol. Alp. Mémoire*, 29, 1–306.

Fudral, S., Deville, E., Marthaler, M. (1987). Distinction de trois ensembles d'unités dans les "Schistes lustrés" compris entre la Vanoise et le Val de Suse (Alpes franco-italiennes septentrionales). *C. R. Acad. Sci. Paris*, 305(2), 467–472.

Ganne, J. (2003). Les dômes de socle HP-BT dans le domaine Pennique des Alpes nord-occidentales (massifs d'Ambin et Vanoise Sud) : modalités de leur exhumation. Thesis, Université de Savoie, Chambéry.

Ganne, J., Bussy, F., Vidal, O. (2003). Multi-stage garnet in the internal Brianconnais Basement (Ambin Massif, Savoy); new petrological constraints on the blueschist-facies metamorphism in the Western Alps and tectonic implications. *Journal of Petrology*, 44(7), 1281–1308.

Geyssant, J. (1973). A propos de l'âge des lambeaux de l'Austro-Alpin supérieur dans les Alpes orientales centrales : essai sur la paléogéographie jurassique des unités supérieures austro-alpines et des Alpes méridionales. *Géologie Alpine*, 49, 41–50.

Gidon, M. (1998). GéolAlpes [Online]. Available at: http://www.geol-alp.com.

Goffé B. (1975). Etude structurale et pétrographique du versant occidental du massif paléozoïque de Chasseforêt (Vanoise méridionale). Thesis, Université Paris Sud – Paris XI.

Graciansky, P.C., Lemoine, M., Mégard-Galli, J., Richez, M. (1988). Tectonique synsédimentaire triasico-jurasique et rifting téthysien dans la nappe briançonnaise de Peyre-Haute (Alpes occidentales). *Bull. Soc. Géol. Fr.*, 8(4), 669–680.

Grandjacquet, C. and Haccard, D. (1977). Position structurale et rôle paléogéographique de l'unité du Bracco au sein du complexe ophiolitique liguro-piémontais. *Bull. Soc. Géol. Fr.*, 7, 18–24.

Handy, M.R., Schmid, S.M., Bousquet, R., Kissling, E., Bernoulli, D. (2010). Reconciling plate-tectonic reconstructions of Alpine Tethys with the geological–geophysical record of spreading and subduction in the Alps. *Earth-Sci. Rev.*, 102, 121–158.

Hässig, M., Rolland, Y., Sahakyan, L., Sosson, M., Galoyan, G., Avagyan, A., Bosch, D., Müller, C. (2015). Multi-stage metamorphism in the South Armenian block during the Late Jurassic to early Cretaceous: Tectonics over south-dipping subduction of Northern branch of Neotethys. *Journal of Asian Earth Sciences*, 102, 4–23.

Haug, E. (1909). Les géosynclinaux de la chaîne des Alpes pendant les temps secondaires. *C. R. Acad. Sci. Paris*, 148, 1637.

Henry, C., Michard, A., Chopin, C. (1993). Geometry and structural evolution of ultra-high-pressure and high-pressure rocks from the Dora-Maira Massif, Western Alps, Italy. *J. Structural Geol.*, 15, 965–981.

Hermann, J., Müntener, O., Trommsdorff, V., Hansmann, W., Piccardo, G.B. (1997) Fossil crust-to-mantle transition, Val Malenco (Italian Alps). *J. Geophys. Res.*, 102, 20123–20132.

Hofer, G., Draganits, E., Wagreich, M., Hofmann, C.C., Reischenbacher, D., Grundtner, M.L., Bottig, M. (2011). Stratigraphy and geochemical characterisation of Upper Cretaceous non-marine-marine cycles (Grünbach Formation, Gosau Group, Austria). *Aust. Journ. Earth Sci.*, 104, 90–107.

Incerpi, N., Martire, L., Manatschal, G., Bernasconi, S.M., Gerdes, A., Czuppon, G., Palcsu, L., Karner, G.D., Johnson, C.A., Figueredo, P.H. (2019). Hydrothermal fluid flow associated to the extensional evolution of the Adriatic rifted margin: Insights from the pre- to post-rift sedimentary sequence (SE Switzerland, N Italy). *Basin Res.*, 00, 1–25.

Ivan, P. (2002). Relics of the Meliata Ocean crust: Geodynamic implications of mineralogical, petrological and geochemical proxies. *Geologica Carpathica*, 53, 245–256.

Jaillard, E. (1985a). Evolutions sédimentaire et paléotectonique de la zone briançonnaise de Vanoise occidentale (Alpes occidentales françaises). *Géologie Alpine*, 61, 85–113.

Jaillard, E. (1985b). La transgression jurassique en Vanoise occidentale (zone briançonnaise, Alpes occidentales françaises). Conséquences paléogéographiques. *C. R. Acad. Sci. Paris*, 301, 633–636.

Jaillard, E. (1988). Une image paléogéographique de la Vanoise briançonnaise. *Eclog. Geol. Helv.*, 81, 553–566.

Jaillard, E. (1989). La transition Briançonnais externe-Briançonnais interne en Savoie. L'Aiguille des Aimes, le Roc du Bourget et le massif d'Ambin. *Géologie Alpine*, 65, 105–134.

Janoschek, W. and Matura, A. (1980). Outline of the geology of Austria. *Abh. Geol. B.-A, 26e CGI, Wien*, 34, 7–98.

Kerckhove, C. (1969). La zone du flysch dans les nappes de l'Embrunais-Ubaye (Alpes occidentales). *Géologie alpine*, 45, 5–204.

Kozur, H. (1991). The evolution of the Meliata-Hallstatt ocean and its significance for the early evolution of the Eastern Alps and Western Carpathians. *Palaeogeogr. Palaeoclimat. Palaeoecol.*, 87, 109–135.

Labesse, B. (1981). A propos de la paléogéographie du domaine Ligure au Crétacé supérieur : le problème de la ride du Bracco (Apennin septentrional, Italie). *C. R. Acad. Sc., Paris*, 293(2), 1087–1090.

Lagabrielle, Y. (1987). Les Ophiolites : marqueurs de l'histoire tectonique des domaines océaniques. Le cas des Alpes franco-italiennes (Queyras, Piémont) ; comparaison avec les ophiolites d'Antalya (Turquie) et du Coast Range de Californie. PhD Thesis, U.B.O., Brest.

Lagabrielle, Y. and Bodinier, J.L. (2008). Submarine reworking of exhumed subcontinental mantle rocks: Field evidence from the Lherz peridotites, French Pyrenees. *Terra Nova*, 20, 11–21.

Lagabrielle, Y. and Cannat, M. (1990). Alpine Jurassic ophiolites resemble the modern central Atlantic basement. *Geology*, 18, 319–322.

Lagabrielle, Y. and Polino, R. (1988). Un schéma structural du domaine des Schistes lustrés ophiolitifères au nord-ouest du massif du Mont Viso (Alpes sud-occidentales) et ses implications. *C.R. Acad. Sci. Paris*, 306(2), 921–928.

Lagabrielle, Y., Asti, R., Duretz, T., Clerc, C., Fourcade, S., Teixell, A., Labaume, P., Corre, B., Saspiturry, N. (2020). A review of cretaceous smooth-slopes extensional basins along the Iberia-Eurasia plate boundary: How pre-rift salt controls the modes of continental rifting and mantle exhumation. *Earth-Sci. Revi.*, 201, 103071.

Laubscher, H.P. and Bernoulli, D. (1977). Mediterranean and Tethys. In *The Ocean Basins and Margins*, Nairn, A.E.M., Stehli, F.G., Malcolm Brown, G. (eds). Plenum, New York.

Le Bayon, B. and Ballèvre, M. (2006). Deformation history of a subducted continental crust (Gran Paradiso, Western Alps): Continuing crustal shortening during exhumation. *Journal of Structural Geology*, 28, 793–815.

Le Guernic, J. (1967). La zone du Roure : contribution à l'étude du Briançonnais interne et du Piémontais en Haute Ubaye. *Trav. Lab. Géol. Grenoble*, 43, 95–127.

Lefèvre, R. (1962). Observations sur le Mésozoïque à faciès briançonnais des environs d'Acceglio (Alpes cottiennes, Italie) ; découverte du Crétacé supérieur fossilifère. *C. R. Acad. Sci, Paris*, 254.

Lefèvre, R. (1968). La structure et le style tectonique de la bande d'Acceglio en Val Maira (Alpes cottiennes italiennes). *Géologie Alpine*, 44, 139–151.

Lefèvre, R. (1982). Les nappes briançonnaises internes et ultrabriançonnaises dans les Alpes cottiennes méridionales. PhD Thesis, Université Paris Sud.

Lefèvre, R. (1984). Les nappes briançonnaises internes et ultrabriançonnaises entre Maira et Grana (Alpes cottiennes méridionales). *Géologie alpine*, 60, 27–36.

Lefèvre, R. and Michard, A. (1976). Les nappes briançonnaises internes et ultrabriançonnaises de la bande d'Acceglio (Alpes franco-italiennes). Une étude structurale et pétrographique dans le faciès des schistes bleus à jadéite. *Bull. Sci. Géol., Strasbourg*, 29, 183–222.

Lemoine, M. (1960). Découverte d'une microfaune du Crétacé supérieur au col du Longet (sources de l'Ubaye, Basses-Alpes) : conséquences tectoniques et paléogéographiques. *C.R. Somm. S.G.F.*, 5(2), 97.

Lemoine, M. (1961). Le Briançonnais interne et le bord de la zone des Schistes lustrés dans les vallées du Guil et de l'Ubaye. *Trav. Lab. Géol. Grenoble*, 37, 97–119.

Lemoine, M. (1971). Données nouvelles sur la série du Gondran près Briançon (Alpes cottiennes). Réflexions sur les problèmes stratigraphique et paléogéographique de la zone piémontaise. *Géologie alpine*, 47, 181–201.

Lemoine, M. (1983). Rifting and early drifting: Mesozoic Central Atlantic and Ligurian Tethys. In *Init. Repts. DSDP, 76*, Sheridan, R.E., Gradstein, F.M. (eds). US Government Printing Office, Washington.

Lemoine, M. (1985). Structuration jurassique des Alpes occidentales et palinspastique de la Téthys ligure. *Bull. Soc. Geol. Fr*, 8(1), 126–137.

Lemoine, M. (1988). Des nappes embryonnaires aux blocs basculés : évolution des idées et des modèles sur l'histoire mésozoïque des Alpes occidentales. *Bull. Soc. Géol. Fr.*, 5, 787–797.

Lemoine, M. (2003). Schistes lustrés from Corsica to Hungary: Back to the original sediments. *Bull. Soc. Géol. Fr.*, 174(3), 197–209.

Lemoine, M. (2006). Une science en crise au milieu du XXe siècle, la géodynamique : des géosynclinaux aux océans disparus. *Travaux du Comité français d'Histoire de la Géologie*, 20, 129–165.

Lemoine, M. and de Graciansky, P.C. (1988). Histoire d'une marge continentale passive : les Alpes occidentales au Mésozoïque, Introduction. *Bull. Soc. Géol. Fr.*, 8(4), 597–600.

Lemoine, M. and Michard, A. (1963). Sur le problème stratigraphique et tectonique des Schistes Lustrés dans les Alpes Cottiennes, *C. R. Acad. Sc. Paris*, 256, 727–732.

Lemoine, M. and Trümpy, R. (1987). Pre-oceanic rifting in the Alps. *Tectonophysics*, 133, 305–320.

Lemoine, M., Steen, D., Vuagnat, M. (1970). Sur le problème stratigraphique des ophiolites piémontaises et des roches sédimentaires associées : observations dans le massif de Chabrière en haute Ubaye (Basses-Alpes, France). *C.R. des Séances, S.P.H.N. Genève*, 5, 44–59.

Lemoine, M., Bourbon, M., Tricart, P. (1978). Le Jurassique et le Crétacé prépiémontais à l'Est de Briançon (Alpes occidentales) et l'évolution de la marge européenne de la Téthys : données nouvelles et conséquences. *C.R. Acad. Sci. Paris*, 286, 1655–1658.

Lemoine, M., Gidon, M., Barfety, J.-C. (1981). Les massifs cristallins externes des Alpes occidentales : d'anciens blocs basculés nés au Lias lors du rifting téthysien. *C.R. Acad. Sci. Paris*, 292(2), 917–920.

Lemoine, M., Marthaler, M., Caron, M., Sartori, M., Amaudric Du Chaffaut, S., Dumont, T., Escher, A., Masson, H., Polino, R., Tricart, P. (1984). Découverte de foraminifères planctoniques du Crétacé supérieur dans les Schistes Lustrés du Queyras (Alpes occidentales). Conséquences paléogéographiques et tectoniques. *C. R. Acad. Sci., Paris*, 299(2), 11.

Lemoine, M., Bas, T., Arnaud-Vanneau, A., Arnaud, H., Dumont, T., Gidon, M., Bourbon, M., de Graciansky, P.C., Rudkiewicz, J.L., Megard-Galli, J., Tricart, P. (1986). The continental margin of the Mesozoic Tethys in the Western Alps. *Mar. Pet. Geol.*, 3, 179–199.

Lemoine, M., Dardeau, G., Delpech, P.Y., Dumont, T., de Graciansky, P.C., Graham, R., Jolivet, L., Roberts, D., Tricart, P. (1989). Extension synrift et failles transformantes jurassiques dans les Alpes occidentales. *C. R. Acad. Sc. Paris*, 309(2), 711–716.

Lemoine, M., de Graciansky, P.C., Tricart, P. (2000). *De l'océan à la chaine de montagne. Tectonique des plaques dans les Alpes*. Gordon and Breach Science Publishers, Paris

Liati, A. and Froitzheim, N. (2006). Assessing the Valais ocean, Western Alps: U-Pb SHRIMP zircon geochronology of eclogite in the Balma unit, on top of the Monte Rosa nappe. *European Journal of Mineralogy*, 18, 299–308.

Liati, A., Froitzheim, N., Fanning, C.M. (2005). Jurassic ophiolites within the Valais domain of the western and central Alps: Geochronological evidence for re-rifting of oceanic crust. *Contrib. Mineral. Petrol.*, 149, 446–461.

Loprieno, A., Bousquet, R., Bucher, S., Ceriani, S., Dalla Torre, F.H., Fügenschuh, B., Schmid, S.M. (2011). The Valais units in Savoy (France): A key area for understanding the paleogeography and the tectonic evolution of the Western Alps. *International Journal of Earth Sciences*, 100, 963–992.

Loubat, H. (1968). Etude pétrographique des ophiolites de la zone du Versoyen, Savoie (France), province d'Aoste (Italie). *Arch. Sci. et C. R. Séan. Soc. Phys. Hist. Nat. Genève*, 21, 265–457

Loubat, H. (1975). La zone du Versoyen : témoin possible d'une intersection entre dorsale océanique et marge continentale. *Arch. Sci. et C. R. Séan. Soc. Phys. Hist. Nat. Genève*, 28, 101–116

Loubat, H. (1984). Considérations préliminaires sur la configuration horizontale de l'édifice subvolcanique du Versoyen (Alpes franco-italiennes) analogue aux basins en distension du type Golfe de Californie. *Géologie alpine*, 60, 37–44.

Loubat, H., and Antoine, P. (1965). Sur l'extension des laves en coussins dans la région du Petit-Saint-Bernard (frontière franco-italienne). *Arch. Sci. et C. R. Séan. Soc. Phys. Hist. Nat. Genève*, 18, 689–692.

Loubat, H. and Delaloye, M. (1984). La zone du Versoyen (Alpes franco-italiennes) : le témoin d'une océanisation mésozoïque circonscrite constituant un milieu hybride, subvolcano-sédimentaire, avec mobilisats et adinoles. *Géologie Alpine*, 60, 45–76.

Lundin, E.R. and Doré, A.G. (2011). Hyperextension, serpentinization and weakening: A new paradigm for rifted margin compressional deformation. *Geology*, 39, 347–350

Manatschal, G. (1999). Fluid- and reaction-assisted low-angle normal faulting: Evidence from rift-related brittle fault rocks in the Alps (Err Nappe, eastern Switzerland). *J. Struct. Geol.*, 21, 777–793.

Manatschal, G. (2004). New models for evolution of magma-poor rifted margins based on a review of data and concepts from West Iberia and the Alps. *Int. J. Earth Sci.*, 93, 432–466.

Manatschal, G. and Bernoulli, D. (1999), Architecture and tectonic evolution of nonvolcanic margins: Present-day Galicia and ancient Adria. *Tectonics*, 18, 1099–1119.

Manatschal, G. and Nievergelt, P. (1997). A continent-ocean transition recorded in the Err and Platta nappes (Eastern Switzerland). *Ecl. Geol. Helv.*, 90, 3–27.

Manatschal, G., Froitzheim, N., Rubenach, M., Turrin, B.D. (2001). The role of detachment faulting in the formation of an ocean-continent transition: Insights from the Iberia Abyssal Plain. *Geol. Soc. Spec. Publ.*, 405–428.

Manatschal, G., Müntener, O., Lavier, L., Minshull, T.A., Péron-Pinvidic G. (2007). Observations from the Alpine Tethys and Iberia-Newfoundland margins pertinent to the interpretation of continental breakup. *Geol. Soc. Spec. Publ.*, 282, 291–324.

Manzotti, P. and Ballèvre, M. (2013). Multistage garnet in high-pressure metasediments: Alpine overgrowths on Variscan detrital grains. *Geology*, 41, 1151–1154.

Manzotti, P., Ballèvre, M., Zucali, M., Robyr, M., Engi, M. (2014a). The tectonometamorphic evolution of the Sesia–Dent Blanche nappes (internal Western Alps): Review and synthesis. *Swiss Journal of Geosciences*, 107, 309–336.

Manzotti, P., Le Carlier de Veslud, C., Le Bayon, B., Ballèvre, M. (2014b). Petro-structural map of the Money Unit (Gran Paradiso Massif, Valnontey valley, Western Alps). *Journal of Maps*, 10, 324–340.

Manzotti, P., Pitra, P., Langlade, J., Ballèvre, M. (2015a). Constraining P-T conditions during thrusting of a higher pressure unit over a lower pressure one (Gran Paradiso, Western Alps). *Journal of Metamorphic Geology*, 33, 981–1002.

Manzotti, P., Poujol, M., Ballèvre, M. (2015b). Detrital zircon geochronology in blueschist-facies meta-conglomerates from the Western Alps: Implications for the late Carboniferous to early Permian paleogeography. *International Journal of Earth Sciences*, 104, 703–721.

Manzotti, P., Ballèvre, M., Poujol, M. (2016). Detrital zircon geochronology in the Dora Maira and Zone Houillère: A record of sediment travel paths in the Carboniferous. *Terra Nova*, 28, 279–288.

Marcoux, J. (1987). Histoire et topologie de la Néo-Tethys. Thesis, Université de Paris.

Masini, E., Manatschal, G., Mohn, G., Ghienne J.F., Lafont, F. (2011). The tectono-sedimentary evolution of a supra-detachment rift basin at a deep-water magma-poor rifted margin: The example of the Samedan Basin preserved in the Err nappe in SE Switzerland. *Basin Res.*, 23, 652–677.

Masson, H. (2002). Ophiolites and other (ultra)basic rocks from the West-Central Alps: New data for a puzzle. *Bull. Soc. Vaud. Sci. Nat.*, 88, 263–276.

Masson, H., Herb R., Steck, A. (1980). Helvetic Alps of Western Swizerland. In *Geology of Switzerland, A Guide-book, Pt. B: Geological Excursions*, Schweizerische Geologische Kommission (ed.). Wepf, Basel.

Masson, H., Bussy, F., Eichenberger, M., Giroud, N., Meilhac, C., Presniakov, S. (2008). Early Carboniferous age of the Versoyen ophiolites and consequences: Non-existence of a "Valais ocean" (Lower Penninic, western Alps). *Bull. Soc. Géol. Fr.*, 179, 337–355.

Mégard-Galli, J. and Baud, A. (1977), Le Trias moyen et supérieur des Alpes nord-occidentales et occidentales : données nouvelles et corrélations stratigraphiques. *Bull. BRGM*, 4(3), 233–250.

Michard, A. (1967). *Etudes géologiques dans les zones internes des Alpes cottiennes*. Editions du CNRS, Paris.

Michard, A. and Henry C. (1988). Les nappes briançonnaises en Haute-Ubaye (Alpes franco-italiennes) ; contribution à la reconstitution paléogéographique du Briançonnais au Mésozoïque. *Bull. Soc. Géol. Fr.*, 8(4), 693–701.

Michard, A., Chopin, C., Henry C. (1993). Compression versus extension in the exhumation of the Dora-Maira coesite-bearing unit, Western Alps, Italy. *Tectonophysics*, 221, 173–193.

Michard, A., Goffé, B., Chopin, C., Henry, C. (1996). Did the Western Alps develop through an Oman-type stage? The geotectonic setting of high-pressure metamorphism in two contrasting Tethyan transects. *Eclogae Geol. Helv.*, 89, 43–80.

Michard, A., Avigad, D., Goffé, B., Chopin, C. (2004). The high-pressure metamorphic front of the south Western Alps (Ubaye-Maira transect, France, Italy). *Schw. Mineral. Petrogr. Mitt.*, 84, 215–235.

Michard, A., Dumont, T., Andreani, L., Loget, N. (2010). Cretaceous folding in the Dévoluy mountains (Subalpine Chains, France): Gravity-driven detachment at the European paleomargin versus compressional event. *Bull. Soc. Géol. France*, 181(6), 565–581.

Missoni, S. and Gawlick, H.-J. (2010). Evidence for Jurassic subduction from the Northern Calcareous Alps (Berchtesgaden; Austroalpine, Germany). *Int. J. Earth Sci.*, 100, 1605–1631.

Mohn, G. (2010). L'évolution tectono-sédimentaire des marges de la Téthys alpine au cours de l'amincissement lithosphérique. PhD Thesis, Université de Strasbourg.

Mohn, G., Manatschal, G., Müntener, O., Beltrando, M., Masini, E. (2010). Unravelling the interaction between tectonic and sedimentary processes during lithospheric thinning in the Alpine Tethys margins, *Int. J. Earth Sci.*, 99, 75–101.

Mohn, G., Manatschal, G., Masini, E., Müntener, O. (2011). Rift-related inheritance in orogens: A case study from the Austroalpine nappes in Central Alps (SE-Switzerland and N-Italy*). Int. J. Earth Sci.*, 100, 937–961.

Mohn, G., Manatschal, G., Beltrando, M., Masini, E., Kusznir, N. (2012). Necking of continental crust in magma-poor rifted margins: Evidence from the fossil Alpine Tethys margins. *Tectonics*, 31, TC1012.

Molli, G. (2008). Northern Apennine–Corsica orogenic system: An updated overview. In *Tectonic Aspects of the Alpine-Dinaride-Carpathian System*, Siegesmund, S., Fügenschuh, B., Froitzheim, N. (eds). The Geological Society, London.

Molli, G. and Malavieille, J. (2010). Orogenic processes and the Alps/Apennines geodynamic evolution: Insights from Taiwan. *International Journal of Earth Sciences (Geol Rundsch)* [Online]. Available at: DOI-10.1007/s00531-010-0598-y.

Mugnier, J.L., Cannic, S., Lapierre, H. (2008). The tholeiites of the Valaisan domain (Versoyen, western Alps): A Carboniferous magma emplaced in a small oceanic basin. *Bull. Soc. Géol. France*, 179, 357–368.

Müntener, O., Pettke, T., Desmurs, L., Meier, M., Schaltegger, U. (2004). Refertilization of mantle peridotite in embryonic ocean basins: Trace element and Nd isotopic evidence and implications for crust-mantle relationships. *Earth Planet. Sci. Lett.*, 221, 293–308

Neubauer, F., Friedl, G., Genser, J., Handler, R., Mader, D., Schneider, D. (2007). Origin and tectonic evolution of the Eastern Alps deduced from dating of detrital white micas: A review. *Aus. J. Earth Sci.*, 100, 8–23.

Olivet, J.L. (1996). La cinématique de la plaque ibérique. *Bulletin des Centres de Recherches Exploration – Production Elf-Aquitaine*, 20(1), 131–195.

Pantet, A., Epard, J.L., Masson, H. (2020). Mimicking Alpine thrusts by passive deformation of synsedimentary normal faults: A record of the Jurassic extension of the European margin (Mont Fort nappe, Pennine Alps). *Swiss Journal of Geosciences*, 113, 13.

Péron-Pinvidic, G. and Manatschal, G. (2009). The final rifting evolution at deep magma-poor passive margins from Iberia-Newfoundland: A new point of view. *Int. J. Earth Sci.*, 98, 1581–1597.

Philippot, P. (1990). Opposite vergence of nappes and crustal extension in the French–Italian Western Alps. *Tectonics*, 9, 1143–1164.

Platt, J.P. and Lister, G.S. (1985a). Structural history of high-pressure metamorphic rocks in the southern Vanoise massif, French Alps, and their relation to Alpine tectonic events. *Journal of Structural Geology*, 7, 19–35.

Platt, J.P. and Lister, G.S. (1985b). Structural evolution of a nappe complex, southern Vanoise massif, French Penninic Alps. *Journal of Structural Geology*, 7, 145–160.

Platt, J.P., Lister, G.S., Cunningham, P., Weston, P., Peel, F., Baudin, T., Dondey, H. (1989). Thrusting and backthrusting in the Briançonnais domain of the western Alps. *Alpine Tectonics. Geol. Soc. London, Spec. Publ.*, 45, 135–152.

Puga, E., Fanning, M., Díaz de Federico, A., Nieto, J.M., Beccaluva, L., Bianchini, G., Díaz Puga, M.A. (2011). Petrology, geochemistry and U–Pb geochronology of the Betic Ophiolites: Inferences for Pangaea break-up and birth of the westernmost Tethys Ocean. *Lithos*, 124, 255–272.

Puga, E., Díaz de Federico, A., Fanning, M., Nieto, J.M., Rodríguez Martínez-Conde, J.A., Díaz Puga, M.A., Lozano, J.A., Bianchini, G., Natali, C., Beccaluva, L. (2017). The Betic ophiolites and the Mesozoic evolution of the Western Tethys. *Geosciences*, 7, 31.

Ribes, C., Manatschal, G., Ghienne, J.F., Karner, G.D., Johnson, C.A., Figueredo, P.H., Incerpi, N., Epin, M.E. (2019). The syn-rift stratigraphic record across a fossil hyper-extended rifted margin: the example of the northwestern Adriatic margin exposed in the Central Alps. *Int. J. Earth Sci.*, 108, 2071–2095.

Ribes, C., Ghienne, J.F., Manatschal, G., Dall'asta, N., Stockli, D., Galster, F., Gillard, M., Karner, G.D. (2020). The "Grès Singuliers" of the Mont Blanc region (France and Switzerland): Stratigraphic response to rifting and crustal necking in the Alpine Tethys. *International Journal of Earth Sciences*, 109(7), 2325–2352.

Ricou, L.E. (1984). Les Alpes occidentales : chaîne de décrochement. *Bull. Soc. Géol. France*, 861–874.

Ricou, L.E. (1988). The Tethyan oceanic gates: A tectonic approach to major sedimentary changes within Tethys. *Geodinamica Acta*, 4, 225–232.

Ricou, L.E. (1994). Tethys reconstructed: Plates, continental fragments and their boundaries since 260 Ma from Central America to South-eastern Asia. *Geodinamica Acta*, 7, 169–218.

Ricou, L.E. (1996). The plate tectonic history of the past Tethys Ocean. In *The Ocean Basin and Margins, V. 8: The Tethys Ocean*, Nairn, A.E.M., Ricou, L.E., Vrielynck, B., Dercourt, J. (eds). Springer-Verlag, New York.

Ricou, L.E., Zonenshain, P., Dercourt, J., Kazmin, V.G., Le Pichon, X., Knipper, L., Grandjacquet, C., Sborshchikov, I.M., Geyssant, J., Lepvrier, C. et al. (1985). Méthodes pour l'établissement de neuf cartes paléographiques de l'Atlantique au Pamir depuis le Lias. *Bull. Soc. Géol. Fr.*, 5, 625–635.

Ricou, L.E., Dercourt, J., Geyssant, J., Grandjacquet, C., Lepvrier, C., Biju- Duval, B. (1986). Geological constraints on the alpine evolution of the Mediterranean Tethys. *Tectonophysics*, 123, 83–122.

Ringgenberg, Y., Tomassi, A., Stampfli, G.M. (2001). La série jurassique de la nappe du Niesen dans la région Le Sépey-La Forclaz (Suisse) : témoin du rifting piémontais dans le domaine paléogéographique de l'Helvétique. *Bull. Soc. Vaud. Sc. Nat.*, 87(4), 353–372.

Robert, C. and Bousquet, R. (2013). *Géosciences, la dynamique du système Terre*. Belin, Paris.

Robertson, A.H.F., Clift, P.D., Degnan, P.J., Jones, G. (1991). Paleogeographic and paleotectonic evolution of the Eastern Mediterranean Neotethys. *Palaeogeography, Palaeoclimatology, Palaeoecology*, 87, 289–343.

Rolland, Y., Perincek, D., Kaymakci, N., Sosson, M., Barrier, E., Avagyan, A. (2012). Evidence for ~80–75 Ma subduction jump during Anatolide–Tauride–Armenian block accretion and ~48 Ma Arabia–Eurasia collision in Lesser Caucasus–East Anatolia. *Journal of Geodynamics*, 56, 76–85.

Sandrone, R., Cadoppi, P., Sacchi, R., Vialon, P. (1993). The Dora–Maira Massif. In *Pre-Mesozoic Geology of the Western Alps*, von Raumer, J.F. and Neubauer, F. (eds). Springer-Verlag, Berlin.

Sartori, M. (1990). L'unité du Barrhorn (Zone pennique, Valais, Suisse). *Mémoire de Géologie (Lausanne)*, 6.

Schaltegger, U., Desmurs, L., Manatschal, G., Müntener, O., Meier, M., Frank, M., Bernoulli, D. (2002). The transition from rifting to sea-floor spreading within a magma-poor rifted margin: Field and isotopic constraints. *Terra Nova*, 14, 156–162.

Schärer, U., Cannic, S., Lapierre, H. (2000). Preliminary evidence for a Hercynian age of the Versoyen complex, western Alps. *C. R. Acad. Sc. Paris*, 330(5), 325–332.

Schmid, S.M., Fügenschuh, B., Kissling, E., Schuster, R. (2004). Tectonic map and overall architecture of the Alpine orogen. *Eclog. Geol. Helv.*, 97, 93–117.

Schmid, S.M., Kissling, E., Diehl, T., van Hinsbergen, D.J.J., Molli, G. (2017). Ivrea mantle wedge, arc of the Western Alps, and kinematic evolution of the Alps–Apennines orogenic system. *Swiss J. Geosci.*, 110, 581–612.

Schwartz, S. (2000). La zone piémontaise des Alpes occidentales : un paléocomplexe de subduction. Arguments métamorphiques, géochronologiques et structuraux. PhD Thesis, Université de Lyon I.

Sengör, A.M.C. (1979). Mid-Mesozoic closure of Permo-Triassic Tethys and its implications. *Nature*, 279, 590–593.

Sengör, A.M.C. and Yilmaz, Y. (1981). Tethyan evolution of Turkey: A plate tectonic approach. *Tectonophysics*, 75, 181–241.

Sosson, M., Rolland, Y., Müller, C., Danelian, T., Melkonyan, R., Kekelia, S., Adamia, S., Babazadeh, V., Kangarli, T., Avagyan, A. et al. (2010). Subductions, obduction and collision in the Lesser Caucasus (Armenia, Azerbaijan, Georgia), new insights. *Geol. Soc., London, Sp. Pub.*, 340, 329–352.

Stampfli, G.M. (1993). Le Briançonnais, terrain exotique dans les Alpes ? *Eclog. Geol. Helv.*, 86, 1–45.

Stampfli, G.M. and Borel, G.D. (2002). A plate tectonic model for the Paleozoic and Mesozoic constrained by dynamic plate boundaries and restored synthetic oceanic isochrons. *Earth and Planetary Science Letters*, 196, 17–33.

Stampfli, G.M. and Hochard, C. (2009). Plate tectonics of the Alpine realm. In *Ancient Orogens and Modern Analogues*, Murphy, J.B., Keppie, J.D., Hynes, A. (eds). The Geological Society, London.

Stampfli, G.M., Borel, G.D., Cavazza, W., Mosar, J., Ziegler, P.A. (eds) (2001). *The Paleotectonic Atlas of the Peritethyan Domain*. European Geophysical Society, Kathlenburg-Lindau.

Stampfli, G.M., Borel, G.D., Marchant, R., Mosar, J. (2002). Western Alps geological constraints on western Tethyan reconstructions. *J. Virtual Explorer*, 8, 77–106.

Steck, A. (2008). Tectonics of the Simplon massif and Lepontine gneiss dome: Deformation structures due to collision between the underthrusting European plate and the Adriatic indenter. *Swiss J. Geosci.*, 101, 515–546.

Steck, A., Masson, H., Robyr, M. (2015). Tectonics of the Monte Rosa and surrounding nappes (Switzerland and Italy): Tertiary phases of subduction, thrusting and folding in the Pennine Alps. *Swiss Journal of Geosciences*, 108, 3–34.

Stern, G., Draganits, E., Wagreich, M., Hofmann, C.C., Reischenbacher, D., Grundtner, M.L., Bottig, M. (2011). Stratigraphy and geochemical characterisation of Upper Cretaceous non-marine-marine cycles (Grünbach Formation, Gosau Group, Austria). *Aust. J. Earth Sci.*, 104, 90–107.

Sutra, E., Manatschal, G., Mohn, G., Unternehr, P. (2013). Quantification and restoration of extensional deformation along the Western Iberia and Newfoundland rifted margins. *Geochem. Geophys. Geosyst.*, 14, 2575–2597.

Tanner, D.C., Behrman, J.H., Dresman, H. (2003). Three-dimensional retro-deformation of the Lechtal Nappe, Northern Calcareous Alps. *Journal of Structural Geology*, 25(5), 737–748.

Trümpy, R. (1955). Remarques sur la corrélation des unités penniques externes entre la Savoie et le Valais, et sur l'origine des Nappes préalpines. *Bull. Soc. Géol. Fr.*, 6(5), 217–231.

Trümpy, R. (1980). *Geology of Switzerland – A Guidebook. Part A: An Outline of the Geology of Switzerland*. Wepf & Co., Basel.

Van Hinsbergen, D.J.J., Torsvik, T.H., Schmid, S.M., Maţenco, L.C., Maffione, M., Vissers, R.L.M., Gürer, D., Spakman, W. (2020). Orogenic architecture of the Mediterranean region and kinematic reconstruction of its tectonic evolution since the Triassic. *Gondwana Research*, 81, 79–229.

Vatin-Pérignon, N., Juteau, T., Le Fort, P. (1972). Les filons du massif du Pelvoux (Alpes occidentales françaises). *Géologie Alpine*, 48, 207–227.

Vialon, P. (1966). Etude géologique du massif cristallin Dora-Maira Alpes Cottiennes internes (Italie). PhD Thesis, Université de Grenoble.

Wagreich, M. and Faupl, P. (1993). Paleogeography and geodynamic evolution of the Gosau Group of the Northern Calcareous Alps (Late Cretaceous, Eastern Alps, Austria). *Palaeogeogr. Palaeoclimat. Palaeoecol.*, 110, 235–254.

Weidmann, M. (1963). Un nouveau lambeau de la nappe de la Simme dans les Préalpes du Chablais. *Bull. lab. Géol. Lausanne.*, 65, 57–73.

Weidmann, M. (1972). Le front de la Brèche du Chablais dans le secteur de Saint-Jean-d'Aulph (Haute-Savoie), quelques remarques générales sur la nappe de la Brèche. *Géologie alpine*, 48, 229–246.

Winkler, W. (1996). The tectono-metamorphic evolution of the Cretaceous northern Adriatic margin as recorded by sedimentary series (western part of the Eastern Alps). *Eclog. Geol. Helv.*, 89, 1.

Ziegler, P.A. and Stampfli, G.M. (2001). Late Paleozoic–Early Mesozoic plate boundary reorganisation: Collapse of the Variscan orogen and opening of Neotethys. In *The Continental Permian of the Southern Alps and Sardinia (Italy) Regional Reports and General Correlations*, Cassinis, R. (ed.). Annali Museo Civico Science Naturali, Brescia.

Ziegler, P.A., Cavazza, W., Robertson, A.H.F., Crasquin-Soleau, S. (eds) (2001). *PeriTethys Memoir 6: Peritethyan Rift/Wrench Basins and Passive Margins, IGCP369*. Mémoires du Museum National d'Histoire Naturelle, Paris.

3

One View on the Petrology and Geochemistry of the Mantle and Oceanic Crustal Remnants in the Alps, Reconciling Rifts and Ultra-slow Seafloor Spreading

Othmar MÜNTENER[1] and Anders MCCARTHY[2,3]

[1]*Institute of Earth Sciences, University of Lausanne, Switzerland*

[2]*CY Cergy Paris Université, France*

[3]*Institute of Geochemistry and Petrology, ETH Zürich, Switzerland*

3.1. Introduction

Ophiolites are considered first-order witnesses in the formation of orogenic belts, with some of them delineating ancient suture zones amalgamating continental crustal blocks or plates. Essentially, they demonstrate that the former seafloor was incorporated into mountain belts. However, the extent to which the architecture of ophiolitic remnants in the Western Alps represent the original seafloor is debated and complicated by their tectonic history from the seafloor into mountain belts. Historically, the complexity of the ophiolitic architecture has been explained by compressional deformation that affected the ophiolitic rock series, but the discovery of broad areas of exhumed mantle rocks along present-day rifted margins and

Geodynamics of the Alps 2,
coordinated by Claudio L. ROSENBERG and Nicolas BELLAHSEN.

ultra-slow spreading ridges (e.g. Boillot et al. 1989; Lagabrielle and Cannat 1990; Beslier et al. 2004; Sauter et al. 2013; Cannat et al. 2019) forced the research community to re-visit the architecture of Western Alpine ophiolites in the context of rifting processes at magma-poor margins and eventual rupture, resulting in the formation of complex, segmented plate boundaries and sea-floor architecture similar to ultra-slow spreading ridges.

In this chapter, we summarize the key findings on petrological research of Alpine-Apennine ophiolites over the last 30 years, while the tectonic and sedimentary rift history is discussed in Chapter 5. Modern analytical studies added essential data and enabled the field-based hypotheses that were developed in a plate tectonic framework to be quantified. Here we focus on aspects of petrology and geochemistry, and argue that the processes recorded are intimately linked to the pre-orogenic spatial distribution of crustal and mantle rocks. Basalts, gabbros and mantle peridotites record processes that are known from modern ultra-slow spreading systems. However, these processes are strongly controlled by inheritance and favor a model for the formation of Alpine ophiolites that combine aspects known from rifted magma-poor continental margins and ultra-slow seafloor spreading. Collectively the data show that the Alpine-Apennine ophiolites are dominated by (serpentinized) peridotites, and comparatively minor gabbroic and basaltic rocks. The chemistry and the distribution of the latter provide key information about the thermal structure of the Western Tethys during the Late Jurassic and Early Cretaceous.

In contrast to many other ophiolites worldwide, there is no evidence for a supra-subduction zone or mantle wedge origin of the Western Tethys Alpine ophiolites. This makes them an excellent analog for understanding processes that are linked to continental breakup, formation of magma-poor passive margins and ultra-slow sea-floor spreading. We will review existing petrological data and add new data and observations within the framework of a paleogeographic map, which emerges as a template for rifts that evolve into ultra-slow sea floor spreadings, such as the Laptev Sea – Gakkel Ridge area in the Arctic or ultra-slow seafloor spreading at the Southwest Indian Ridge (Sauter et al. 2013).

3.2. Geological constraints on the Alpine Tethys seafloor

The architecture of the Alpine-Apennine ophiolites has long been known to be different from other ophiolites around the world (Steinmann 1927; Decandia and Elter 1969; Bonatti 1971). Western Tethys ophiolites are dismembered slivers dominated by variably serpentinized mantle rocks with minor basaltic dykes and flows and small gabbro intrusions. The major difference between Penrose type ophiolites is the lack of sheeted dike complexes and the generally minor volumes of gabbroic rocks. Ophiolitic sandstones, arenites, gabbro and basaltic breccias and olistoliths testify to tectonic activity on the seafloor (e.g. Tricart and Lemoine 1991). Radiolarites and pelagic limestones overlie serpentinized mantle rocks, ophicarbonates, exhumed gabbros and lava flows, pillow basalts and pillow breccias covering mantle rocks (e.g. Lagabrielle et al. 1984). Carbonated ultramafic rocks, known as "ophicalcites" or "ophicarbonates" were recognized early on (Lemoine et al. 1987; Weissert and Bernoulli 1985) and document low-temperature hydrothermal processes similar to those found in the Lost City on the Mid-Atlantic ridge (e.g. Kelley et al. 2001). Combined with stable isotope studies and sedimentary structures, such carbonated ultramafic rocks provide a strong argument for seafloor alteration (e.g. Lagabrielle et al. 1984; Bernoulli and Weissert 1985).

Plate separation evolved from rifted passive margins to ultra-slow seafloor spreading. Early on it was stressed that deformation processes during rifting and ultra-slow seafloor spreading are fundamental to explaining some of the Western Alps ophiolites (e.g. Tricart and Lemoine 1986). Since the discovery of exhumed peridotites at the Iberia margin (e.g. Boillot et al. 1989) the simple comparison of Alpine-Apennine ophiolites with a ridge setting (e.g. Lagabrielle and Cannat 1990) has been challenged, and the focus moved to the identification of ocean-continent transitions preserved in the Alps, in addition to studies that compared the Alpine ophiolites with (ultra-) slow-spreading ridges. The results are schematically illustrated in Figure 3.1. It follows that in some places an ocean-continent transition setting is well preserved. This is certainly the case in the Eastern Central Alps, where field relations show that thinned continental crust is directly in contact with exhumed subcontinental mantle rocks. Faulted lower and upper crustal rocks, and mantle rocks are sealed by deep-sea sediments with complex 3-D geometries (Figure 3.1, e.g. Florineth and Froitzheim 1994; Manatschal and Nievergelt 1997; Beltrando et al. 2014; Epin et al. 2019; Ribes et al. 2020). The petrology and geochemistry of crustal and mantle rocks are strikingly similar to the Iberia-Newfoundland transect in the Cretaceous and document the transition from continental to more oceanic settings (e.g. Müntener and Manatschal 2006; Müntener et al. 2010). However, in other areas, field geometries and reconstructions revealed or inferred oceanic core complexes (OCCs) (Figure 3.1, e.g. Manatschal et al. 2011; Lagabrielle et al. 2015; Decrausaz et al. 2021), or even nearly amagmatic ultramafic seafloor areas covered with ultramafic breccias and limited magmatic products (e.g. Manatschal and Müntener, 2009). These studies reinforced the similarity to ultra-slow spreading settings, such as the Southwest Indian Ridge (SWIR) (Figure 3.1, e.g. Sauter et al. 2013; Cannat et al. 2019; Bickert et al. 2020). The emerging picture for Alpine-Apennine ophiolites is that low angle detachment faults and high-angle normal faults accommodate a significant amount of plate separation and interact with periods of magmatic additions. The ratio between magmatic and tectonic plate separation (M, as defined in numerical modeling studies (Buck et al. 2005; Tucholke et al. 2008)) varies both in space and time, and leads to strongly segmented seafloor with smooth or rugged topography (Figure 3.1). The M-value may range from 0.5 and nearly 0, thus providing a conceptual framework for a system that oscillates between magma-starved and magma-rich periods (Figure 3.1). This interplay between extension-controlled and magma-controlled plate separation is intrinsic to two geodynamic domains – ocean-continent transition and ultra-slow spreading. We will show that the excellent field context of the Alps allows for rigorous testing of petrological processes in time and space, which control the chemical signature of crustal and mantle rocks. Alpine ophiolites do not only allow for identifying processes for mid-ocean ridge basalt (MORB) generation and mantle melting, but, importantly, provide the necessary temporal and spatial resolution to quantify the changes of the ultramafic seafloor and the magmatic products in an evolving system from continental breakup to ultra-slow seafloor spreading. In the following paragraphs, we focus on geochemical data and petrology to decipher these processes.

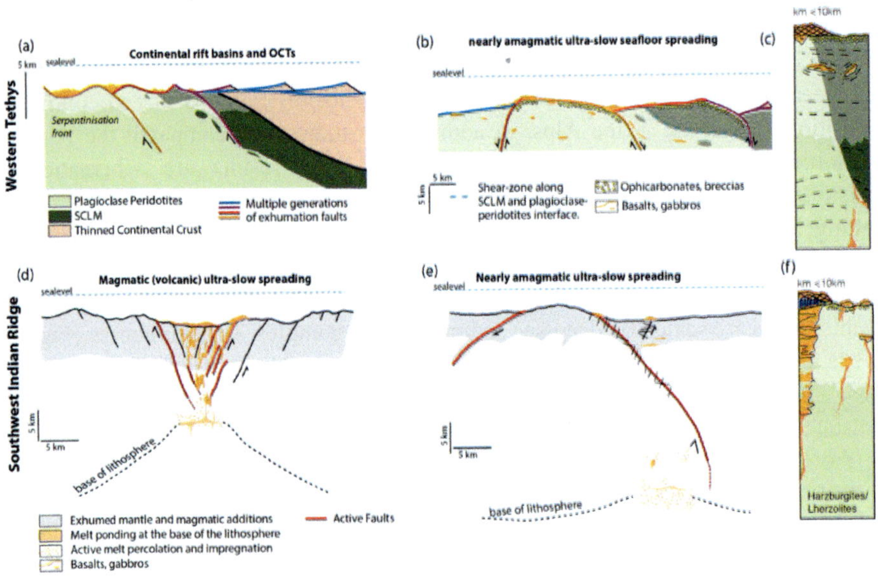

Figure 3.1. *Seafloor types in the Western Tethys*

COMMENTS ON FIGURE 3.1.– *(a) Continental rift basins and ocean-continent transitions zones. In the proximal parts, the subcontinental mantle is exhumed on the seafloor, with occasional remnants of the continental upper or lower crust. Laterally, the mantle rocks change from subcontinental to "re-fertilized" mantle rocks. Magmatic additions are mostly restricted to the re-fertilized domains (modified from Epin et al. (2019)). (b) Oceanic core complexes and smooth seafloor with very little magmatism, typical of ultra-slow spreading ridges that vary between moderate magmatic additions to nearly amagmatic, tectonic seafloor spreading. The volume of gabbroic bodies is variable. Type localities for the Alps: OCC' Aiguilles Rouges d'Arolla; OCT (Tasna-Platta) smooth seafloor Upper Platta. (c) Schematic illustration of Tethyan oceanic crust, representing both subcontinental and re-fertilized mantle types, limited magmatism and coverage of seafloor with ophicarbonates and occasional MOR-type basalts and flows. Note that sheeted dikes are absent. (d) Magmatic spreading and (e) nearly amagmatic spreading ("flip flop detachment spreading" on the Southwest Indian Ridge (figures redrawn from Cannat et al. (2019)). (f) Schematic illustration of southwest Indian oceanic crust, which varies between nearly amagmatic spreading and smooth seafloor and more robust magmatic spreading. (c) and (f) redrawn from McCarthy et al. (2020).*

To limit the length of this chapter, this contribution will not treat the so-called orogenic peridotites for which the Alps are also famous. They are incorporated in the continental crust, but lack evidence of being exhumed on the seafloor, or being associated with MOR-type gabbros and basalts. They resemble (or are identical to) the inherited subcontinental mantle described below, yet their tectonic framework with respect to the opening of the Western Tethys is different. For a more complete geochemical view of orogenic peridotites and mantle rocks and xenoliths in the context of the Western Tethys, we refer to the excellent review papers (e.g. Bodinier and Godard 2013, Picazo et al. 2016).

3.3. Alpine-Apennine peridotites: fingerprints of (ancient) melting and syn-rift re-fertilization

3.3.1. *Three main groups of peridotites*

In this section, we compare the petrological and geochemical characteristics of mantle peridotites from the Alpine-Apennine area with abyssal peridotites. Although in general strongly serpentinized, we distinguish three groups of peridotites – (i) subcontinental lithospheric mantle or "inherited mantle", (ii) plagioclase peridotites or "re-fertilized" mantle, and (iii) refractory peridotites. Although peridotites in the subcontinental group range from lherzolite to harzburgite and dunite, we restrict the use of refractory peridotites to samples that are closely related to plagioclase peridotites.

The subcontinental mantle is heterogeneous, with pyroxenites, harzburgites and dunites, garnet pyroxenites, rare phlogopite-bearing or Ti-hornblende-bearing metasomatic veins. In places, these rocks are associated with lower crustal continental rocks, hence their classification (Florineth and Froitzheim 1994; Müntener et al. 2000; Pelletier and Müntener 2006). Basaltic dikes and gabbro bodies of oceanic affinity in these peridotites are rare or missing. Examples can be found in Balangero (Lanzo), Malenco, Tasna-Nauders, Totalp, Upper Platta and Suvero in the External Ligurides (Figure 3.2). Thermobarometry using equilibria among pyroxenes, olivine, spinel and/or garnet indicate temperatures between about 850 and 950°C, indicating long-term cooling approaching a continental geotherm (e.g. Müntener et al. 2010).

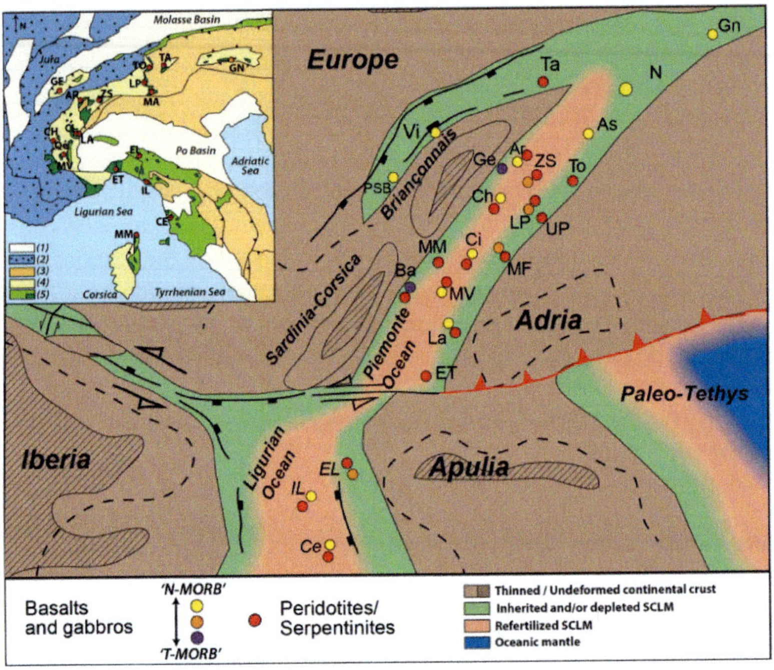

Figure 3.2. *Overview map of key mantle and oceanic crust localities (Ophiolite paleogeography) distributed over the Alps and Apennines*

COMMENTS ON FIGURE 3.2.– *Inset: Tectonic overview and location of main ophiolites in the Alpine–Apennine domain, modified from Schaltegger et al. (2002)). (1) Tertiary basins, (2) European domain, (3) Adriatic domains, (4) Briançonnais and Valais units and (5) Ophiolites and Peridotite bodies. Red dots: peridotite localities, orange dots: basalt/gabbro localities. Schematic paleogeographic map of the Western Tethys at ~110 Ma, adapted from Picazo et al. (2016). Alpine MOR basalt chemistry is represented by yellow, orange, and purple dots, ranging from*

N-MORB to T-MORB signatures. The approximate location is based on its geological context and the geochemical data of basalts and mantle peridotites. Note that oceanic crust (basalts, mantle rocks) with high-pressure metamorphism was located north of the large transcurrent fault zone linking the Paleothetys to the Pyrenees: As: Arosa, Ar: Aiguilles Rouges, Ba: Balagne, Ce: Cecina Valley, Ch: Chenaillet, Ci: Civrari, EL: External Ligurides, ET: Erro-Tobbio, Ge: Gets Nappe, Gn: Glockner nappe, IL: Internal Ligurides, La: Lanzo, LP: Lower Platta, MF: Malenco-Forno, MM: Monte Maggiore, MV: Monte-Viso, PSB: Petit St. Bernard, Qe: Queyras, N: Northern Calcareous Nappes, UP: Upper Platta, Ta: Tasna, To: Totalp, Vi: Visp, ZS: Zermatt-Saas.

Plagioclase peridotites include rather homogeneous plagioclase lherzolites and a few plagioclase spinel pyroxenites. Most plagioclase peridotites record reactive porous flow processes, cotectic intergrowth of orthopyroxene and plagioclase replacing clinopyroxene, and microgabbro dike formation, although some plagioclase may be formed by subsolidus reactions (Rampone et al. 1993). Locally, discordant dunite and olivine-rich troctolites are preserved. Examples can be found in Lanzo, Chenaillet, Queyras, Lower Platta, Erro Tobbio, Internal Ligurides, Corsica and Tuscany. Thermometry based on major and trace element equilibria among plagioclase-bearing assemblages yield much higher temperatures ranging from ~1100 to >1200°C, indicating near-solidus temperatures prior to exhumation. Cooling of the plagioclase peridotites has to be rapid after melt-rock reaction to freeze disequilibrium textures such as orthopyroxene-plagioclase symplectites replacing mantle clinopyroxene and has been modeled via orthopyroxene Cr-Al diffusion profiles (Jollands and Müntener 2019). Time scales retrieved from simple modeling of Cr-diffusion in orthopyroxene range between ~5–15 Ma for cooling to about 700°C (Jollands and Müntener 2019).

Refractory peridotites in Alpine-Apennine ophiolites are rare, with only a few unequivocal findings as lenses within plagioclase peridotites, identified so far only in the Lower Platta, Civrari, Internal Ligurides, Corsica and Tuscany (Figure 3.2). There is no particular field or microtextural criteria to identify them and therefore proper identification needs geochemical data (see below).

3.3.2. *Bulk rock chemistry: fertile versus re-fertilized*

First, we examine bulk rock major element variations of peridotites from the Alpine ophiolites and compare them with abyssal peridotites that were calculated from mineral compositions and modes (Warren 2016). Alpine-Apennine peridotites are generally enriched in Al_2O_3 and TiO_2 relative to abyssal peridotites (Figure 3.3).

This is particularly clear when comparing the TiO_2 content of spinel peridotites from the Alps and abyssal peridotites (Figure 3.3(b)). TiO_2 is sensitive to fractional melting of the mantle and this is recorded in most abyssal peridotites with $TiO_2<0.1$ wt%. The distribution of TiO_2 for Alpine–Apennine peridotites covers a much broader range up to ~0.25 wt% TiO_2. Although some of the Alpine–Apennine data have high loss on ignition indicating serpentinization, the large range in TiO_2 reflects the heterogeneous composition of the subcontinental mantle. Alpine–Apennine peridotites also show a bimodal distribution in Al_2O_3. Considering only spinel peridotites, similar distribution as for TiO_2 is apparent, with a long tail to high Al_2O_3 contents. A few plagioclase peridotites in our dataset have $Al_2O_3 < 2.5$ wt%, but most of them display much higher Al_2O_3 contents clearly different from that of typical abyssal peridotites. To explain this, it has been suggested that this difference arises from re-fertilization processes in the thermal boundary layer between the lithosphere and asthenosphere, as documented by reaction textures of clinopyroxene and symplectitic intergrowth of orthopyroxene and plagioclase (e.g. Müntener et al. 2004; Müntener and Piccardo 2003; Rampone et al. 2008). Bulk rock Al_2O_3 and TiO_2, and more incompatible elements rise as more melt reacts with mantle rocks and remains trapped (Rampone et al. 1997). Accumulation of melt locally produces peridotite compositions that are enriched in fusible components relative to estimates of the primitive mantle (e.g. Kaczmarek and Müntener 2010).

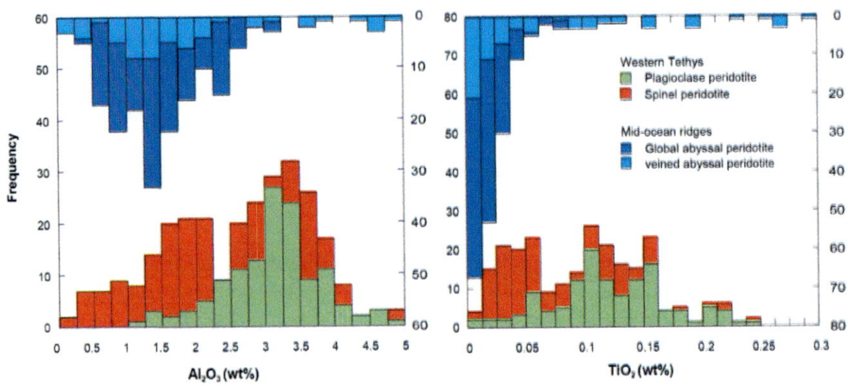

Figure 3.3. *Distribution of bulk rock Al_2O_3 (a) and TiO_2 (b) in spinel and plagioclase peridotites from the Alpine–Apennine mountain belt, compared to residual abyssal peridotites and "veined" abyssal peridotites for the global mid-ocean ridge system (data from Warren (2016)). Note that the plagioclase peridotites are much more fertile than spinel peridotites. Note also that refractory peridotites are not distinguished. Data sources: Bodinier 1988; Chalot-Prat et al. 2003; Ernst 1978; Kaczmarek and Müntener 2010; Melcher et al. 2002; McCarthy and Müntener 2019; Müntener et al. 2010; Ottonello et al. 1984; Piccardo et al. 2007; Rampone et al. 1995, 2005, 2008*

3.3.3. *Mineral chemistry: inherited versus re-fertilized*

We turn now to the chemical characteristics of peridotite minerals. We restrict our discussion to clinopyroxene and spinel, as they have been shown to be most sensitive to distinguish ancient subcontinental mantle peridotites from plagioclase peridotite and refractory spinel peridotites. We base our discussion on the major element chemistry on the data compilation of Picazo et al. (2016), supplemented by data cited in McCarthy and Müntener (2019) and our unpublished data. Although clinopyroxene in spinel and plagioclase peridotites from the Alps overlap to some degree (Figure 3.4(a)) in terms of Cr# (that is molar Cr/Cr+Al) versus TiO_2, if plotted against Na_2O they clearly discriminate between clinopyroxene from subcontinental spinel peridotite and plagioclase peridotite (Figure 3.4(b)). The refractory peridotites show the lowest TiO_2 and Na_2O, and refer to samples that are rarely found within plagioclase peridotites, but show exceptionally low clinopyroxene REE contents that are (so far) only reported for a few localities (Internal Ligurides, Mte Maggiore, Platta, Civrari, see Figure 3.2 for locations).

Spinel peridotites of subcontinental origin have Na_2O contents exceeding 1 wt%, while plagioclase peridotites generally have $Na_2O < 1$ wt%. It is evident from the data distribution on bulk rock chemistry discussed above (Figure 3.3) that the chemical variation in clinopyroxene does not solely arise from subsolidus reactions, although 2-pyroxene solvus thermometry may indicate equilibration temperatures below 1000°C. Metamorphic reactions of clinopyroxene, orthopyroxene and spinel to olivine + plagioclase would not change the bulk rock composition. The wide range of clinopyroxene TiO_2 contents with generally increasing Cr# rather suggest reactive porous flow processes with MORB-type liquids. This is also supported by the overall higher TiO_2 in clinopyroxene from plagioclase peridotites relative to abyssal peridotites (Warren 2016), but this is less clear for Na_2O, which is partitioned between clinopyroxene and plagioclase. Na_2O varies substantially in abyssal peridotites, and this could be related to Na-metasomatism, such as re-fertilization of peridotites by migrating basaltic melt and/or diffusive infiltration processes (e.g. Lundstrom 2000; Hellebrand and Snow 2003), or to late-stage alteration. The refractory spinel peridotites from the Alps plot at the lower end of the plagioclase peridotites in terms of Na_2O and TiO_2. The lowest TiO_2 and Na_2O are from the Civrari peridotites (Italy) and together with samples from Mte Maggiore, Internal Ligurides and Platta nappe in the Eastern Central Alps (Figure 3.1), represent residues of partial melting similar to abyssal peridotites. Major and trace element modeling showed that the Civrari peridotites are an almost perfect fit to a simple fractional melting model starting in the garnet peridotite field (McCarthy and Müntener 2015).

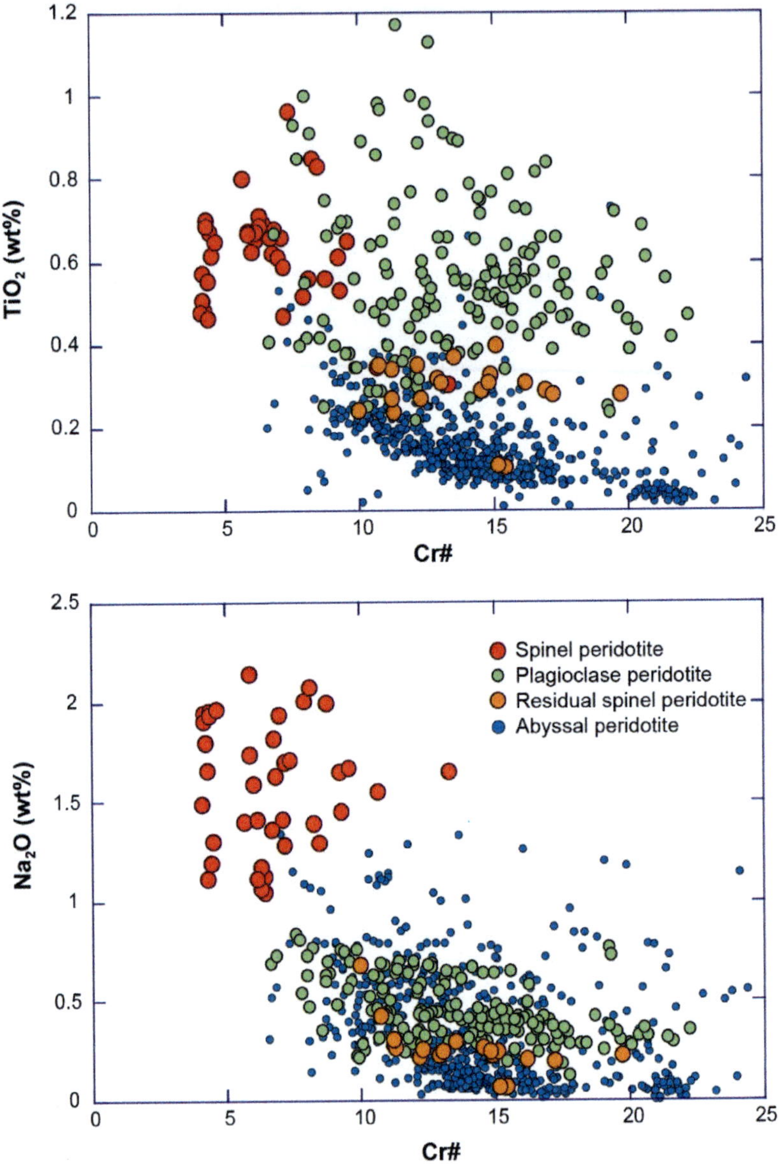

Figure 3.4. *Major element composition of clinopyroxene from mantle peridotites of the Alpine-Apennine ophiolites*

COMMENTS ON FIGURE 3.4.– *Variability of clinopyroxene illustrated as variation diagrams in terms of Cr# versus TiO_2 (a) and Cr# versus Na_2O. Clinopyroxene from subcontinental peridotites is much higher in Na_2O and has slightly lower Cr# than plagioclase peridotites and residual peridotites. TiO_2 in clinopyroxene from residual spinel peridotites resemble depleted abyssal peridotites. TiO_2 contents of clinopyroxene from plagioclase peridotites scatter widely but are mostly >0.4 wt%, with variable Cr#. Residual spinel peridotites with refractory isotopic compositions plot at the lower end of the plagioclase peridotite. Data sources are compiled in Picazo et al. (2016). Abyssal peridotite dataset from Warren (2016).*

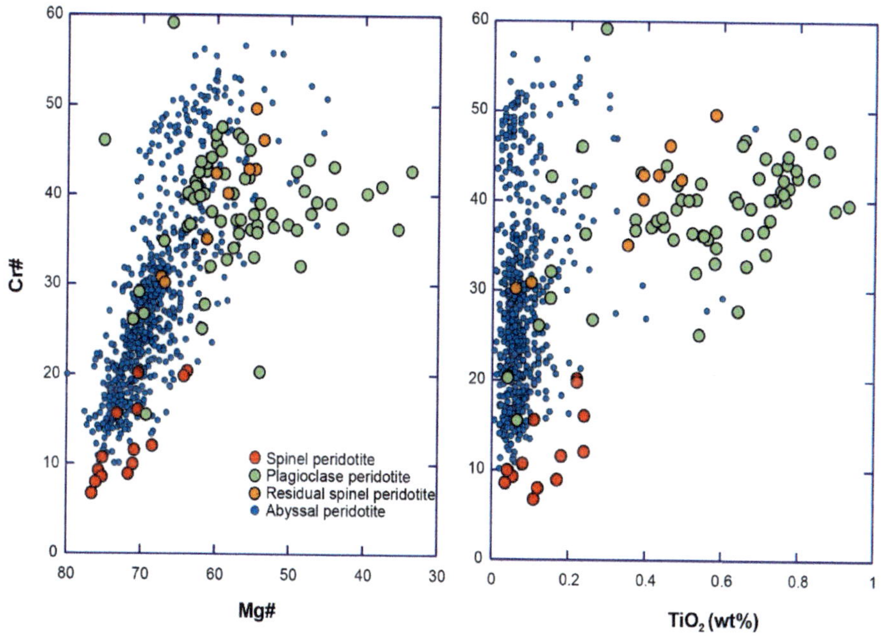

Figure 3.5. *Variation in spinel Cr# as a function of Mg# (a) and TiO_2 (b). Residual abyssal peridotites are plotted as a background dataset (Warren 2016). Spinel from plagioclase peridotite display elevated TiO_2 content. Note that most of the residual spinel peridotite also indicate elevated TiO_2 content, consistent with their occurrence in the field as lenses within plagioclase peridotite*

The compositional variation of spinel is illustrated in Figure 3.5 and shows a clear distinction between spinel peridotites and plagioclase peridotites. During melting, the Cr# of spinel increases and the Mg# of spinel decreases, as shown by

Dick (1989) and Hellebrand et al. (2001). Subcontinental spinel peridotites from the Alps generally show Cr# < 20, while the refractory spinel peridotites (with a depleted Nd-isotopic composition, see below) display higher Cr# between 30 and 50, similar to abyssal peridotites. Although there is some overlap, most plagioclase peridotites from the Alpine–Apennine ophiolites plot at TiO_2 contents exceeding 0.2 wt%, while only a few abyssal peridotites from the global ridge system have TiO_2 > 0.2 wt%. Titanium enrichment in spinel from plagioclase peridotites is taken as an indication of equilibration with a percolating melt, consistent with the interpretation that plagioclase peridotites represent the product of re-fertilization processes. Interestingly, most of the spinel from refractory spinel peridotites display elevated TiO_2 consistent with their association with plagioclase peridotites.

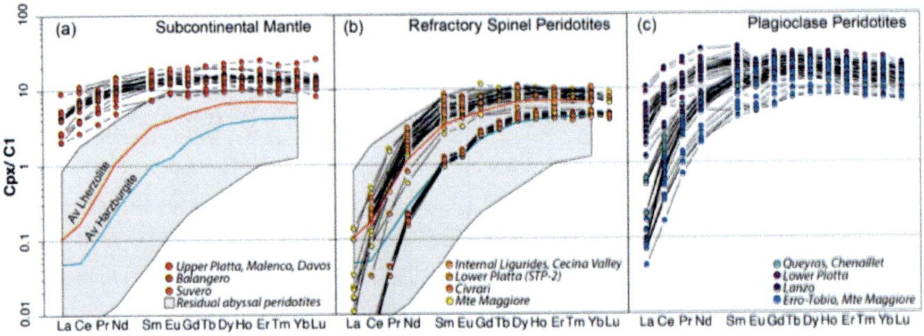

Figure 3.6. *Clinopyroxene chondrite normalized REE compositions, grouped after their occurrence in various Alpine peridotites (a) REE compositions of clinopyroxene from subcontinental peridotites. Note the moderate Ce_N/Yb_N ratios. (b) Refractory spinel peridotite from four different ophiolites. Note the extremely low Ce_N/Yb_N ratios, similar to abyssal peridotites. (c) Plagioclase peridotites with variably fractionated Ce_N/Yb_N, flat to positive Gd_N/Yb_N ratios and weak to moderate Eu negative anomalies. Data sources: subcontinental mantle: (Rampone et al. 1995; Müntener et al. 2010; Müntener, unpublished data (Balangero)). Refractory spinel peridotites: (Rampone et al. 1996; Müntener et al. 2004; Tribuzio et al. 2004; Rampone et al. 2008; McCarthy and Müntener 2015). Plagioclase peridotites: (Rampone et al. 1995; Müntener et al. 2004; Rampone et al. 2004; Borghini et al. 2007; Müntener et al. 2010; Kaczmarek and Müntener 2010; Basch et al. 2018; Müntener unpublished data (Queyras, Chenaillet)). Average lherzolite and harzburgite, and the field of abyssal peridotite is from Warren (2016)*

In Figure 3.6, we have compiled representative REE clinopyroxene patterns from the Alps and Apennines, split into three groups, representing compositions from the subcontinental mantle (a), clinopyroxene from residual spinel peridotites (b), and clinopyroxene from plagioclase peridotites (c). Clinopyroxene from subcontinental peridotites show generally high Ce_N/Yb_N and no Eu anomalies, which result in rather flat REE spectra (Figure 3.6(a)). Clinopyroxene REE spectra from plagioclase peridotites vary much more with respect to La_N/Sm_N and Ce_N/Vb_N ratios, but are remarkably similar in displaying a "humped shaped" REE pattern, with Sm_N/Yb_N >1 and a small, distinct Eu anomaly. Quantitative modeling of re-fertilization indicates that the compositions of plagioclase peridotites can best be reproduced by crystallizing 4–12% of a MORB-type melt that equilibrated with peridotite in the stability field of plagioclase (Müntener et al. 2010). More recent studies in Corsica and the Internal Ligurides support these models of melt entrapment in the shallow mantle lithosphere. Melt-rock reactions consuming olivine and precipitating plagioclase and orthopyroxene indicate that melt stagnation and crystallization in dunite will transform them into (olivine-rich) troctolite (Sanfilippo and Tribuzio 2011; Basch et al. 2018). Somewhat to our surprise, there are subtle but systematic differences for clinopyroxene from the Western Alps such as Queyras, Chenaillet, Corsica and Erro Tobbio, compared to those from Lanzo and Platta (Figure 3.6(c)). From the geological context, Platta and Lanzo are peridotites that were close to thinned continental crust, while Chenaillet, Queyras, Corsica and Erro Tobbio represent "more distal" parts of the mantle (see Figure 3.2 for locations). The La_N/Sm_N ratio of clinopyroxene is much lower for the samples from the Western Alps and Corsica. This could either represent a higher fraction of trapped melt in the Ligurides and Platta (Müntener et al. 2004) or, alternatively, the reactive melts in the Chenaillet, Queyras, Corsica and Erro-Tobbio peridotites are more depleted than those in Lanzo and Platta (e.g. Rampone et al. 1997, 2005). Lanzo clinopyroxene illustrates a transitional signature, consistent with the interpretation of the Lanzo peridotites located at the transition from more continental to oceanic settings. Pressure estimates based on the equilibrium Forsterite-Anorthite-Ca-Tschermak-Enstatite indicate that most plagioclase peridotites equilibrated at pressures between 3–7 kbar (Fumagalli et al. 2017). The crust and the mantle lithosphere, therefore, were thinned prior to melt-rock reaction and re-fertilization in the plagioclase peridotite facies. This indicates that the earliest signatures of magmatism in Alpine-Apennine ophiolites are recorded in peridotites and expressed by the reactive migration of asthenosphere-derived melt in exhuming peridotites (see also Rampone and Sanfilippo (2021)).

Refractory spinel peridotites have Ce_N/Yb_N much less than 0.02 and Gd_N/Yb_N <1, similar to abyssal peridotites. However, clinopyroxene from these residual peridotites sometimes shows weak Eu anomalies (e.g. Müntener et al. 2004; McCarthy and Müntener 2015), consistent with partial equilibration with plagioclase. Because all these refractory peridotites occur within areas of plagioclase peridotites, we interpret them to be only weakly affected by re-equilibration with percolating melts, thereby providing information about the mantle prior to re-fertilization. Perhaps the best example of this process is preserved in the Lanzo peridotites, where infiltration and percolative fractional crystallization of a melt along a thermal gradient is recorded in some of the depleted peridotites. For the case of the Civrari peridotite, light REE in reactive clinopyroxene may vary by two orders of magnitude from core to rim. Migrating melts reacted with the refractory clinopyroxene and major and trace elements increased approaching compositions of microgabbros or clinopyroxene from plagioclase peridotites (McCarthy and Müntener 2019). The solidification processes of "reactive liquids" involving some dissolution of pre-existing peridotite minerals are most spectacularly shown by the symplectitic growth of orthopyroxene and spinel corroding clinopyroxene, and are widespread in plagioclase peridotites (e.g. Müntener and Piccardo 2003; Rampone et al. 2008). These melt-rock reaction processes in a thermal gradient are associated with a significant evolution of melt composition from originally basaltic to progressively volatile-enriched, possibly culminating in low-temperature small-volume apatite saturated melts (McCarthy and Müntener 2019). It therefore seems clear that along with the fertile bulk rock compositions (Figure 3.3), the major and trace element compositions of spinel and clinopyroxene are indicators of re-fertilization processes in the plagioclase peridotite stability field. This "fractional fertilization" mechanism results in peridotite compositions that significantly depart from melting trends of abyssal peridotites, and locally produces compositions in major and trace elements that are more fertile than that defined for the primitive upper mantle (Kaczmarek and Müntener 2010).

In summary, Alpine peridotites preserve systematic changes in bulk rock and mineral chemical composition, and show variations that are spatially controlled, mainly by the interaction of migrating melt within extending mantle lithosphere. Subcontinental mantle compositions with Na-rich clinopyroxene are exclusively preserved in "cold" peridotites from ocean-continent transitions zones and these peridotites preserve characteristic banding of spinel and garnet pyroxenites, hallmarks of the subcontinental mantle. Refractory spinel peridotites, although rare, are found within domains of re-fertilized plagioclase peridotites that show typical textures of (disequilibrium) melt-rock reactions. Based on the spatial distribution, we conclude that most plagioclase peridotites are produced by melt-rock reaction and crystallization of trapped melt near the base of the extending lithosphere. They are most similar to peridotites found in domains of ultra-slow seafloor spreading.

3.4. Gabbroic rocks: small volumes of crystallization – differentiation, mostly in the mantle

Gabbros associated with Alpine-Apennine ophiolites encompass a large range of compositions, from olivine-rich troctolites, olivine-gabbros, gabbros, gabbronorites to Fe-Ti gabbros, hornblende-gabbros and plagiogranites. Clinopyroxene-rich gabbros are the most common rock type, whereas olivine-gabbros and troctolites remain comparatively sparse. Coherent gabbro bodies exceeding ~1 km are rare and are restricted to a few areas in the Western Alps, such as Saas-Zermatt, Monte Viso and Aiguilles Rouges, and in the Apennines and Corsica (Figure 3.2).

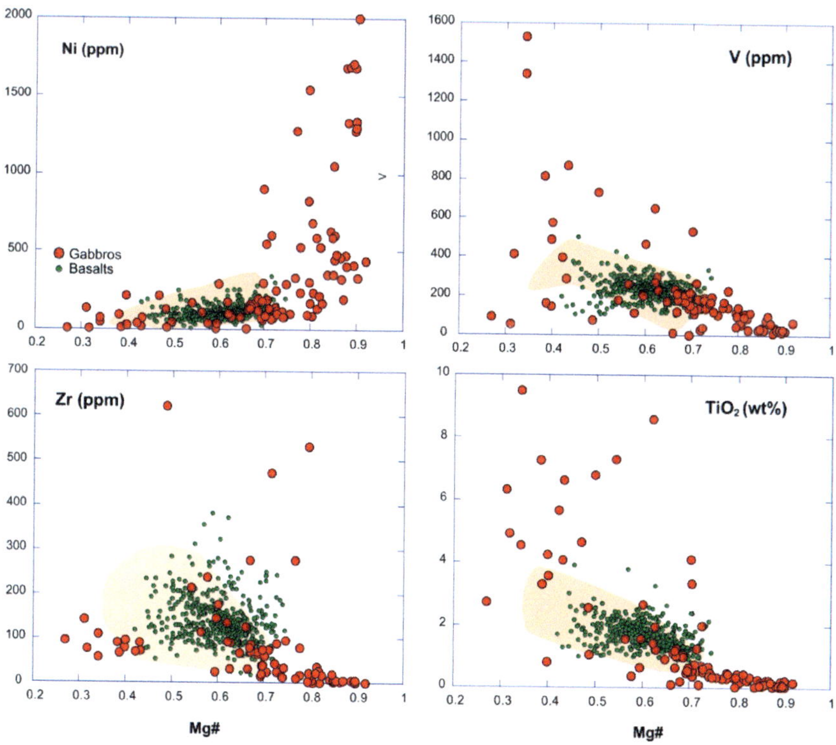

Figure 3.7. *Mg# versus major and trace elements for gabbros and basalts from the Alps and Apennines. a) Mg# versus Ni, b) Mg# versus V, c) Mg# versus Zr, d) Mg# versus TiO_2. The orange shaded field covers the area of MOR basalts (Gale et al. 2013). The compositional variation of gabbros is much larger that the basalts. Primitive gabbro cumulates (troctolite, olivine gabbros) with Mg# approaching 0.9 and Ni contents approaching mantle olivine indicate that they may be residues from which some of the basalts were extracted*

All gabbroic intrusions locally show weak modal or grain size layering, often subparallel to the sill-like structures of the gabbros, however, larger sill-like bodies common to fast-spreading ridges (e.g. Oman) are not found. There have been no attempts to distinguish the gabbroic sequences into (lower) layered gabbros and (upper) isotropic gabbros, as commonly proposed for Penrose type crustal sections of fast-spreading ridges (e.g. Oman and Nicolas 1989), simply because such a systematic pattern cannot be identified in exposed gabbros from the Alps and Apennines. On the contrary, these sparsely distributed and isolated gabbroic rocks display a remarkable variety of compositions at a local scale and show intrusive contacts into peridotites (e.g. Müntener and Piccardo 2003; Sanfilippo and Tribuzio 2011; Sanfilippo and Tribuzio 2013). Gradational transitions from re-fertilized plagioclase peridotites to micro-gabbros, plagioclase-enriched dunites and plagioclase-bearing troctolites are locally observed, indicating a common origin of plagioclase peridotites and gabbros (Higgie and Tommasi 2014). A second remarkable feature is that highly evolved gabbros, such as Fe-Ti-P gabbros, hornblende gabbros and plagiogranites are found in most gabbro bodies. These evolved gabbros either form small crosscutting dikes, often strongly deformed at high temperatures (e.g. Kaczmarek et al. 2008), or small patches and lenses, indicating in-situ fractionation processes (e.g. Desmurs et al. 2002; Sanfilippo and Tribuzio 2013). All these characteristics are remarkably similar to gabbros from (ultra-) slow-spreading ridges (e.g. Cannat et al. 1997; Dick et al. 2008).

The bulk rock composition of gabbros was compiled from various places from the Northern Apennines to the Eastern Central Alps (Figure 3.2), and from various sources summarized in Figures 3.7 and 3.8. As shown in Figure 3.7, average MORB from the Alpine-Apennine ophiolites forms a narrow field with an average Mg# of 0.61, TiO_2 of 1.62 wt%, Zr of 138 ppm and Ni of about 110 ppm (Table 3.1), indicating that they are not in equilibrium with mantle peridotite. Gabbroic rocks, however, show extreme compositional variability. Primitive olivine-rich gabbroic rocks and olivine-rich troctolites reach more than 300 ppm Ni, and less than 30 ppm Zr and $TiO_2 < 0.2$ wt% (Figure 3.7). In contrast, many of the evolved ferrogabbros and plagiogranites have Mg# lower than any basalt and some with extremely enriched incompatible element compositions (Figures 3.7(c) and (d)). Only the primitive olivine-rich troctolites and the olivine gabbros have appropriate high Mg# and correspondingly low Zr and TiO_2 to represent cumulates complementary to the evolved composition of the basalts.

Table 1. Average composition of basaltic rocks from Alpine-Apennine ophiolites compared to mid-ocean ridge basalts (Gale et al. 2013 and additional data for SWIR (Paquier et al. 2006)

	Mid Ocean ridges			Basalts Alps		N-MORB*
	MAR	SWIR	Gakkel	Subset*	All	
n**	*1138*	*154*	*326*	*330*	*558*	
SiO_2	50.77	50.62	50.42	51.12	50.91	50.77
TiO_2	1.40	1.42	1.52	1.62	1.62	1.54
Al_2O_3	15.04	16.66	16.49	16.70	16.80	15.24
FeO_{TOT}	10.09	8.83	8.95	9.03	9.06	9.88
MnO	0.17	0.17	0.16	0.16	0.17	0.17
MgO	8.00	7.82	7.71	7.76	7.42	7.81
CaO	11.61	10.48	10.97	8.83	9.12	11.43
Na_2O	2.56	3.60	3.33	4.30	4.24	2.85
K_2O	0.21	0.22	0.28	0.26	0.42	0.14
P_2O_5	0.16	0.18	0.18	0.23	0.23	0.17
Total	100.00	100.00	100.00	100.00	100.00	100.00
Mg#	0.611	0.637	0.630	0.630	0.619	0.610
Trace elements						
Li	5.74	6.56	5.70	n.a.	n.a.	6.10
Be	0.49	0.75	0.72	n.a.	n.a.	0.64
Sc	40.56	37.60	35.60	33.97	33.91	38.4
V	286.88	234.03	247.50	233.74	229.84	280
Cr	292.11	284.41	302.90	243.64	246.66	263.00
Cu	84.77	69.54	75.49	72.875	65.99	73.00
Co	44.52	41.82	41.37	38.87	39.86	42.30
Ni	111.61	132.61	132.29	109.13	110.39	100
Zn	85.85	78.78	77.01	75.844	76.28	85.30
Rb	4.11	2.38	5.70	*7.37*	11.15	1.84
Sr	130.39	176.99	181.19	206.54	255.46	128
Y	29.44	31.99	30.57	33.05	31.32	33.2
Zr	87.51	116.90	111.30	139.03	136.85	101.9
Nb	6.61	4.95	6.58	3.75	3.75	3.62
Cs	0.06	0.05	0.09	*0.34*	0.29	0.02
Ba	51.55	25.48	61.82	*32.79*	46.65	19.6
La	5.41	5.66	5.71	5.26	5.11	4.19
Ce	13.45	15.51	15.05	15.14	14.99	12.42
Pr	1.97	2.49	2.31	2.36	2.33	1.98
Nd	10.09	12.31	11.67	12.86	12.58	10.66
Sm	3.16	3.55	3.54	4.02	3.93	3.48
Eu	1.15	1.31	1.28	1.41	1.39	1.26
Gd	4.11	4.62	4.77	4.85	4.74	4.55
Tb	0.73	0.83	0.83	0.85	0.83	0.82
Dy	4.91	5.30	5.18	5.34	5.20	5.5
Ho	1.06	1.14	1.15	1.13	1.11	1.18
Er	3.08	3.29	3.19	3.17	3.09	3.42
Tm	n.a.	n.a.	n.a.	0.49	0.48	0.52
Yb	2.96	3.17	3.00	3.17	3.10	3.28
Lu	0.44	0.48	0.48	0.47	0.46	0.48
Hf	2.15	2.60	2.73	3.34	3.24	2.46
Ta	0.43	0.24	0.43	0.26	0.25	0.24
Pb	0.57	0.76	0.70	3.73	3.12	0.51
Th	0.51	0.29	0.52	0.28	0.32	0.252
U	0.15	0.10	0.15	0.13	0.14	0.083
Ga	16.70	18.48	16.87	*16.50*	*16.25*	17.00
Tl	0.02	*0.01*	0.02	n.a.	n.a.	0.02
Mo	0.51	*0.70*	0.41	n.a.	n.a.	0.36
Sn	0.77	*1.34*	1.10	*1.89*	*1.81*	0.80
Trace element ratios						
Zr/Hf	40.70	44.91	40.77	41.59	42.17	41.42
Zr/Nd	8.67	9.50	9.54	10.81	10.88	9.56
Nb/Ta	15.37	20.27	15.30	14.14	14.83	15.08
(La/Yb)N***	1.24	1.21	1.29	1.13	1.12	0.99
(Ce/Sm)N***	1.03	1.05	1.03	0.91	0.92	0.86
(Sm/Yb)N***	1.16	1.22	1.28	1.38	1.38	1.15
Sr/Y	4.43	5.53	5.93	6.25	8.16	3.86
Ti/V	29.31	36.46	36.78	41.59	42.29	33.02
Zr/Y	2.97	3.65	3.64	4.21	4.37	3.07
Zn/Fe (*10exp4)	12.17	12.76	12.30	12.01	12.04	12.35
V/Sc	7.07	6.22	6.95	6.88	6.78	7.29
Nb/La	1.22	0.87	1.15	0.71	0.73	0.86
U/Th	1.19	1.38	1.17	1.84	1.78	1.33

*subset is calculated for bulk rocks with $0.02 < K2O < 1.00$, and $CaO < 13$ and $45 < SiO2 < 56$

**n refers to major element analysis, trace elements are based on less analysis than major elements

***chondrite-normalized ratios

numbers in italics (average based on less than 40 samples)

Table 3.1. *Average composition of Alpine-Apennine basalts, relative to compositions of N-MORB and averages of the Mid-Atlantic Ridge (MAR), the Southwest Indian Ridge (SWIR) and the Gakkel Ridge (data from Gale et al. (2013)), complemented with data from Paquet et al. (2016) for the SWIR. Average composition of Alpine basalts based on data from (Ferrara et al. 1976; Lewis and Smewing 1980; Venturelli et al. 1981; Pognante et al. 1985; Bodinier et al. 1986; Bertrand et al. 1987; Dürr et al. 1993; Vannucci et al. 1993; Frisch et al. 1994; Steinmann 1994; Durand Delga et al. 1997; Rampone et al. 1998; Steinmann and Stille 1999; Bill et al. 2000; Puschnig 2000; Padoa et al. 2001; Desmurs et al. 2002; Chalot-Prat et al. 2003; Kramer et al. 2003; Chalot-Prat 2005; Kaczmarek et al. 2008; Montanini et al. 2008; Saccani et al. 2008; Renna et al. 2018; Amann et al. 2019; McCarthy and Müntener 2019; Decrausaz et al. 2021; Viereck 2006)*

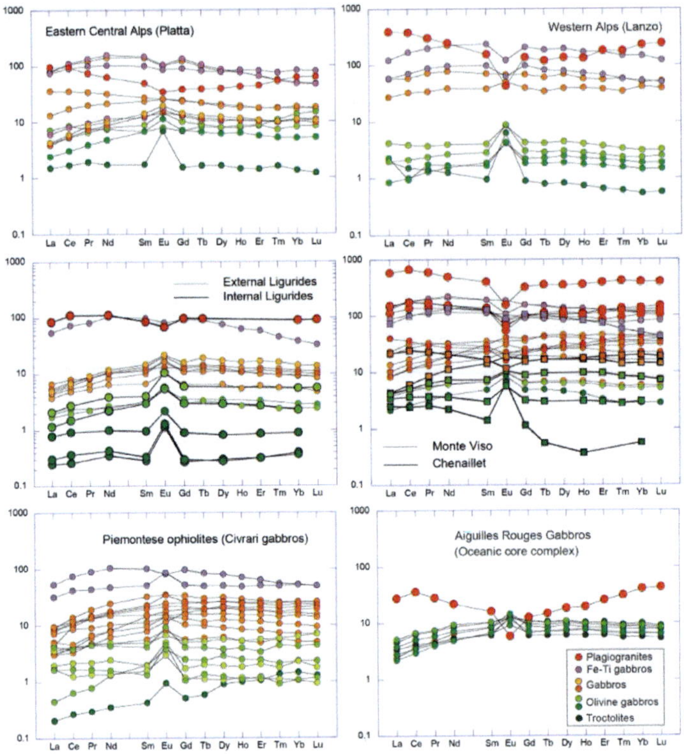

Figure 3.8. *Chondrite normalized REE diagrams for various gabbros from the Alps and Apennines. Most gabbros record differentiation from troctolites and olivine gabbros (greenish colors) to gabbros (orange colors), Fe-Ti gabbros (violet) and plagiogranites (red), with more than two orders of magnitude variation in REE*

COMMENTS ON FIGURE 3.8.– *Note that the gabbros from the oceanic core complex in the Combin zone (Aiguilles rouges gabbros) are much more homogeneous. For the location of gabbros, see Figure 3.1. Plagiogranites display the highest REE with a V-shaped REE pattern and strongly negative Eu anomalies (data sources: Bertrand et al. 1987; Borsi et al. 1996; Rampone et al. 1998; Messiga et al. 1999; Costa and Caby 2001; Desmurs et al. 2002; Borghini and Rampone 2007; Castelli and Lombardo 2007; Kaczmarek et al. 2008; Montanini et al. 2008; Sanfilippo and Tribuzio 2011; McCarthy and Müntener 2019; Locatelli et al. 2019; Decrausaz et al. 2021).*

In Figure 3.8, we illustrate the evolution of the chondrite normalized REE pattern from primitive troctolites to oxide gabbros and plagiogranites in Alpine gabbros. This is done for six representative localities from the Eastern Central Alps (panel A) to the northern Apennines (Panel B) and variably metamorphosed gabbros from the Western Alps (Panels C–F). The data for all these gabbros are remarkably consistent and display variations from primitive troctolites and olivine gabbros (chondrite normalized REE < 5) to gabbros (chondrite normalized REE between 5 and 20), to oxide gabbros (chondrite normalized REE up to 100) and plagiogranites with characteristic V-shaped REE and negative Eu anomalies, representing >90% of crystallization from a primitive liquid. The most notable exception is the Aiguilles rouges gabbros with a rather homogeneous composition. This is one of the largest coherent gabbro bodies in the Alpine ophiolites and has some characteristics of oceanic core complexes (Decrausaz et al. 2021). Why are these rather homogeneous gabbros so rare? From field relations, we argue that most gabbros show sharp contacts and intrude into relatively "cold" mantle peridotites. This is crucial with respect to crystallization differentiation processes, which are clearly shown by the REE evolution (Figure 3.8). Most oxide gabbros crystallized interstitial hornblende, which is only stable below about 950–1000°C and 3–4 wt% H_2O. We infer that the temperature of the host peridotite during intrusion of the gabbros is not substantially higher than hornblende stability, otherwise, crystal fractionation processes as shown by the incompatible element enrichment such as REE, Zr and TiO_2 are unlikely. Many of the oxide gabbros are so evolved that they have no volcanic equivalents among the Alpine basalts (Figure 3.7), consistent with the idea that much of the liquids are differentiated and crystallized as isolated bodies in the mantle. This is typical for ocean-continent transition zones (e.g. Desmurs et al. 2002) and (ultra)-slow-spreading ridges (Cannat et al. 2007; Dick et al. 2000).

3.5. Basalts: records of low-degree melting of garnet-bearing sources

Basalt rocks from the Alps occur as hyaloclastites, lava flows, pillow lavas and breccias up to a maximum thickness of about 300 m, but of limited lateral extent. They are often directly extruded on top of exhumed mantle rocks, and/or associated with high angle normal faults, indicating that these structures acted as feeding systems cutting exhumed mantle rocks (e.g. Chalot-Prat 2005; Manatschal et al. 2011). Most basalts are plagioclase-phyric and some of them show pseudomorphs of olivine.

The basalt data from the Alpine ophiolites are compiled from weakly metamorphosed ophiolites and areas that have been affected by greenschist or higher grade metamorphism. Data used in this section ($n = 617$ basalt analysis unfiltered for alteration) are compiled from various sources (Figure 3.9). While this list is not exhaustive, we consider the dataset sufficiently large and representative to explore the primary characteristics of basalts from the Piemonte-Liguria ocean. We compare these data to basalts from the Mid-Atlantic Ridge (MAR), the Gakkel Ridge and the Southwest Indian Ridge (SWIR), compiled and analyzed by Gale et al. (2013), and supplemented with glass analysis of Gakkel Ridge MORB from Shaw et al. (2010), Wanless et al. (2014) and Lissenberg et al. (2019).

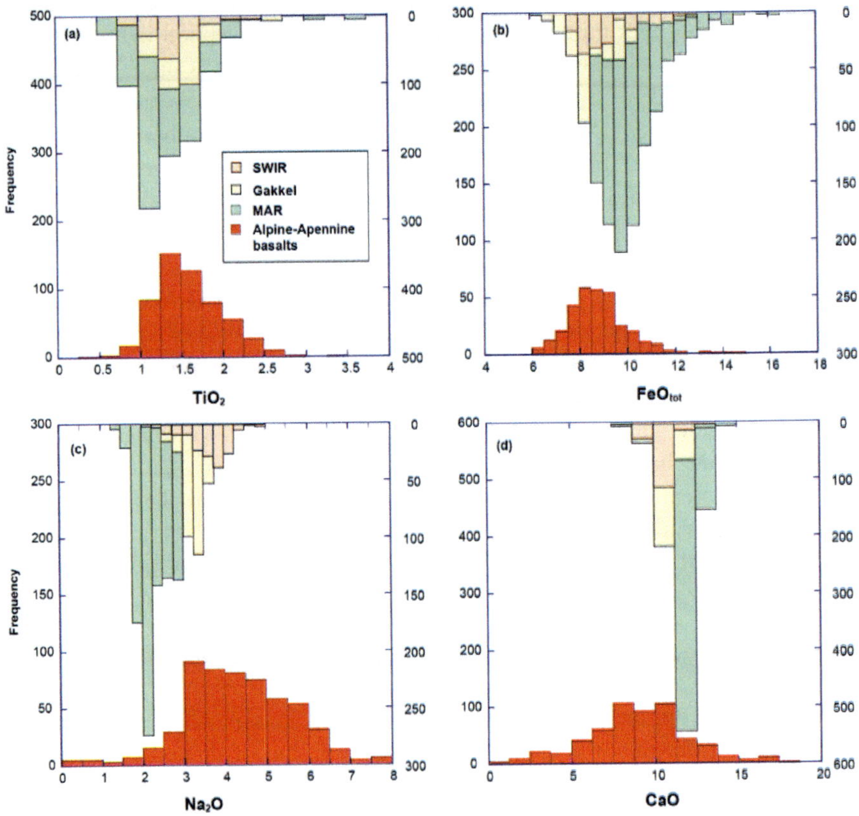

Figure 3.9. *Comparative histograms of basalts of the Alps, relative to global mid-ocean ridge data (MAR, Gakkel and SWIR). The distribution is most similar to Gakkel Ridge basalts in terms of TiO_2 and FeO_{tot}. Note that the variability in Na_2O and CaO is much greater for basalts from the Alps than for mid-ocean ridge basalts, illustrating the combined effects of alteration and metamorphism*

In comparing Alpine-Apennine basalts (for locations, see Figure 3.2) with fresh MORB from the global ridge system, it is important to keep in mind that (i) some of the major and trace elements such as Na_2O and CaO simply cannot be used to infer primary processes. In fact, the bulk rock chemistry of basalts has been affected by alteration, either on the seafloor of the Piemonte Ligurian Ocean or during Alpine metamorphism. This is illustrated in Table 3.1 and histograms of compiled bulk rock major element analyses of basalts from Alpine ophiolites (Figure 3.9) compared to fresh MORB from the MAR, the Gakkel Ridge and the SWIR (compiled by Gale et al. (2013)). While the Na_2O distribution for fresh MORB from global mid-ocean-ridges covers a narrow range between 1.5 and 3.5 wt%, the basalts from Alpine ophiolites vary between 0 and 8 wt%. Similar observations can be made for CaO. This indicates either loss or gain of Na (spilitization) or Ca (rodingitization) during sea-floor metasomatism or metamorphism. While this is certainly an important aspect of basalt processing, it is not the topic of this chapter and details on the metamorphic evolution of ophiolites are discussed by Agard in Chapter 4 and Manzotti and Ballèvre in Chapter 5. (ii) Many of the Alpine-Apennine basalts have very high large ion lithophile (LIL) element (Cs, Rb and Pb) abundances, about five times as high as N-MORB, which equally reflect the combined effects of alteration and/or metamorphic processes. Finally, Alpine basalts show higher Al_2O_3 contents than N-MORB or basalts from the MAR. Whether or not this is associated with the accumulation of plagioclase is an open question, however, we stress that basalts from the ultra-slow spreading Gakkel Ridge and SWIR are similarly high in their Al_2O_3 contents. It is thus likely that this is a primary igneous feature, potentially related to a lower degree of melting and/or higher pressures of fractionation and greater extraction depth related to ultra-slow spreading rates (Villiger et al. 2007).

Some major elements and notably the more incompatible trace elements are similar to global data sets (Figure 3.10). Numerous studies since the 1970s (e.g. Pearce 1982) have shown that incompatible elements reliably constrain formation processes and tectonic provenance, despite LIL elements being modified to a variable degree. Because Ti, Y, Zr, Nb, Th, Ta and La are among the most immobile during low-temperature alteration and metamorphism of basalts, trace element ratios such as Nb/La, Zr/Y or U/Th have been used to constrain basalt provenance (e.g. Pearce 1982). The first observation is that Alpine-Apennine basalts are remarkably similar to the global MORB data with respect to Mg# and transition elements (Ni, V, Cr, Co and Cu), indicating similar degrees of fractionation from primitive melts or cumulates. Overall, this indicates similar liquid lines of descent for most Alpine basalts. Average incompatible element compositions are reported in Table 3.1 and compared to basalts from the MAR, Gakkel Ridge and SWIR (Gale et al. 2013). Although the overall trace element patterns are comparable, there are some notable similarities and also some remarkable differences between the different datasets.

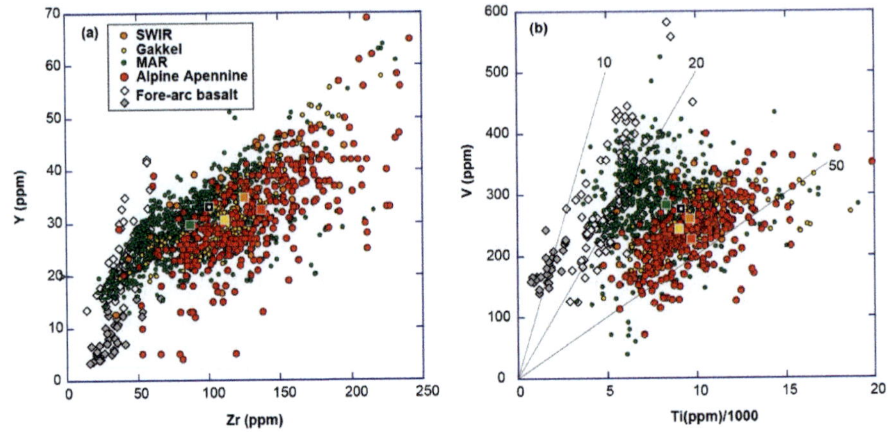

Figure 3.10. *Comparison of basalts from the Alps with datasets from global mid-ocean ridges (MAR, Gakkel and SWIR) and forearc basalts from the Izu Bonin-Mariana arc. Averages are indicated with squares*

COMMENTS ON FIGURE 3.10.– *(a) Zr versus Y abundances in basalts noting that high Zr/Y is indicative of an enriched component in the mantle. Note that average Zr/Y is highest for Alpine basalts, followed by Gakkel and SWIR. The lowest averages are for the MAR basalts and N-MORB. (b) V versus Ti abundance in basalts. Alpine basalts show the highest Ti(ppm)/1000/V(ppm), and all but a few MAR basalts have Ti (ppm)/1000/V (ppm) > 20, which separates MORB from tholeiites in island arcs and fore-arcs (data sources: MAR, Gakkel, SWIR: Gale et al. 2013. Fore-arc basalts: Arculus et al. 2015. Basalts from the Alps: Ferrara et al. 1976; Venturelli et al. 1979; Lewis and Smewing 1980; Venturelli et al. 1981; Pognante et al. 1985; Bodinier et al. 1986; Bertrand et al. 1987; Dürr et al. 1993; Vannucci et al. 1993; Frisch et al. 1994; Steinmann 1994; Durand Delga et al. 1997; Marroni et al. 1998; Rampone et al. 1998; Steinmann and Stille 1999; Bill et al. 2000; Puschnig 2000; Padoa et al. 2001, Desmurs et al. 2002; Chalot-Prat et al. 2003, Kramer et al. 2003; Chalot-Prat 2005; Viereck 2006; Kaczmarek et al. 2008; Montanini et al. 2008; Saccani et al. 2008; Renna et al. 2018; Amann et al. 2019; McCarthy and Müntener 2019; Decrausaz et al. 2021).*

Basalts from ophiolitic sequences in the Alps and Apennines correspond globally to N-MORB to T-MORB on the basis of chondrite normalized REE contents (Figure 3.11). Using the definition of Gale et al. (2013) for E-MORB, no more than 4% of basalts from the Piemonte-Liguria ocean are E-MORB, while all the others range between N-MORB and T-MORB. However, the critical distinctive characteristics of the Alpine–Apennine basalts compared to the MORB compilation of Gale et al.

(2013) are their elevated Ti/V and Zr/Y ratios (Figure 3.10). These characteristics are compatible with an overall lower degree of melting relative to global mid-ocean-ridge basalts and those that formed along the slowest spreading ridges, such as the SWIR and Gakkel Ridge. One of the recurrent questions is whether the Alpine-Apennine basalts carry any subduction signature, which is so common among global ophiolites. A comparison of the Ti-V and Zr-Y systematics from fore-arc basalts and boninites as compiled by Arculus et al. (2015) also indicate that this is not the case (Figure 3.9). The basalts originating from the Piemonte Liguria ocean are different than lavas from fore-arcs – the latter are derived from mantle sources that are more depleted than those from mid-ocean ridges.

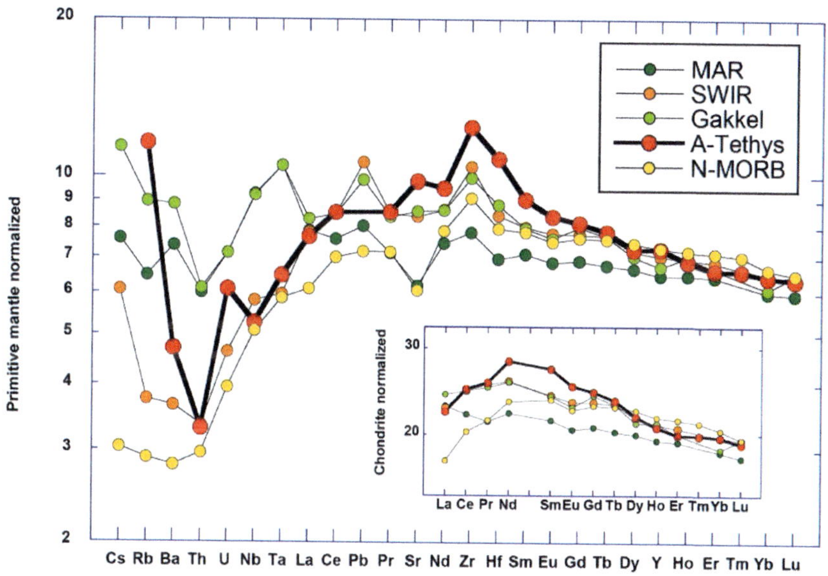

Figure 3.11. *Primitive mantle normalized spider diagrams for basalts from the Piemonte-Liguria ocean, compared to N-MORB (Gale et al. 2013), and slow to ultra-slow spreading ridges (MAR, Gakkel and SWIR). The basaltic rocks from the Alps have the highest Sm_N/Yb_N, the highest Zr/Y and the lowest Nb/La ratios on average. They probably record partial melting of the depleted asthenospheric mantle with a significant proportion of garnet pyroxenites (see text for details). Inset shows average chondrite normalized REE concentrations for mid-ocean ridges (from Gale et al. (2013)) compared to Alpine Apennine basalts. C1 Chondrite values: McDonough and Sun (1995); Primitive mantle values: Sun and McDonough (1989)*

All average basalts compiled from the global mid-ocean ridge system show U/Th > 1, yet the alpine basalts are the only ones with U/Nb > 1 (Figure 3.10), which may be due to the combined effects of alteration and metamorphism, as can be inferred by comparing the entire Alpine dataset (U/Th ~1.78) with those data filtered for alteration (U/Th ~1.38). The U/Th ratio of this subset, however, is similar to SWIR basalts and N-MORB (Figure 3.10), so we prefer the interpretation that the U/Th ratios of Alpine basalts are near-primary. Ba/La and Th/La are low and comparable to global MORB (Figure 3.10), indicating that the Alpine-Apennine basalts do not contain any substantial enrichment related to the recent Alpine collisional history.

A remarkable feature is that average Alpine-Apennine basalts have the lowest Nb/La ratio compared to the data compiled from different mid-ocean ridges, similar to some basalts erupted at the SWIR (e.g. Paquet et al. 2016; Table 3.1, Figure 3.12), but lower than the recent estimate of N-MORB (Gale et al. 2013). This low Nb/La ratio is coupled with an elevated average Sm/Yb ratio. Average Sm/Yb increases from the slow-spreading MAR to the ultra-slow spreading SWIR and Gakkel Ridges, and is highest in Alpine-Apennine basalts. This cannot be solely related to the degree of melting. This feature is rather attributed to a higher garnet-bearing component in the mantle source of MORB (e.g. Hellebrand et al. 2002; Montanini et al. 2008).

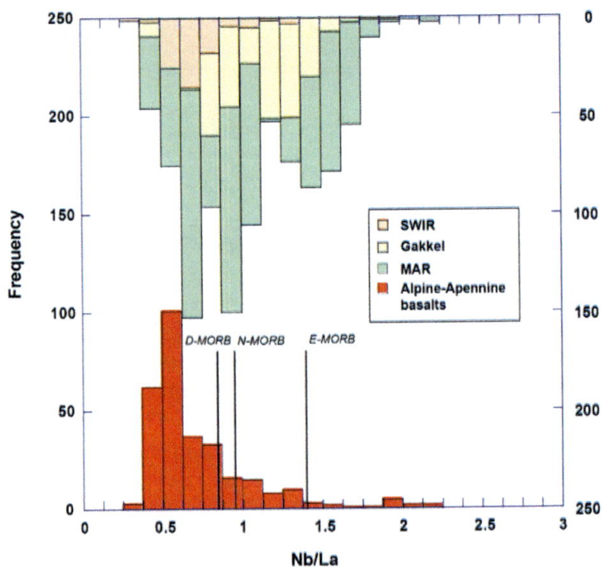

Figure 3.12. *Nb/La ratios of basalts. Histograms for Alpine basalts, MAR, Gakkel and SWIR. Most basalts from the Alps show Nb/La substantially lower than depleted MOR basalts. Gakkel Ridge basalts and MAR show two peaks in their distribution, closely corresponding to N-MORB and E-MORB. Data sources as in Figure 3.11*

Trace element modeling (e.g. Montanini et al. 2008; Saccani and Principi 2016; Amann et al. 2019) indicates that these ratios are likely to be controlled by the amount of a garnet-bearing component in the mantle source, most probably garnet-pyroxenites. Intriguingly, there is a large subset of Alpine basalts (>60%) that display Nb/La < 0.7 (Figure 3.12), which cannot be modeled by low to moderate degrees of partial melting of a normal DMM source. Indeed, these basalts probably result from the melting of a source that is more enriched than global asthenospheric DMM. If so, the source of Alpine–Apennine basalts is probably a combination of depleted peridotite with variable proportions of enriched components (e.g. garnet-pyroxenite), a hallmark of (ancient) continental peridotites. A pyroxenite component is not uncommon in the MOR subaxial mantle, but is hidden in most basalts as a result of higher degrees of melting (e.g. Paquet et al. (2016) and many others). The Nd-isotopic data of Alpine-Apennine basalts are consistent with this interpretation. Refractory peridotite mixed with (garnet-) pyroxenite is typically formed in arc-related mantle wedge settings. While it is tempting to relate these signatures to processes of the Variscan and post-variscan evolution of the European mantle, this is a hypothesis that needs further work. Interestingly, the hypothesis of ancient depleted peridotite is consistent with some of the Nd-isotopic systematics of mantle clinopyroxene from Alpine peridotites (see below).

With the caveat of alteration, it follows that the main characteristics of the Alpine-Apennine basalts are different from the global MORB dataset in two respects: (1) Some trace element indicators (e.g. Nb/La, Ce/Sm) call for low degree melting of a source that is more refractory than those beneath mid-ocean ridges, but unlike melt depleted sources as invoked for fore-arc basalts and boninites. (2) Some trace element indicators (e.g. Sm/Yb, Nb/La) point to contributions of garnet–rich lithologies in the mantle source of Alpine Apennine basalts. The combination of a rather refractory mantle source containing some garnet pyroxenites is a hallmark of subcontinental peridotites. If so, this could explain why the volumes of MORB-type magmatism in the Alpine Apennine area are small. Most of the source seems depleted and the melt productivity over the entire lifetime of the Piemonte-Liguria ocean is low.

3.6. Timing of magmatism of Alpine-Apennine ophiolites: a short-lived event

The timing of magmatism in the Piemonte-Liguria ocean has primarily been addressed by two different methods – Sm-Nd mineral isochrons on plagioclase peridotites and olivine gabbros, and U-Pb dating on zircons from evolved Fe-Ti gabbros and plagiogranites. In the last 30 years, high-precision U/Pb dates on zircon has constrained the crystallization age of MOR-type gabbros from the Alpine-Apennine ophiolites (e.g. Borsi et al. 1996; Rubatto et al. 1998; Schaltegger et al. 2002; Lombardo et al. 2002; Kaczmarek et al. 2008; Tribuzio et al. 2016), with implications for the timescales of magmatism in the Piemonte-Liguria ocean. More recent studies also reported Hf and oxygen isotopes on zircon that provide further constraints on the magma sources. However, here we need to be brief and cannot provide a comprehensive discussion on possible discrepancies in the interpretations. Since absolute age data on ophiolitic gabbros are of considerable interest for other chapters covered in this book, we summarize our understanding of recent work on this topic.

Several recent studies noted that there is a discrepancy between isochron ages (mostly Sm-Nd isochron ages determined on mineral separates from gabbros) and U-Pb dates on zircons (e.g. Rampone et al. 2014; Tribuzio et al. 2016). These Sm-Nd isochron ages reflect the closure temperature for the Sm-Nd system at about 800°C (Mezger et al. 1992) and are therefore interpreted as a minimum age for melt migration and re-fertilization of peridotite and crystallization of olivine gabbros in peridotites.

Both gabbros from Erro Tobbio and the external and internal Ligurides are older than 170 Ma (Figure 3.13). Sm-Nd ages around 170 Ma (170 ±13 and 173.5 ±4.8 Ma, initial εNd = 8.5–8.9; Tribuzio et al. 2004) were also documented in olivine gabbros from Tuscany. Notably, the above ages are overall older than the oldest pelagic sediments (Bathonian, ~165 Ma; 165 Ma; Bill et al. 1997, 2001; O'Dogherty et al. 2005), which track continental breakup and oceanization of the Piemonte–Liguria ocean. This was originally explained by the fact that most mineral isochrones were determined on zircon-free primitive Mg-gabbros, while U-Pb on zircon was restricted to highly differentiated Fe-Ti gabbros and plagiogranites. Accordingly, ages as old as ~180 Ma were taken as an argument for a syn-rift origin of these gabbros (e.g. Tribuzio et al. 2004; Rampone et al. 2014). In contrast, a dataset on the external and internal Ligurides gabbros convincingly showed that the U-Pb ages of zircon are restricted to a much narrower range of 166–158 Ma (Tribuzio et al. 2016). Combined with U-Pb ages from the northern part of the Piemonte-Liguria ocean and the Valais branch (Figure 3.13), U-Pb-ages on zircon provide a narrow age range of 165 ±2 to 155 ±2 Ma.

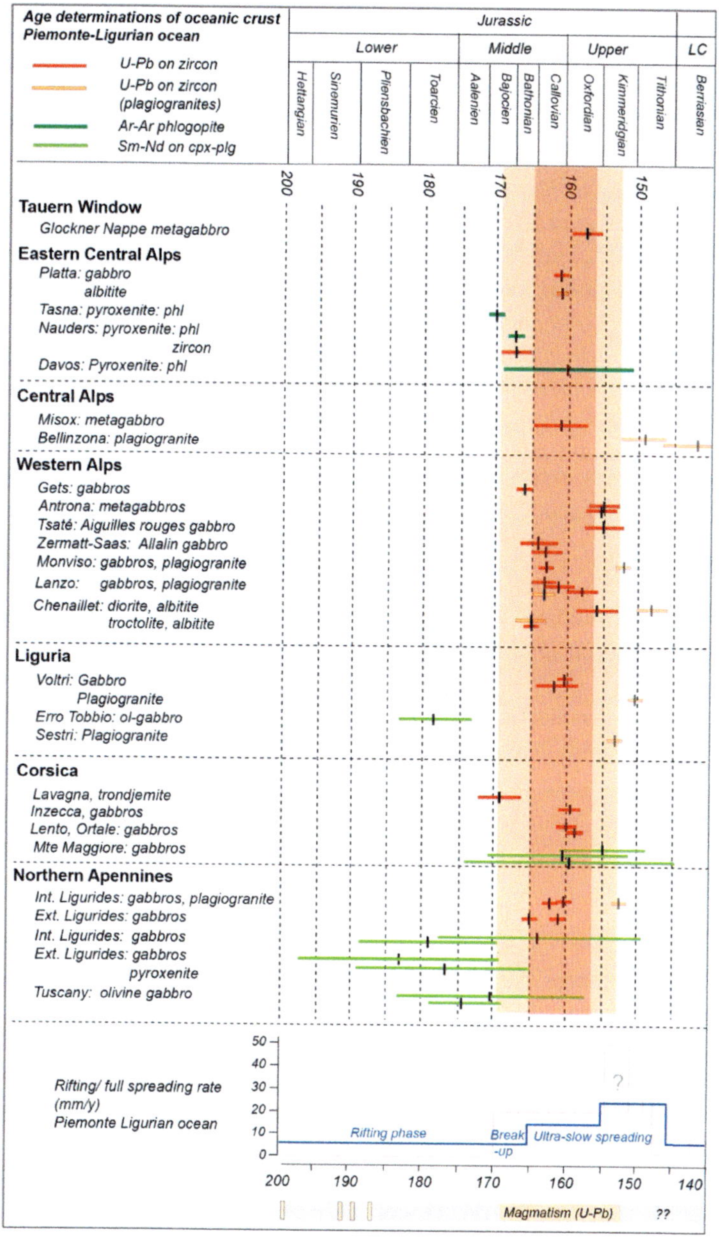

Figure 3.13. *Ar-Ar ages on phlogopite pyroxenite, Sm-Nd isochrones on clinopyroxene-plagioclase pairs, and U-Pb zircon dates of gabbros and plagiogranites from the Alpine-Apennine ophiolites*

COMMENTS ON FIGURE 3.13.– *Most U-Pb zircon dates from gabbros fall in an age range between 167 and 155 Ma, while many Sm-Nd isochrons are subject to larger errors and often provide older ages than U-Pb on zircon. The interpretation of the Sm-Nd ages is ambiguous. Zircon ages younger than 155 Ma are all derived from plagiogranites likely affected by recrystallization (see text for further discussion). Age data are compiled in Manatschal and Müntener (2009) and supplemented by more recent studies (Li et al. 2013; Rampone et al. 2014; Li et al. 2015; Tribuzio et al. 2016; Decrausaz et al. 2021; Gleißner et al. 2021). (b) Full spreading rates of the Piemonte-Liguria ocean, red line from Vissers et al. (2013), blue line from Le Breton et al. (2021). Punctuated magmatism (U-Pb on zircon from syenites and carbonatites) during rifting as recorded in the Ivrea lower crust by Schaltegger et al. (2015) and Galli et al. (2019).*

A few U-Pb studies on zircon and Ar-Ar on phlogopite have provided slightly older ages of 167 ±1 Ma (Hauser 2011), consistent with the idea that the earliest magmatism initiated in ocean-continent transitions. The volume of magmatism, however, is extremely small and is related to zircon-bearing phlogopite-orthopyroxene veins in a subcontinental mantle of the Nauders-Tasna area (Hauser, 2011). The Nauders-Tasna area is now known as a classical example of an exhumed ocean-continent transition (e.g. Florineth and Froitzheim 1994; Ribes et al. 2020), suggesting that these ages represent the earliest intrusions in the exhumed subcontinental mantle, rather than magmatism that is related to the formation of ultra-slow spreading crust. Note that these ages are all substantially younger than alkaline intrusions and associated carbonatites in the lower crust of the Ivrea Zone, which range from ~225–178 Ma (Stähle et al. 1990; Schaltegger et al. 2015; Galli et al. 2019).

Ages considerably younger than the above range defined by gabbros were found for plagiogranites in the Chenaillet, some Ligurian units, Monte Viso and the Central Alps (for references see Figure 3.13). Ages obtained on plagiogranite dikes from these units range from 142–155 Ma, and often define a time gap relative to the ages of associated gabbros (e.g. Central Alps and Tuscany, see Figure 3.13). The major unknown with respect to U-Pb ages of zircon is those of plagiogranites, which tend to be younger than zircons from gabbros, especially for samples dated prior to 2005 and those that are overprinted by higher grade metamorphism (e.g. Stucki et al. 2003). In some ophiolite units (e.g. Platta, Lanzo, Inzecca (Corsica), Aiguilles rouges (Combin zone) and recent dating in Chenaillet; see Figure 3.13), plagiogranites yield ages identical within error of the associated gabbros. Field relations indicate that they are contemporaneous to MOR type gabbros (Kaczmarek et al. 2008; Ohnenstetter et al. 1981; Rubatto et al. 2008; Decrausaz et al. 2021), and derived from a depleted, MORB-like mantle source (Schaltegger et al. 2002; Li et al. 2015; Decrausaz et al. 2021). These plagiogranites are interpreted to represent cathodo-crystallization products of late-stage differentiated melts. Detailed luminescence studies have shown that zircons from plagiogranites are partial to being completely replaced by "spongy" textures (e.g. Lombardo et al. 2002; Rubatto et al. 2008; Kaczmarek et al. 2008; Li et al. 2013; Tribuzio et al. 2016; Li et al. 2013) indicative of dissolution-recrystallization processes, coupled with a high common Pb component related to hydrothermal alteration. It is possible that early studies on zircons in plagiogranites are affected by hydrothermal alteration, explaining the younger ages of some of the plagiogranites (e.g. Borsi et al. 1996; Costa and Caby 2001; Stucki et al. 2003). More recent studies have confirmed hydrothermal alteration of zircons from plagiogranites by analyzing in-situ oxygen isotopes in zircon (Li et al. 2015).

If all the plagiogranite zircons that are younger than associated gabbros were affected by hydrothermal alteration, we would propose that the main magmatic activity for the Piemonte-Liguria ocean ranges from 165 ±2 to 153 ±2 Ma, indicating about 10 ±4 Ma of oceanic magmatism. Chronostratigraphic studies using radiolarites covering continental basement and serpentinites and/or basaltic rocks range from about 169 to 154 Ma, in excellent agreement with U-Pb studies on zircon (Lombardo et al. 2002; Cordey and Bailly 2007).

The present understanding of the Alpine ophiolites is that some of them represent parts of an ocean-continent transition (e.g. Manatschal and Müntener 2009), while others originated from an ultra-slow spreading ridge and oceanic core complexes (e.g. Lagabrielle et al. 2015; Decrausaz et al. 2021). If the entire magmatism in the Piemonte-Liguria ocean were the result of ultra-slow spreading, then the preserved duration of "oceanic magmatism" of 10 ±3 Ma is probably a maximum. This is considerably shorter than previous estimates of the magmatic activity of the Piemonte-Liguria ocean (~25 Ma (Rampone et al. 2014) to ~40 Ma (Handy et al. 2010)). A remarkable feature is that most studies analyzing the paleomagnetic data and reconstructions between Europe, Iberia, Africa, Adria and the North Atlantic consider 20 mm/year as a maximum upper bound for Piemonte-Liguria spreading rates (Rosenbaum and Lister 2002; Vissers et al. 2013; LeBreton et al. 2021), with small excursions between 155 and 145 that are possibly slightly higher (e.g. Vissers et al. 2013; LeBreton et al. 2021). The duration of magmatic activity coupled with estimates of spreading rates is, of course, crucial for estimates of the width of the Piemonte-Liguria ocean.

If we assume that the U/Pb ages on zircon from MOR gabbros date the accretion of the oceanic crust, that remnants of the entire magmatic activity are preserved and there were plate separation rates of ±1.5 cm/year as a maximum between 165 and 155 Ma (Le Breton et al. 2021), the Alpine ophiolites were sampled within a zone about 150 ±20 km wide, which corresponds approximately to the width of the present-day OCT in the Iberia–Newfoundland margins. Considering the possibility of complete subduction of ultra-slow spreading crust formed between 154 and 145 Ma, an additional ~200 km could be added. Such an estimate (~350 km) is at the lower end relative to previous estimates and has important implications for shortening rates, subduction depth, potential slabs identified on seismic tomography, and the overall construction of the Alpine orogeny. For an overview of potential consequences, the reader is referred to McCarthy et al. (2020a, 2021).

3.7. A brief overview of isotopic (dis)equilibrium between magmatic and mantle rocks

It has been shown that Sr and Nd isotopic compositions of basalts from the global mid-ocean ridges are much more homogeneous than associated mantle rocks (e.g. Stracke et al. 2011) and the results from the Alpine-Apennine system are no exception (see Rampone and Hofmann (2012), for a review on these aspects). Compiled $^{143}Nd/^{144}Nd$ data for basalts from the Alpine-Apennines show a narrow compositional field ranging from 0.5126 to 0.5130 (at 160 Ma) for more than 95% of the data and covers a similar range as the global MORB data (as shown by Warren et al. (2009)). However, there is a tendency of Alpine-Apennine basalts towards more enriched compositions.

The discovery of extreme variations of Hf isotopic compositions from depleted peridotites raised an important question of whether ultra-depleted domains of the oceanic mantle are represented in the global dataset of basalts and gabbros of the oceanic crust (e.g. Liu et al. 2008; Byerly and Lassiter 2014). The emerging idea is that the ultra-depleted mantle is the volumetrically dominant portion of the depleted mantle, but contributes little to the incompatible budget of the oceanic crust (e.g. Stracke et al. 2019). Recent studies on Hf isotopes from the Lanzo peridotite in the Western Alps were interpreted as infiltration of melts derived from an ancient depleted source (Sanfilippo et al. 2019). Here we add recent isotopic studies, notably those on the Civrari peridotites and mafic rocks in the Western Alps (McCarthy and Müntener 2015, 2019), to previous compilations (Rampone and Hofmann 2012). In our compilation of Nd isotopic analysis of clinopyroxene separates from the Alpine-Apennine peridotites (Hf isotopic data are too sparse at present), most clinopyroxene show $^{147}Sm/^{144}Nd < 0.3$ close to the depleted MORB mantle, with significant variations in $^{143}Nd/^{144}Nd$ (Figure 3.14). This isotopic variability has been interpreted as recording Proterozoic melting events and/or heterogeneity of the mantle source, as reflected by their variable $^{143}Nd/^{144}Nd$ (e.g. Bodinier et al. 1991; Rampone et al. 1995). However, almost all clinopyroxene with elevated $^{147}Sm/^{144}Nd$, that is clinopyroxene with strongly fractionated LREE, show increasing $^{143}Nd/^{144}Nd$ (Figure 3.14). Even after correction for radiogenic ingrowth to 160 Ma, the average age of the oceanic crust as inferred from U-Pb dates on zircon (see 6.6 above), some clinopyroxene from peridotites have extremely high ^{143}Nd /^{144}Nd ratios that are unknown from studies on mantle clinopyroxene from the global mid-ocean ridge system (e.g. Warren et al. 2009), calling for ancient depletion event(s) in the mantle sources of Alpine Apennine peridotites. The timing of ancient depletion is difficult to determine due to the general lack of isochron relationships for individual rock samples. Taking the LREE depleted mantle clinopyroxene from Tuscany, the Internal Ligurides, the Platta nappe and the Civrari peridotite, (i.e. those with the highest $^{143}Nd/^{144}Nd$ isotopic ratios), an "errorchon" can be calculated to 273 ±24 Ma, assuming that these clinopyroxenes are derived from the depleted mantle (McCarthy and Müntener 2015). Using the latter as a potential source is justified as illustrated in Figure 3.14(b), where the isotopic composition of the Alpine Apennine basalts and gabbros are plotted along with the depleted mantle. Ages derived from plagioclase-clinopyroxene Sm-Nd isochrones on gabbros and re-fertilized plagioclase peridotites are generally in agreement with U-Pb dates on zircons and give Jurassic ages between 150 and 180 My. These data suggest that the MORB-type gabbros and basalts are not in isotopic equilibrium with the extremely LREE depleted clinopyroxene of refractory mantle peridotites (Rampone et al. 1998; Müntener et al. 2004).

One possibility is that the peridotites with highly radiogenic Nd isotopic compositions represent the residues of Permian magmatism (e.g. McCarthy and Müntener 2015). An alternative suggests that they are infiltrated by melts with an ultra-depleted signature during the Jurassic formation of the oceanic crust (e.g. Sanfilippo et al. 2019). It is, however, remarkable that the errorchron age calculated from the residual clinopyroxene over a regional scale of at least 500 km is in agreement with a period of massive transcrustal magmatism during the Early Permian (e.g. Schuster and Stüwe 2008), in the area where the future Piemonte-Liguria and Valais oceans formed. Given the general scarcity of magmatic activity during the formation of oceanic crust in the Alpine-Apennine area, such inherited mantle domains, which escaped re-equilibration and re-fertilization by migrating melt, may have greater chances to be preserved and found in mantle regions characterized by a lower degree of melting.

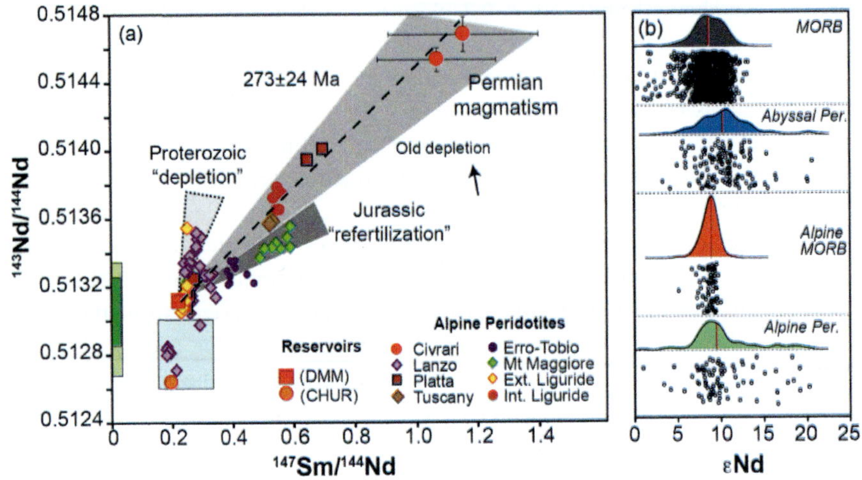

Figure 3.14. *The isotopic composition of Alpine peridotite, basalt and gabbros*

COMMENTS ON FIGURE 3.14.– *(a) Present-day ^{143}Nd/^{144}Nd versus ^{147}Sm/^{144}Nd values of Alpine-Apennine peridotites, modified from McCarthy and Müntener (2015). Depleted mantle and CHUR values from Faure (1986). Blue field represents ^{143}Nd/^{144}Nd versus ^{147}Sm/^{144}Nd values for clinopyroxene, plagioclase, bulk rock gabbros and basalts (at 160 Ma) from the Alps and Apennines. Green field indicates the range of ^{143}Nd/^{144}Nd for global MORB (at 0 Ma) and the dark-green field contains >90 % of the data (from Warren et al. 2009). Note that the Alpine-Apennine basalts are less depleted than global MORB in terms of Nd-isotopes. (b) Ridge-line and scatter plots showing the density distribution of εNd for present-day MORB and Abyssal peridotites (Abyssal Per.), and the Alpine Apennine ophiolites (Alpine MORB) and Alpine peridotites (red line is the median). For the Alpine Apennine ophiolites, εNd values are corrected to 160 Ma. MORB data compilation is from Gale et al. (2013). For Abyssal peridotites, the data compilation is from Warren (2016). For Alpine MORB and mantle rocks, data is from Bodinier et al. (1991), Borsi et al. (1996), Rampone et al. (1995, 1996, 1998, 2005, 2009), Costa and Caby (2001), Schaltegger et al. (2002), Müntener et al. (2004), Tribuzio et al. (2004), Mugnier et al. (2008), Borghini et al. (2013), McCarthy and Müntener (2015, 2019) and Sanfilippo et al. (2019).*

It has long been inferred that refractory subcontinental mantle fragments may be stranded in the Mid-Atlantic ocean, even close to the ridge (e.g. Bonatti 1971), but these ideas found little currency in studies about the isotopic evolution of the mantle.

One of the key assumptions in interpreting isotope data from peridotites along a mid-ocean ridge system is that they cooled from the asthenosphere over a maximum of 20 Ma so radiogenic ingrowth for the Nd isotopic system is negligible (e.g. Warren et al. 2009). However, contamination of asthenospheric mantle by subcontinental or arc mantle bodies has been inferred at slow- and ultraslow ridge environments (Müntener and Manatschal 2006; Urann et al. 2020) and from very depleted Os isotopic signatures documented in some MAR peridotites (Alard et al. 2005; Harvey et al. 2006). These data have been interpreted as a hallmark of depleted arc(-hean) mantle slices incorporated in the asthenosphere, but it could also be that lithospheric mantle bodies simply represent "rafts" from the recent rifting history, stranded as continental mantle allochthons among the global mid-ocean ridge system. The isotopic geochemical signature of western Gakkel Ridge MOR basalts was interpreted as evidence of a delaminated continental mantle incorporated into the Arctic convective asthenosphere (Goldstein et al. 2008). The extremely depleted Os and Hf isotopic values found in some Gakkel Ridge peridotites (Liu et al. 2008; Stracke et al. 2011) could also be interpreted as evidence of incorporation of the subcontinental lithosphere in the sub-ridge Gakkel mantle, or simply as isolated mantle fragments during extension of the continental lithosphere. In view of the geochemistry of the Alpine Apennine basalts, and the widespread evidence for massive mantle melting during the Permian, combined with the field evidence of a successive overprint of the continental mantle by migrating melts (e.g. Müntener and Piccardo 2003; Müntener et al. 2004; Müntener et al. 2010), we prefer the hypothesis that the most depleted isotopic signature of some Alpine-Apennine peridotites results from peridotites that record Permian depletion events. They are locally preserved from the almost pervasive nature of melt-rock reaction processes as recorded by plagioclase peridotite during the opening of the Piemonte-Liguria ocean.

3.8. Conclusion and outlook

Alpine ophiolites are dramatically different from most ophiolites worldwide. Together with peridotites from passive margins and ultra-slow spreading ridges, Alpine ophiolites exemplify an endmember that is dominated by exhumed and serpentinized peridotites, with very minor and local MORB basalts and gabbros. They do not represent supra-subduction zone ophiolites. Despite some selective sampling related to the collisional history of the Alps, the results presented in this chapter are robust. There are fresh peridotites over the entire ~500 km long Alpine-Apennine mountain belt and these show petrological and geochemical signatures that are spatially arranged in a surprisingly coherent manner. In most respects, Alpine peridotites are similar to dredged and drilled peridotites from passive margins and ultra-slow spreading ridges, reinforcing the hypothesis that they represent distal margins and ultra-slow spreading oceanic lithosphere.

Heterogeneous spinel peridotites with garnet pyroxenites, websterites, dunites and locally metasomatic, phlogopite-amphibole bearing peridotites are the results of long-term interaction of (ancient) depleted peridotites and migrating melts beneath the continental crust. Thermal constraints on these peridotites are unlike geotherms from mid-ocean ridges, indicating long-term cooling to a continental geotherm. They are associated with lower crustal granulites and mafic intrusions, which are best explained by an origin as mantle lithosphere beneath continental crust. These peridotites are typically associated with continental slivers and are locally covered by sediments derived from the continental margin. They are most similar to those dredged from the Galicia Bank (e.g. Chazot et al. 2005) and the Australia-Antarctic margins (McCarthy et al. 2020b).

Plagioclase peridotites, with spinel-plagioclase pyroxenites and local harzburgites and dunites, are more homogeneous and chemically fertile, approaching primitive upper mantle compositions. They show characteristic textures of melt-rock reactions, are associated with rare gabbroic dikes and are sometimes locally covered by MOR-basalts, indicating an oceanic origin. Reactive textures between pyroxenes and plagioclase are largely determined by the nature of the Si-rich reactive melts and indicate the addition of MOR tholeiite to previously depleted peridotites. The preservation of reaction textures and Cr-Al diffusion profiles in orthopyroxene indicates rapid exhumation and cooling of plagioclase peridotite along with a system of detachment faults.

Rare, LREE-depleted clinopyroxene resembling abyssal peridotite clinopyroxene derived from near-fractional melting is preserved as small lenses within plagioclase peridotites. Their Nd-isotopic signature is elevated and much higher than the upper mantle at ~160 Ma. They all align at a pseudo-isochron of ~273 ±24 Ma, which coincides with a period of widespread, transcrustal post-Variscan magmatic activity (Schuster and Stüwe, 2008; McCarthy and Müntener 2015). It is possible that these peridotites are remnants of ancient, near-fractional melting processes during the Permian, unaffected by the Jurassic processes in the Western Tethys. Alternatively, they represent peridotites that were affected by melts derived from ultra-depleted parts of the asthenospheric mantle. In any case, despite being possible residues of the formation of gabbros and basalts of the Western Tethys as indicated by trace element models (e.g. McCarthy and Müntener 2015), they are in isotopic disequilibrium with the melts that formed the Jurassic oceanic crust of the Western Tethys. The only possible exceptions are the spinel and plagioclase peridotites from

Mte Maggiore (Corsica), which indicate isotopic equilibrium between peridotite and gabbroic rocks (Rampone et al. 2009).

The compositional variation of MOR-type gabbros ranges from troctolite to Fe-Ti gabbros and plagiogranites. Zircon ages from oceanic gabbros cover a surprisingly small range between ~167 and ~155 Ma, favoring the hypothesis that the magmatic activity in this oceanic basin was short-lived and volumetrically minor. Locally, plagioclase peridotites are replaced by reactive harzburgites and dunite channels and cut by gabbroic dikes that testify to melt extraction and magmatism during the formation of the Piemonte-Liguria ocean. The volumes, however, are comparatively minor and testify to a segmented oceanic lithosphere similar to the ultra-slow spreading SWIR (Cannat et al. 2019). The basalt chemistry resembles MORB from ultra-slow spreading ridges (SWIR, Gakkel Ridge) in terms of REE elements, but depleted in Nb/La at elevated Sm/Yb. Possible avenues in the future should address in more detail the surprisingly depleted composition of the basaltic rocks, despite being formed in an ultra-slow spreading system. We argue that they represent a hallmark of ancient (Variscan) history. It is possible that the magmatic system that formed the Alpine-Apennine ophiolites was short-lived because once enriched mantle domains (amphibole peridotites, pyroxenites) were molten, the mantle source "ran out of fusible components", presumably as a consequence of previous orogenies and the overall depleted nature of the mantle lithosphere. Limited magmatism and the predominance of depleted MORB type basalts indicate that active asthenospheric upwelling was probably limited and "large scale, active steady-state" plate separation such as in the Atlantic never initiated. Finally, the tectonic and magmatic inheritance produced by a short-lived episode of continental rifting and ultra-slow spreading has consequences for the mechanisms of convergence and subduction leading to the formation of the Alpine orogen (McCarthy et al. 2018, 2020, 2021). Most notably, any interpretation of the deep structure of the Alps needs to be compatible with a rather small oceanic domain of the Piemonte Ligurian and Valain ocean (<350 km), with important consequences for the interpretation of tomographic data and the volume and size of "slabs" beneath the present days Alps.

3.9. Acknowledgments

We thank Yves Lagabrielle, Nicolas Bellahsen and Claudio L. Rosenberg for inviting us to contribute to this book and for their patience with respect to deadlines. An informal review of Peter Ulmer was very helpful in sharpening the arguments. Mathilde Cannat and Elisabetta Rampone provided stimulating and very helpful reviews. OM was supported in part by the Swiss National Science

Foundation (Research grant nr. 200021_162666). AMC was supported by IODP France and a Postdoc Mobility grant by the Swiss National Science Foundation (P400P2_180794).

3.10. References

Alard, O., Luguet, A., Pearson N.J., Griffin W.L., Lorand J.-P., Gannoun, A., Burton K.W., O'Reilly, S.Y. (2005). In-situ Os isotopes in abyssal peridotites bridge the isotopic gap between MORBs and their source mantle. *Nature,* 436, 1005–1008.

Amann, M., Ulrich, M., Manatschal, G., Pelt, E., Epin, M.-E., Autin, J., Sauter, D. (2019). Geochemical characteristics of basalts related to incipient oceanization: The example from the Alpine-Tethys OCTs. *Terra Nova,* 32, 75–88.

Arculus, R.J., Ishizuka, O., Bogus, K.A., Gurnis, M., Hickey-Vargas, R., Aljahdali, M.H., Bandini-Maeder, A.N., Barth, A.P., Brandl, P.A., Drab et al. (2015). A record of spontaneous subduction initiation in the Izu-Bonin-Mariana arc. *Nature Geoscience,* 8, 728–733.

Basch, V., Rampone, E., Crispini, L., Ferrando, C., Ildefonse, B., Godard, M. (2018). From mantle peridotites to hybrid troctolites: Textural and chemical evolution during melt-rock interaction history (Mt. Maggiore, Corsica, France). *Lithos,* 323, 4–23.

Beltrando, M., Manatschal, G., Mohn, G., Dal Piaz, G.V., Vitale Brovarone, A., Masini, E. (2014). Recognizing remnants of magma-poor rifted margins in high-pressure orogenic belts: The Alpine case study. *Earth-Science Reviews,* 131, 88–115.

Bernoulli, D. and Weissert, H. (1985). Sedimentary fabrics in Alpine ophicalcites, South Pennine Arosa zone, Switzerland. *Geology,* 13, 755–758.

Bertrand, J., Dietrich, V., Nievergelt, P., Vuagnat, M. (1987). Comparative major and trace element geochemistry of gabbroic and volcanic rock sequences, Montgenèvre ophiolite, Western Alps. *Schweizerische Mineralogische und Petrographische Mitteilungen,* 67, 147–169.

Beslier, M.O., Royer, J.-Y., Girardeau, J., Hill, P.J., Boeuf, E., Buchanan, C., Chatin, F., Jacovetti, G., Moreau, A., Munschy, M. et al. (2004). Une large transition continent-océan en pied de marge sud-ouest australienne : premiers résultats de la campagne MARGAU/MD110. *Bulletin de la Societé Géologique de France,* 175, 629–641.

Bickert, M., Lavier, L.L., Cannat, M. (2020). How do detachment faults form at ultraslow mid-ocean ridges in a thick axial lithosphere? *Earth and Planetary Science Letters,* 533, 116048.

Bill, M., Bussy, F., Cosca, M., Masson, H., Hunziker, J.C. (1997). High-precision U-Pb and Ar-40/Ar-39 dating of an Alpine ophiolite (Gets nappe, French Alps). *Eclogae Geologicae Helvetiae,* 90, 43–54.

Bill, M., Nägler, T., Masson, H. (2000). Major, minor and trace element, Sm-Nd and Sr isotope composition of mafic rocks from the earliest oceanic crust of the Alpine Tethys. *Schweizerische Mineralogische und Petrographische Mitteilungen,* 80, 131–146.

Bill, M., O'Dogherty, L., Guex, J., Baumgartner, P.O., Masson, H. (2001). Radiolarite ages in Alpine-Mediterranean ophiolites: Constraints on the oceanic spreading and the Tethys-Atlantic connection. *Geological Society of America Bulletin,* 113, 129–143.

Bodinier, J.L. (1988). Geochemistry and petrogenesis of the Lanzo peridotite body, Western Alps. *Tectonophysics,* 149, 67–88.

Bodinier, J.L. and Godard, M. (2003). Orogenic, ophiolitic, and abyssal peridotites. In *Treatise on Geochemistry, Vol. 2: Mantle and Core,* Carlson, R.W. (ed.). Elsevier Science, Amsterdam.

Bodinier, J.L., Guiraud, M., Dupuy, C., Dostal, J. (1986). Geochemistry of basic dikes in the Lanzo massif (Western Alps): Petrogenetic and geodynamic implications. *Tectonophysics,* 128, 75-95.

Bodinier, J.L., Menzies, M.A., Thirlwall, M.F. (1991). Continental to oceanic mantle transition: REE and Sr-Nd isotopic geochemistry of the Lanzo Lherzolite Massif. *Journal of Petrology,* 191–210.

Boillot, G., Féraud, G., Recq, M., Girardeau, J. (1989). Undercrusting by serpentine beneath rifted margins. *Nature,* 341, 523–525.

Bonatti, E. (1971). Ancient continental mantle beneath oceanic ridges. *Journal of Geophysical Research,* 76, 3825–3831.

Borghini, G., Rampone, E., Crispini, L., De Ferrari, R., Godard, M. (2007). Origin and emplacement of ultramafic-mafic intrusions in the Erro-Tobbio mantle peridotite (Ligurian Alps, Italy). *Lithos,* 94, 210–229.

Borghini, G., Rampone, E., Zanetti, A., Class, C., Cipriani, A., Hofmann, A.W., Goldstein, S.L. (2013). Meter-scale Nd isotopic heterogeneity in pyroxenite-bearing Ligurian peridotites encompasses global-scale upper mantle variability. *Geology,* 41, 1055–1058.

Borsi, L., Schärer, U., Gaggero, L., Crispini, L. (1996). Age, origin and geodynamic significance of plagiogranites in lherzolites and gabbros of the Piedmont-Ligurian ocean basin. *Earth and Planetary Science Letters,* 140, 227–241.

Buck, W.R., Lavier, L.L., Poliakov, A.N.B. (2005). Modes of faulting at mid-ocean ridges. *Nature,* 434, 719–723.

Cannat, M., Sauter, D., Lavier, L.L., Bickert, M., Momoh, E., Leroy, S. (2019). On spreading modes and magma supply at slow and ultraslow mid-ocean ridges. *Earth and Planetary Science Letters,* 519, 223–233.

Castelli, D. and Lombardo, B. (2007). The plagiogranite – Fe-Ti gabbro association of Verne (Monviso metamorphic ophiolite, Western Alps). *Ofioliti,* 32, 1–14.

Chalot-Prat, F. (2005). An undeformed ophiolite in the Alps: Field and geochemical evidence for a link between volcanism and shallow plate tectonic processes. In *Plates, Plumes, and Paradigms,* Foulger, G.R., Natland, J.H., Presnall, D.C., Anderson, D.L. (eds). Geological Society of America Special Publication, Washington DC.

Chalot-Prat, F., Ganne, J., Lombard, A. (2003). No significant element transfer from the oceanic plate to the mantle wedge during subduction and exhumation of the Tethys lithosphere (Western Alps). *Lithos,* 69, 69–103.

Cordey, F. and Bailly, A. (2007). Alpine ocean seafloor spreading and onset of pelagic sedimentation: New radiolarian data from the Chenaillet-Montgenèvre ophiolite (French-Italian Alps). *Geodinamica Acta,* 20, 131–138.

Costa, S. and Caby, R. (2001). Evolution of the Ligurian Tethys in the Western Alps: Sm/Nd and U/Pb geochronology and rare-earth element geochemistry of the Montgenèvre ophiolite (France). *Chemical Geology,* 175, 449–466.

Decandia, F.A. and Elter, P. (1969). Riflessioni sul problema delle ofioliti nell'Appennino settentrionale, nota preliminare. *Atti della Società Toscana di Scienze Naturali,* 75, 1–9.

Decrausaz, T., Müntener, O., Manzotti, P., Lafay, R., Spandler, C. (2020). Fossil oceanic core complexes in the Alps. New constraints from the Tethyan Aiguilles Rouges ophiolite (Val d'Hérens, Western Alps, Switzerland). *Swiss Journal of Geosciences,* 114, 3.

Desmurs, L., Müntener, O., Manatschal, G. (2002). Onset of magmatic accretion within magma-poor passive margins: A case study from the Err-Platta ocean-continent transition, Eastern Switzerland. *Contributions to Mineralogy and Petrology,* 144, 365–382.

Dick, H.J.B. (1989). Abyssal peridotites, very slow-spreading ridges and ocean ridge magmatism. In *Magmatism in the Ocean Basins,* Saunders, A.D. and Norry, M.J. (eds). Geological Society of London Special Publication, Washington DC.

Durand-Delga, M., Peybernès, B., Rossi, P. (1997). Arguments en faveur de la position, au Jurassique, des ophiolites de Balagne (Haute-Corse, France) au voisinage de la marge continentale européenne. *Comptes rendus de l'Académie des Sciences série II*, 325, 973–981.

Dürr, S.B., Ring, U., Frisch, W. (1993). Geochemistry and geodynamic significance of North Penninic ophiolites from the Central Alps. *Schweizerische Mineralogische Petrographische Mitteilungen,* 73, 407–419.

Epin, M.-E., Manatschal, G., Amann, M., Ribes, C., Clausse, A., Guffon, T., Lescanne, M. (2019). Polyphase tectono-magmatic evolution during mantle exhumation in an ultra-distal, magma-poor rift domain: Example of the fossil Platta ophiolite, SE Switzerland. *International Journal of Earth Sciences,* 108, 2443–2467.

Ernst, W.G. (1978). Petrochemical study of lherzolitic rocks from the Western Alps. *Journal of Petrology,* 19, 341–392.

Faure, G. (1986). *Principles of Isotope Geology.* John Wiley and Sons, London, New York.

Ferrara, G., Innocenti, F., Ricci, C.A., Serri, G. (1976). Ocean-floor affinity of basalts from North Appennine Ophiolites: Geochemical evidence. *Chemical Geology,* 17, 101–111.

Florineth, D. and Froitzheim, N. (1994). Transition from continental to oceanic basement in the Tasna nappe (Engadine window, Graubünden, Switzerland): Evidence for Early Cretaceous opening of the Valais ocean. *Schweizerische Mineralogische und Petrographische Mitteilungen,* 74, 437–448.

Frisch, W., Ring, U., Dürr, S., Borchert, S., Biehler, D. (1994). The Arosa zone and Platta nappe ophiolites (Eastern Swiss Alps): Geochemical characteristics and their meaning for the evolution of the Penninic Ocean. *Jahrbuch der Geologischen Bundesanstalt Wien,* 137, 19–33.

Fumagalli, P., Borghini, G., Rampone, E., Poli, S. (2017). Experimental calibration of Forsterite–Anorthite–Ca-Tschermak–Enstatite (FACE) geobarometer for mantle peridotites. *Contributions to Mineralogy and Petrology,* 172, 38.

Gale, A., Dalton, C.A., Langmuir, C.H., Su, Y., Schilling, J.-G. (2013). The mean composition of ocean ridge basalts. *Geochemistry, Geophysics, Geosystems,* 14, 489–518.

Galli, A., Grassi, D., Sartori, G., Gianola, O., Burg, J.P., Schmidt, M.W. (2019). Jurassic carbonatite and alkaline magmatism in the Ivrea Zone (European Alps) related to the breakup of Pangea. *Geology,* 47, 199–202.

Gleißner, P., Franz, G., Frei, D. (2021). Contemporaneous opening of the Alpine Tethys in the Eastern and Western Alps: Constraints from a Late Jurassic gabbro intrusion age in the Glockner nappe, Tauern Window, Austria. *International Journal of Earth Sciences,* 110, 2705–2724.

Handy, M.R., Schmid, S.M., Bousquet, R., Kissling, E., Bernoulli, D. (2010). Reconciling plate-tectonic reconstructions of Alpine Tethys with the geological–geophysical record of spreading and subduction in the Alps. *Earth-Science Reviews,* 102, 121–158.

Harvey, J., Gannoun, A., Burton, K.W., Rogers, N.W., Alard, O., Parkinson, I.J. (2006). Ancient melt extraction from the oceanic upper mantle revealed by Re-Os isotopes in abyssal peridotites from the Mid-Atlantic Ridge. *Earth and Planetary Science Letters,* 244, 606–621.

Hellebrand, E. and Snow, J.E. (2003). Deep melting and sodic metasomatism underneath the highly oblique-spreading Lena Trough (Arctic Ocean). *Earth and Planetary Science Letters,* 216, 283–299.

Hellebrand, E., Snow, J.E., Dick, H.J.B., Hofmann, A.W. (2001). Coupled major and trace elements as indicators of the extent of melting in mid-ocean-ridge peridotites. *Nature,* 410, 677–681.

Higgie, K. and Tommasi, A. (2014). Deformation in a partially molten mantle: Constraints from plagioclase lherzolites from Lanzo, western Alps. *Tectonophysics,* 615, 167–181.

Jollands, M. and Müntener, O. (2019). Testing orthopyroxene diffusion chronometry on rocks from the Lanzo peridotite. *Journal of Geophysical Research-Solid Earth,* 124, 7822–7841.

Kaczmarek, M.-A. and Müntener, O. (2008). Juxtaposition of melt impregnation and high-temperature shear zones in the upper mantle; Field and petrological constraints from the Lanzo Peridotite (Northern Italy). *Journal of Petrology,* 49, 2187–2220.

Kaczmarek, M.-A. and Müntener, O. (2010). The variability of peridotite composition across a mantle shear zone (Lanzo massif, Italy): Interplay of melt focusing and deformation. *Contributions to Mineralogy and Petrology,* 160, 663–679.

Kaczmarek, M.-A., Müntener, O., Rubatto, D. (2008). Trace element chemistry and U-Pb dating of zircons oceanic gabbros and their relationship with whole rock composition (Lanzo, Italian Alps). *Contributions to Mineralogy and Petrology,* 155, 295–312.

Kelley, D.S., Karson, J.A., Blackman, D.K., Früh-Green, G.L., Butterfield, D.A., Lilley, M.D., Olson, E.J., Schrenk, M.O., Roe, K.K., Lebon, G.T., Rivizzigno, P., Party, A.-S. (2001). An off-axis hydrothermal vent field near the Mid-Atlantic Ridge at 30°N. *Nature,* 412, 145–149.

Kramer, J., Abart, R., Müntener, O., Schmid, S.M., Stern, W.B. (2003). Geochemistry of metabasalts from ophiolitic and adjacent distal continental margin units: Evidence from the Monte Rosa region (Swiss and Italian Alps). *Schweizerische Mineralogische Petrographische Mitteilungen,* 83, 1–23.

Lagabrielle, Y. and Cannat, M. (1990). Alpine Jurassic ophiolites resemble the modern central Atlantic Basement. *Geology,* 18, 319–322.

Lagabrielle, Y., Polino, R., Auzende, J.M., Blanchet, R., Caby, R., Fudral, S., Lemoine, M., Mevel, C., Ohnenstetter, M., Robert, D., Tricart, P. (1984). Les témoins d'une tectonique intra-océanique dans le domaine téthysien : analyse des rapports entre les ophiolites et leur couverture métasédimentaire dans la zone piémontaise des Alpes franco-italiennes. *Ofioliti,* 9, 67–88.

Lagabrielle, Y., Vitale Brovarone, A., Ildefonse, B. (2015). Fossil oceanic core complexes recognized in the blueschist metaophiolites of Western Alps and Corsica. *Earth-Science Reviews,* 141, 1–26.

Le Breton, E., Brune, S., Ustazewski, K., Zahirovic, S., Seton, M., Müller, R.D. (2021). Kinematics and extent of the Piemont-Liguria basin – Implications for subduction processes in the Alps. *Solid Earth,* 12, 885–913.

Lemoine, M., Tricart, P., Boillot, G. (1987). Ultramafic and gabbroic ocean floor of the Ligurian Tethys (Alps, Corsica, Apennines). In search of a genetic model. *Geology,* 15, 622–625.

Lewis, A.D. and Smewing, J.D. (1980). The Montgenevre ophiolite (Hautes Alpes, France): Metamorphism and trace element geochemistry of the volcanic sequence. *Chemical Geology,* 28, 291–306.

Li, X.H., Faure, M., Lin, W., Manatschal, G. (2013). New isotopic constraints on age and magma genesis of an embryonic oceanic crust: The Chenaillet Ophiolite in the Western Alps. *Lithos,* 160, 283–291.

Li, X.-H., Faure, M., Rossi, P., Lin, W., Lahondère, D. (2015). Age of Alpine Corsica ophiolites revisited: Insights from in situ zircon U–Pb age and O–Hf isotopes. *Lithos,* 220, 179–190.

Lissenberg, J., Macleod, C.J., Bennett, E.N. (2019). Consequences of a crystal mush-dominated magma plumbing system: A mid-ocean ridge perspective. *Philosophical Transactions of the Royal Society of London, Series A,* 377, 20180014.

Liu, C.-Z., Snow, J.E., Hellebrand, E., Brügmann, G., Von Der Handt, A., Büchl, A., Hofmann, A.W. (2008). Ancient, highly heterogeneous mantle beneath Gakkel Ridge, Arctic Ocean. *Nature,* 452, 311–316.

Locatelli, M., Verlaguet, A., Agard, P., Pettke, T., Federico, L. (2019). Fluid pulses during stepwise brecciation at intermediate subduction depths (Monviso Eclogites, W. Alps): First internally then externally sourced. *Geochemistry, Geophysics, Geosystems,* 20. doi: 10.1029/2019GC008549.

Lombardo, B., Rubatto, D., Castelli, D. (2002). Ion microprobe U-Pb dating of zircon from a Monviso metaplagiogranite: Implications for the evolution of the Piedmont-Liguria Tethys in the Western Alps. *Ofioliti,* 27, 109–117.

Lundstrom, C. (2003). Rapid diffusive infiltration of sodium into partially molten peridotite *Nature,* 403, 527–530.

Manatschal, G. and Müntener, O. (2009). A type sequence across an ancient magma-poor ocean-continent transition: The example of the Western Alpine Tethys ophiolites. *Tectonophysics,* 473, 4–19.

Manatschal, G. and Nievergelt, P. (1997). A continent-ocean transition recorded in the Err and Platta nappes (Eastern Switzerland). *Eclogae Geologicae Helvetiae,* 90, 3–27.

Manatschal, G., Sauter, D., Karpoff, A.M., Masini, E., Mohn, G., Lagabrielle, Y. (2011). The Chenaillet Ophiolite in the French/Italian Alps: An ancient analogue for an Oceanic Core Complex? *Lithos,* 124, 169–184.

Marroni, M., Molli, G., Montanini, A., Tribuzio, R. (1998). The association of continental crust rocks with ophiolites in the Northern Apennines (Italy): Implications for the continent-ocean transition in the Western Tethys. *Tectonophysics,* 292, 43–66.

McCarthy, A. and Müntener, O. (2015). Ancient depletion and mantle heterogeneity: Revisiting the Permian-Jurassic paradox of Alpine peridotites. *Geology,* 43, 255–258.

McCarthy, A. and Müntener, O. (2019). Evidence for ancient fractional melting, cryptic re-fertilization and rapid exhumation of Tethyan mantle (Civrari Ophiolite, NW Italy). *Contributions to Mineralogy and Petrology,* 174, 69.

McCarthy, A., Chelle-Michou, C., Müntener, O., Arculus, R.J., Blundy, J.D. (2018). Subduction initiation without magmatism: The case of the missing Alpine magmatic arc. *Geology,* 46, 1059–1062.

McCarthy, A., Tugend, J., Mohn, G., Candioti, L., Chelle-Michou, C., Arculus, R., Schmalholz, S.M., Müntener, O. (2020a). A case for Ampferer-type subduction and consequences for the Alps and the Pyrenees. *American Journal of Science,* 320, 313–372.

McCarthy, A., Falloon, T.J., Sauermilch, I., Whittaker, J.M., Niida, K., Green, D.H. (2020b). Revisiting the Australian-Antarctic Ocean-Continent Transition Zone using petrological and geophysical characterization of exhumed subcontinental mantle. *Geochemistry, Geophysics, Geosystems,* 21, e2020GC009040.

McCarthy, A., Tugend, J., Mohn, G. (2021). Formation of the Alpine orogen by amagmatic convergence and assembly of previously rifted lithosphere. *Elements,* 17, 29–34.

McDonough, W.F. and Sun, S.-S. (1995). The composition of the Earth. *Chemical Geology,* 120, 223–253.

Melcher, F., Meisel, T., Puhl, J., Koller, F. (2002). Petrogenesis and geotectonic setting of ultramafic rocks in the Eastern Alps: Constraints from geochemistry. *Lithos,* 65, 69–112.

Messiga, B., Kienast, J.R., Rebay, G., Riccardi, M.P., Tribuzio, R. (1999). Cr-rich magnesiochloritoid eclogites from the Monviso ophiolites (Western Alps, Italy). *Journal of Metamorphic Geology,* 17, 287–299.

Mezger, K., Essene, E.J. Halliday, A.N. (1992). Closure temperatures of the Sm–Nd system in metamorphic garnets. *Earth and Planetary Science Letters,* 113(3), 397–409. doi: 10.1016/0012-821X(92)90141-H.

Montanini, A., Tribuzio, R., Vernia, L. (2008). Petrogenesis of basalts and gabbros from an ancient continent-ocean transition (External Liguride ophiolites, Northern Italy). *Lithos,* 101, 453–479.

Müntener, O. and Manatschal, G. (2006). High degrees of melting recorded by spinel harzburgites on the Newfoundland margin: Role of inheritance and consequences for the evolution of the southern North Atlantic. *Earth and Planetary Science Letters,* 252, 437–452.

Müntener, O. and Piccardo, G.B. (2003). Melt migration in ophiolitic peridotites: The message from Alpine-Apennine peridotites and implications for embryonic ocean basins. In *Ophiolites in Earth History,* Dilek, Y., Robinson, P.T. (eds). The Geological Society, London.

Müntener, O., Hermann, J., Trommsdorff, V. (2000). Cooling history and exhumation of lower crustal granulites and upper mantle (Malenco, Eastern Central Alps). *Journal of Petrology,* 41, 175–200.

Müntener, O., Pettke, T., Desmurs, L., Meier, M., Schaltegger, U. (2004). Refertilization of mantle peridotite in embryonic ocean basins: Trace element and Nd-isotopic evidence and implications for crust-mantle relationships. *Earth and Planetary Science Letters,* 221, 293–308.

Müntener, O., Manatschal, G., Desmurs, L., Pettke, T. (2010). Plagioclase peridotites in ocean-continent transitions: Re-fertilized mantle domains generated by melt stagnation in the shallow mantle lithosphere. *Journal of Petrology,* 51, 255–294.

Nicolas, A. (1989). *Structures of Ophiolites and Dynamics of Oceanic Lithosphere.* Kluwer Academic Publishers, Dordrecht.

Ohnenstetter, M., Ohnenstetter, D., Vidal, P., Cornichet, J., Hermitte, D., Mace, J. (1981). Crystallization and age of zircon from Corsican ophiolitic albitites: Consequences for oceanic expansion in Jurassic times. *Earth and Planetary Science Letters,* 54, 397–408.

Ottonello, G., Joron, J.L., Piccardo, G.B. (1984). Rare earth and 3d transition element geochemistry of peridotitic rocks II: Ligurian peridotites and associated basalts. *Journal of Petrology,* 25, 373–393.

Padoa, E. and Durand-Delga, M. (2001). L'unité ophiolitique du Rio Magno en Corse alpine, élément des Ligurides de l'Apennin septentrional. *Comptes rendus de l'Académie des Sciences series: Earth and Planetary Science,* 333, 285–293.

Paquet, M., Cannat, M., Brunelli, D., Hamelin, C., Humler, E. (2016). Effect of melt/mantle interactions on MORB chemistry at the easternmost Southwest Indian Ridge (61–67°E). *Geochemistry Geophysics Geosystems,* 17, 4605–4650.

Pearce, J.A. (1982). Trace element characteristics of lavas from destructive plate boundaries. In *Andesites,* Thorpe, R.S. (ed.). John Wiley & Sons, New York.

Pelletier, L. and Müntener, O. (2006). High-pressure metamorphism of the Lanzo peridotite and its oceanic cover, and some consequences for the Sesia-Lanzo zone (Northwestern Italian Alps). *Lithos,* 90, 111–130.

Picazo, S., Müntener, O., Manatschal, G., Bauville, A., Karner, G.D., Johnson, C.A. (2016). Mapping the nature of mantle domains in Western and Central Europe based on clinopyroxene and spinel chemistry: Evidence for mantle modification during an extensional cycle. *Lithos,* 266, 233–263.

Piccardo, G.B., Zanetti, A., Müntener, O. (2007). Melt/peridotite interaction in the Lanzo South peridotite: Field, textural and geochemical evidence. *Lithos,* 94, 181–209.

Pognante, U., Rösli, U., Toscani, L. (1985). Petrology of ultramafic and mafic rocks from the Lanzo peridotite body (Western Alps). *Lithos,* 18, 201–214.

Puschnig, A.R. (2000). The oceanic Forno unit (Rhetic Alps): Field relations, geochemistry and paleogeographic setting. *Eclogae Geologicae Helvetiae,* 93, 103–124.

Rampone, E. and Hofmann, A.W. (2012). A global overview of isotopic heterogeneities in the oceanic mantle. *Lithos,* 148, 247–261.

Rampone, E. and Sanfilippo, A. (2021). The heterogeneous Tethyan oceanic lithosphere of the Alpine Ophiolite. *Elements,* 17, 23–28.

Rampone, E., Piccardo, G.B., Vannucci, R., Bottazzi, P., Ottolini, L. (1993). Subsolidus reactions monitored by trace element partitioning: The spinel- to plagioclase-facies transition in mantle peridotites. *Contributions to Mineralogy and Petrology,* 115, 1–17.

Rampone, E., Hofmann, A.W., Piccardo, G.B., Vannucci, R., Bottazzi, P., Ottolini, L. (1995). Petrology, mineral and isotope geochemistry of the External Liguride Peridotites (Northern Apennines, Italy). *Journal of Petrology,* 36, 81–105.

Rampone, E., Hofmann, A.W., Piccardo, G.B., Vannucci, R., Bottazzi, P., Ottolini, L. (1996). Trace element and isotope geochemistry of depleted peridotites from an N-Morb type ophiolite (Internal Liguride, N-Italy). *Contributions to Mineralogy and Petrology,* 123, 61–76.

Rampone, E., Piccardo, G.B., Vannucci, R., Bottazzi, P. (1997). Chemistry and origin of trapped melts in ophiolitic peridotites. *Geochimica et Cosmochimica Acta,* 61, 4557–4569.

Rampone, E., Hofmann, A.W., Raczek, I. (1998). Isotopic contrasts within the Internal Liguride ophiolite (N. Italy): The lack of a genetic mantle-crust link. *Earth and Planetary Science Letters,* 163, 175–189.

Rampone, E., Romairone, A., Hofmann, A.W. (2004). Contrasting bulk and mineral chemistry in depleted mantle peridotites: Evidence for reactive porous flow. *Earth and Planetary Science Letters,* 218, 491–506.

Rampone, E., Piccardo, G.B., Hofmann, A.W. (2008). Multi-stage melt-rock interaction in the Mt. Maggiore (Corsica, France), ophiolitic peridotites: Microstructural and geochemical evidence. *Contributions to Mineralogy and Petrology,* 166, 453–475.

Rampone, E., Hofmann, A.W., Raczek, I. (2009). Isotopic equilibrium between mantle peridotite and melt: Evidence from the Corsica ophiolite. *Earth and Planetary Science Letters,* 288, 601–610.

Rampone, E., Borghini, G., Romairone, A., Abouchami, W., Class, C., Goldstein, S.L. (2014). Sm-Nd geochronology of the Erro-Tobbio gabbros (Ligurian Alps, Italy): Insights into the evolution of the Alpine Tethys. *Lithos,* 205, 236–246.

Renna, M.R., Tribuzio, R., Sanfilippo, A., Thirlwall, M. (2018). Role of melting process and melt–rock reaction in the formation of Jurassic MORB-type basalts (Alpine ophiolites). *Contributions to Mineralogy and Petrology,* 173, 31.

Ribes, C., Petri, B., Ghienne, J.-F., Manatschal, G., Galster, F., Karner, G.D., Figueredo, P.H., Johnson, C.A., Karpoff, A.M. (2020). Tectono-sedimentary evolution of a fossil ocean-continent transition: Tasna nappe, central Alps (SE Switzerland). *Geological Society of America Bulletin,* 132, 1427–1446.

Rosenbaum, G., Lister, G.S., Duboz, C. (2002). Relative motions of Africa, Iberia and Europe during Alpine orogeny. *Tectonophysics,* 359, 117–129.

Rubatto, D., Gebauer, D., Fanning, M. (1998). Jurassic formation and Eocene subduction of the Zermatt – Saas-Fee ophiolites: Implications for the geodynamic evolution of the Central and Western Alps. *Contributions to Mineralogy and Petrology,* 132, 269–287.

Rubatto, D., Müntener, O., Barnhoorn, A., Gregory, C.J. (2008). Dissolution-precipitation of zircon at low-temperature, high-pressure conditions (Lanzo Massif, Italy). *American Mineralogist,* 93, 1519–1529.

Saccani, E. and Principi, G. (2016). Petrological and tectono-magmatic significance of ophiolitic basalts from the Elba Island within the Alpine Corsica-Northern Apennine system. *Mineralogy and Petrology,* 110, 713–730.

Saccani, E., Principi, G., Garfagnoli, F., Menna, F. (2008). Corsica ophiolites: Geochemistry and petrogenesis of basaltic and metabasaltic rocks. *Ofioliti,* 33, 187–207.

Sanfilippo, A. and Tribuzio, R. (2011). Melt transport and deformation history in a nonvolcanic ophiolitic section, northern Apennines, Italy: Implications for crustal accretion at slow-spreading settings. *Geochemistry, Geophysics, Geosystems,* 12, 1–34.

Sanfilippo, A., Salters, V.J.M., Tribuzio, R., Zanetti, A. (2019). Role of ancient, ultra-depleted mantle in Mid-Ocean-Ridge magmatism. *Earth and Planetary Science Letters,* 511, 89–98.

Sauter, D., Cannat, M., Rouméjon, S., Andreani, M., Birot, D., Bronner, A., Brunelli, D., Carlut, J., Delacour, A., Guyader, V. et al. (2013). Continuous exhumation of mantle-derived rocks at the Southwest Indian Ridge for 11 million years. *Nature Geoscience,* 6, 314–320.

Schaltegger, U., Desmurs, L., Manatschal, G., Müntener, O., Meier, M., Frank, M., Bernoulli, D. (2002). The transition from rifting to seafloor spreading within a magma-poor rifted margin: Field and isotopic constraints. *Terra Nova,* 14, 156–162.

Schaltegger, U., Ulianov, A., Müntener, O., Ovtcharova, M., Vonlanthen, P., Vennemann, T., Antognini, M., Girlanda, F. (2015). Megacrystic zircon with planar fractures in miaskite-type nepheline pegmatites formed at high pressures in the lower crust (Ivrea Zone, southern Alps, Switzerland). *American Mineralogist,* 100, 84–94.

Schuster, R. and Stüwe, K. (2008). Permian metamorphic event in the Alps. *Geology,* 36, 603–606.

Shaw, A.W., Behn, M.D., Humphries, S.E., Sohn, R.A., Gregg, P.M. (2010). Deep pooling of low degree melts and volatile fluxes at the 85 degrees E segment of the Gakkel Ridge: Evidence from olivine-hosted melt inclusions and glasses. *Earth and Planetary Science Letters,* 289, 311–322.

Stähle, V., Frenzel, G., Kober, B., Michard, A., Puchelt, H., Schneider, W. (1990). Zircon syenite pegmatites in the Finero peridotite (Ivrea zone): Evidence for a syenite from a mantle source. *Earth and Planetary Science Letters,* 101, 196–205.

Steinmann, G. (1927). Die ophiolitischen Zonen in den mediterranen Kettengebirgen. *14th International Geological Congress,* 637–668.

Steinmann, M. (1994). Die nordpenninischen Bündnerschiefer der Zentralalpen Graubündens: Tektonik, Stratigraphie und Beckenentwicklung. PhD Thesis, ETH Zürich.

Steinmann, M. and Stille, P. (1999). Geochemical evidence for the nature of the crust beneath the eastern North Penninic basin of the Mesozoic Tethys ocean. *Geologische Rundschau,* 87, 633–643.

Stracke, A., Snow, J.E., Hellebrand, E., Von Der Handt, A., Bourdon, B., Birbaum, K., Guenther, D. (2011). Abyssal peridotite Hf isotopes identify extreme mantle depletion. *Earth and Planetary Science Letters,* 308, 359–368.

Stucki, A., Rubatto, D., Trommsdorff, V. (2003). Mesozoic ophiolite relics in the Southern Steep Belt of the Central Alps. *Schweizerische Mineralogische und Petrographische Mitteilungen*, 83, 285–299.

Sun, S.-S. and McDonough, W.F. (1989). Chemical and isotopic systematics of oceanic basalts: Implications for mantle composition and processes. In *Magmatism in the Oceanic Basins*, Saunders, A.D. and Norry, M.J. (eds). The Geological Society, London.

Tribuzio, R., Thirlwall, M.F., Vannucci, R. (2004). Origin of the gabbro-peridotite association from the Northern Apennine ophiolites (Italy). *Journal of Petrology*, 45, 1109–1124.

Tribuzio, R., Garzetti, F., Corfu, F., Tiepolo, M., Renna, M.R. (2016). U–Pb zircon geochronology of the Ligurian ophiolites (Northern Apennine, Italy): Implications for continental breakup to slow seafloor spreading. *Tectonophysics*, 666, 220–243.

Tricart, P. and Lemoine, M. (1986). From faulted blocks to megamullions and megaboudins: Tethyan heritage in the structure of the Western Alps. *Tectonics*, 5, 95–118.

Tricart, P. and Lemoine, M. (1991). The Queyras ophiolite west of Monte Viso (Western Alps): Indicator of a peculiar ocean floor in the Mesozoic Tethys. *Journal of Geodynamics*, 13, 163–181.

Tucholke, B.E., Behn, M.D., Buck, W.R., Lin, J. (2008). Role of melt supply in oceanic detachment faulting and formation of megamullions. *Geology*, 36, 455–458.

Urann, B.M., Dick, H.J.B., Parnell-Turner, R., Casey, J.F. (2020). Recycled arc mantle recovered from the Mid-Atlantic Ridge. *Nature Communications*, 11, 3887.

Vannucci, R., Rampone, E., Piccardo, G.B., Ottolini, L., Bottazzi, P. (1993). Ophiolitic magmatism in the Ligurian Tethys: An ion microprobe study of basaltic clinopyroxenes. *Contributions to Mineralogy and Petrology*, 115, 123–137.

Venturelli, G., Capedri, S., Thorpe, R.S., Potts, P.J. (1979). Rare-earth and other element distribution in some ophiolitic metabasalts of Corsica, Western Mediterranean. *Chemical Geology*, 24, 393–353.

Venturelli, G., Thorpe, R.S., Potts, P.J. (1981). Rare earth and trace element characteristics of ophiolitic metabasalts from the Alpine-Apennine belt. *Earth and Planetary Science Letters*, 53, 109–123.

Viereck, L. (2006). Spatial distribution of textural and compositional variations of MORB in the Arosa-Zone. *IV Swiss Geoscience Meeting*. Swiss Academy of National Science, Bern.

Villiger, S., Müntener, O., Ulmer, P. (2007). Crystallisation of MORB in the mantle: Implications from major element variations of glasses from equilibrium and fractional crystallization experiments. *Journal of Geophysical Research*, 112, B01202.

Vissers, R.L.M., Van Hinsbergen, D.J.J., Meijer, P.T., Piccardo, G.B. (2013). Kinematics of Jurassic ultra-slow spreading in the Piemonte Ligurian ocean. *Earth and Planetary Science Letters*, 380, 138–150.

Wanless, V.D., Behn, M.D., Shaw, A.M., Plank, T. (2014). Variations in melting dynamics and mantle compositions along the Eastern Volcanic Zone of the Gakkel Ridge: Insights from olivine-hosted melt inclusions. *Contributions to Mineralogy and Petrology,* 167, 1005.

Warren, J.M. (2016). Global variations in abyssal peridotite compositions. *Lithos,* 248, 193–219.

Warren, J.M., Shimizu, N., Sakaguchi, C., Dick, H.J.B., Nakamura, E. (2009). An assessment of upper mantle heterogeneity based on abyssal peridotite isotopic compositions. *Journal of Geophysical Research,* 114, B12203.

Weissert, H. and Bernoulli, D. (1985). A transform margin in the Mesozoic Tethys: Evidence from the Swiss Alps. *Geologische Rundschau,* 74, 665–679.

4

Oceanic Subduction in the Alps: Lessons from the Rock Record

Philippe AGARD
Sorbonne Université, CNRS-INSU,
Institut des Sciences de la Terre Paris,
ISTeP UMR 7193, France

4.1. Introduction: how important is "oceanic" subduction in the Alps?

Subduction of oceanic lithosphere is a key ingredient of Solid Earth dynamics (Forsyth and Uyeda 1975; Stern 2002; Coltice et al. 2019). Documenting the chemical and mechanical processes operating during "oceanic subduction" is thus essential for elucidating plate tectonics as well as for risk assessment. Active subduction dynamics, however, essentially happen at depths beyond reach for sampling and/or observation, at least with sufficient spatial resolution (e.g. Calahoranno et al. 2008). A complementary approach is to use samples exhumed from fossil subduction zones (e.g. Jolivet et al. 2003; Bebout 2007), in fact, most commonly sourced from the downgoing plate (i.e. the "slab", Agard et al. 2018).

The Alps, with its abundant and well-preserved subducted remnants (blueschists and eclogites), seem an ideal location to study deep subduction processes and reconstruct the initial subduction geometry. While past oceanic subduction was recognized in the Alps shortly after the advent of plate tectonics (Ernst 1971; Dal Piaz 1971; Dal Piaz et al. 1972; Michard 1977), the concept of subduction itself roots deeply into Alpine studies (Ampferer and Hammer 1911; Amstutz 1951;

Geodynamics of the Alps 2,
coordinated by Claudio L. ROSENBERG and Nicolas BELLAHSEN.

White 1970). Two centuries of detailed structural investigations have also shown that major units and paleogeographic domains have not been tremendously offset with respect to former convergence directions (e.g. Schmid et al. 2017). And the Alpine belt is small enough for its entire structure to be envisioned in 3D and tightly connected to geophysical imaging (e.g. Schmid and Kissling 2000; Kästle et al. 2019).

The quest for subducted fragments, i.e. rocks deeply buried yet maintained in relatively cold, high-pressure–low-temperature conditions (HP-LT), reached new heights during the 1980s (e.g. Goffé and Chopin 1986; Pognante and Kienast 1987) and culminated with the discovery of coesite and deep continental subduction (Chopin 1984). The tectono-metamorphic map of the Alps (Handy and Oberhänsli 2004; Oberhänsli and Goffé 2004), together with some later additions (Bousquet et al. 2008; Berger and Bousquet 2008; Agard et al. 2009; Beltrando et al. 2010; Bousquet et al. 2012), emphasized how widespread these fragments are. They once belonged to a slow-spreading ocean (Lemoine et al. 1970, 1984, 1987; Lagabrielle and Cannat 1990) whose margins are relatively well-preserved (Manatschal and Müntener 2009; Mohn et al. 2011). Paleogeographic reconstructions (e.g. Stampfli and Marchant 1997; Vissers et al. 2013) and studies in search of former slabs, tear faults or slab breakoff events at depth (Lippitsch et al. 2003; Handy et al. 2010; Kästle et al. 2019) have also placed important constraints on the disappearance of the Alpine ocean.

Yet, we may wonder: is the Alpine example representative of subduction dynamics worldwide? After all, the vanished Alpine ocean was a relatively small ~500 km wide domain characterized by Atlantic type, slow-spreading oceanic lithosphere, with trench sediments quite different from Peri-Pacific subduction zones today, and furthermore left no volcanic arc behind. In order to answer this question, the present contribution provides an overview of the (mostly petrological) subduction archive of the Alpine ocean. I then discuss what the Alpine record tells us about subduction and/or exhumation dynamics, and evaluate the major differences along the belt and the role of tectonic inheritance. The short-lived period of continental subduction that followed oceanic subduction is only mentioned briefly here (see Manzotti and Ballèvre for details), to the extent it affected the fate of subducted oceanic fragments.

4.2. Petrology and geodynamics

Oceanic fragments bearing diagnostic subducted-related HP-LT mineral assemblages are found within the core of the entire Alpine orogen, from Genova to Vienna, from the Ligurian Alps to the Rechnitz window (Figure 4.1(a)). They are found in the Liguro-Piemont and Valais domains (Upper and Lower Penninic domains, respectively), two former oceanic domains merging to the east but separated in the Western and Central Alps by a continental fragment (Briançonnais, or "Middle-Penninic"). A schematic cross-section (Figure 4.1(b)) illustrates the structural organization of the major paleogeographic domains recognized in the belt, from the Austro-Alpine units on top to the European units below. The detailed architecture of the Alps can be found in many other publications as this is not recalled here (see cross-sections of Figures 4.3 and 4.4; e.g. Debelmas and Lemoine 1970; Froitzheim et al. 1996; Stampfli et al. 1998; Schmid et al. 2004, 2017; Nagel 2008; Manzotti et al. 2014, and references therein).

4.2.1. *Remnants of subducted oceanic lithosphere*

Alpine subducted oceanic fragments are dominated by "Mesozoic calc-schists, greenstones and serpentinized peridotites" (Dal Piaz and Ernst 1978) showing distinctive metamorphic mineralogy upon subduction: e.g. lawsonite, sodic clinopyroxene ("omphacite"), blue amphibole (glaucophane and "crossite"), garnet, in metamorphic rocks; ferro-magnesio-carpholite, lawsonite, chloritoid, garnet, talc, coesite in metasedimentary rocks and antigorite, titano-clinohumite, titano-chondrodite in metaperidotites.

In the case of the Western Alps, it was soon realized that the intensity of HP-LT metamorphism increases towards the southeast (e.g. Kienast and Velde 1970; Dal Piaz et al. 1972; Goffé et al. 1973). There, the distribution of metamorphosed oceanic units, from blueschist to transitional blueschist-eclogite to eclogite facies (Figures 4.1(c) and (d)) advocates for southeast-directed subduction of oceanic lithosphere below the Adriatic plate. The same fossil subduction zone is exposed in the Central and Eastern Alps, indicating subduction below the Austroalpine and South Alpine domains. It is therefore relatively easy to bridge the gap, geometrically, between the present-day architecture and the subduction history (compare Figures 4.1(b) and (e)). Along-strike, the preservation of HP-LT metamorphism varies with the amount of Tertiary collision: it is generally less where the collision was strong (Figure 4.1(f); Schmid et al. 2004; Bousquet et al. 2008).

A distinct record of late Cretaceous intracontinental subduction (~105–90 Ma; e.g. Thöni (1999)), unrelated to the closing of the Alpine ocean, is found in the Eastern Alps. Those volumetrically small HP-LT remnants, in the Saualpe, Koralpe and Porhoje (Figure 4.1(a)), are generally thought to originate from the closure of the westernmost sector of the Neotethys, the Vardar/Meliata/Sava ocean (Schmid et al. 2020 and references therein).

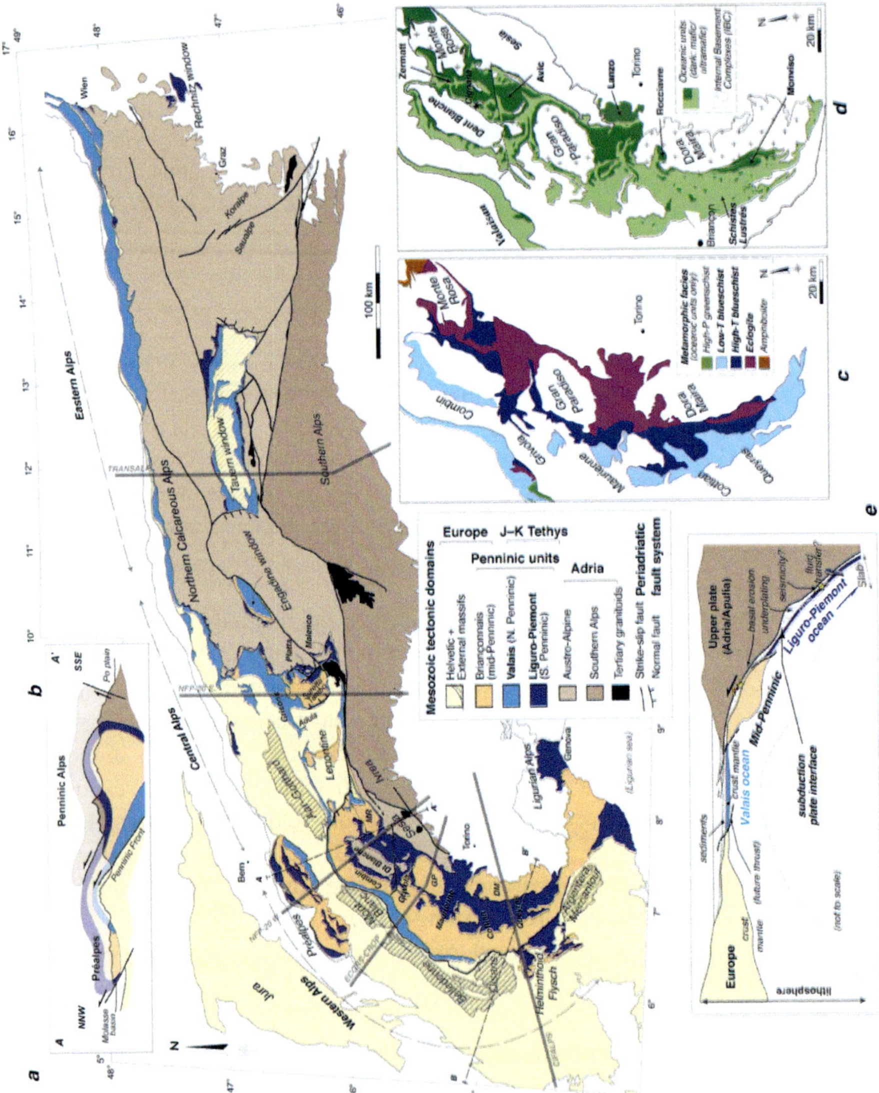
a
b
c
d
e
Mesozoic tectonic domains
Europe
J–K Tethys
Penninic units
Adria
Helvetic + External massifs
Briançonnais (mid-Penninic)
Valais (N. Penninic)
Liguro-Piemont (S. Penninic)
Austro-Alpine
Southern Alps
Tertiary granitoids
Strike-slip fault
Normal fault
Periadriatic fault system
Western Alps
Central Alps
Eastern Alps
Northern Calcareous Alps
Southern Alps
Tauern window
Engadine window
Rechnitz window
Jura
Préalpes
Penninic Alps
Penninic Front
Wien
Graz
Torino
Genova
Bern
100 km
Metamorphic facies (oceanic units only)
High-P greenschist
Low-T blueschist
High-T blueschist
Eclogite
Amphibolite
Monte Rosa
Gran Paradiso
Dora Maira
Combin
Zermatt
Dent Blanche
Sesia
Lanzo
Monviso
Schistes Lustrés
Briançon
Valaisan
Oceanic units (dark: mafic/ultramafic)
Internal Basement Complexes (IBC)
20 km
Europe
Upper plate (Adria/Apulia)
Valais ocean
Liguro-Piemont ocean
Mid-Penninic
subduction plate interface
lithosphere
(not to scale)

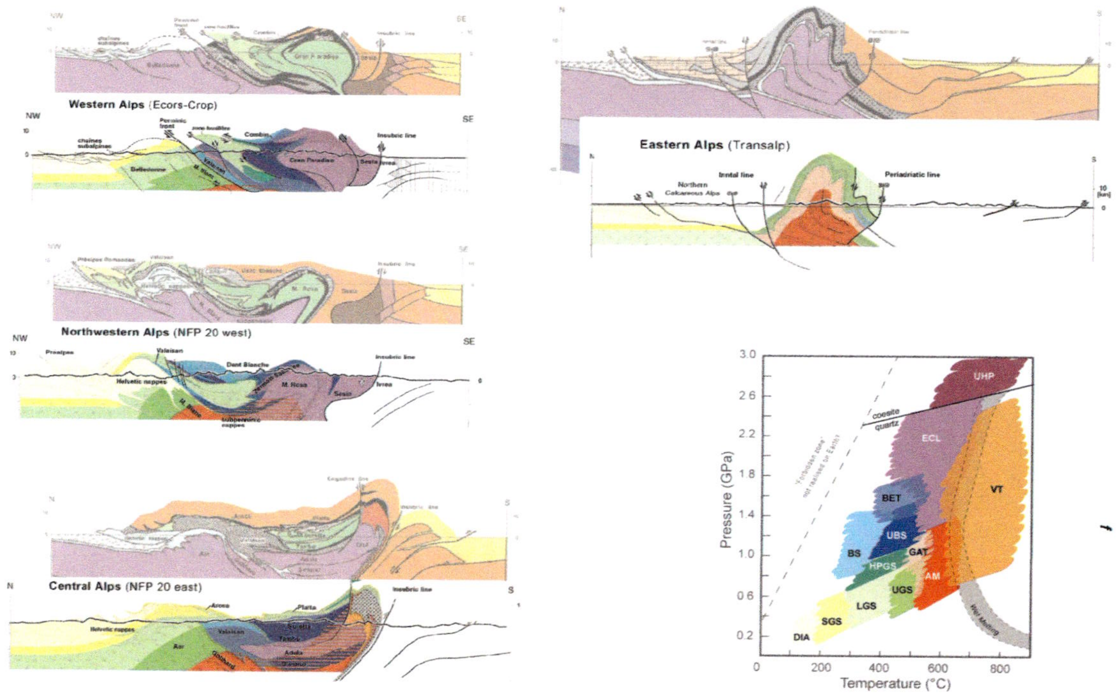

Figure 4.1. *a) General structural map of the Alpine belt (after Handy and Oberhänsli 2004; Schmid et al. 2004). The two branches of the Alpine 'ocean', the Valais and Liguro-Piemont Penninic domains, are outlined in light and dark blue respectively. b) Schematic cross-section along A-A' illustrating the structural organization of the major paleogeographic domains recognized in the belt, from the Austro-Alpine units on top to the European units below (after Agard and Lemoine 2005)*

COMMENTS ON FIGURE 4.1.– *The limits of the Western, Central and Eastern Alps corresponds adopted in the text are shown here. The location of the main geophysical profiles and cross-sections shown in later figures are indicated (B–B': Figure 4.3(b); D–D': Figure 4.4(b)). Abbreviations: AM: Argentera-Mercantour; AG: Aar-Gotthard; BO: Belledonne-Oisans; DM: Dora Maira; GP: Gran Paradiso; MB: Mont Blanc; MR: Monte Rosa. c) Distribution of metamorphic facies (as in Oberhänsli and Goffé 2004) within the ocean-derived units across the same area. d) Distribution of lithologies across the oceanic units separating the sedimentary-dominated ("S") domains from those dominated by oceanic crust and serpentinized peridotites (i.e. mafic and ultramafic: "MUM"). See section 4.3.1 and 4.3.2. Note the marked W-E zoning in c) and d). For the sake of convenience, the main sectors of the Western Alps are sub-divided into the Combin, Grivola, Maurienne, Cottian and Queyras c). The main mafic- and ultramafic-dominated MUM massifs referred to in the text are shown in d). e) Section showing the restored geodynamic subduction setting of the Alps, with emphasis on subduction interface processes. The respective location of the Valais and Liguro-Piemont oceans can be readily deduced from the structural relationships between the European, Penninic and Austro/Southalpine domains (to be compared with Figure 4.1(b)). f) Comparison of metamorphic facies and collisional "maturity": HP-LT metamorphism is much less preserved where Tertiary collision was strong (e.g. western parts of the Central Alps; after Schmid et al. 2004 and Bousquet et al. 2008).*

4.2.2. Geodynamic setting and boundary conditions for oceanic subduction

The overall time frame of the Alpine ocean, from birth to death, is recalled in Figure 4.2(a). Kinematic reconstructions (Dercourt et al. 1986; Dewey et al. 1989; Stampfli et al. 1998; Rosenbaum et al. 2002; Handy et al. 2010; Vissers et al. 2013; van Hinsbergen et al. 2020) indicate that the Alpine Ocean was ~400–700 km wide and comprised, in its southern part (Figure 4.2(b)), the ~300–400 km wide Liguro-Piemont Ocean, the ~100–150 km wide continental fragment of the Briançonnais Domain and, in the southwest, the 50–200 km wide Valais Ocean.

Slow opening of the Liguro–Piemont Ocean (<2 cm/year) occurred between 170 and 120 Ma (Figure 4.2(b); Vissers et al. 2013), with evidence for considerable continental thinning (Figure 4.2(c); e.g. or the Central Alps: Mohn et al. 2011). A schematic reconstruction of the Alpine oceanic domain is shown in Figure 4.2(d) at ~120–110 Ma. The age range of gabbros (~165–150 Ma; Figure 4.2(a); Manatschal and Müntener 2009), which are the remnants of the Liguro–Piemont crust shows that subsequent magmatism was either insignificant and/or that the entire oceanic lithosphere formed between ~150 and 120 Ma was subducted without a trace.

Continental subduction affected the Sesia Zone (75–65 Ma), the Briançonnais Domain, including the Internal Basement Complexes of the Monte Rosa, Gran Paradiso and Dora Maira massifs (~42–35 Ma), and the European margin at 38–35 Ma (e.g. Adula; Nagel 2008). Collision between Europe and Adria started at 34–32 Ma in the Western Alps (Simon-Labric et al. 2009; Beltrando et al. 2009).

Given that 400–700 km of the lithosphere (including the Briançonnais Domain) was subducted within ~60 Ma, the inferred subduction rate from 100–95 to 40–35 Ma was ~1 cm/year. Independent, rough petrological constraints yield 0.5–1 cm/yr (Lapen et al. 2003). A minimum of ~200–300 km of this subduction was truly oceanic in the Liguro–Piemont Ocean (Handy et al. 2010; Manzotti et al. 2014). In the Central and Western Alps, the large detached Alpine slab possibly reached down to the Mantle Transition Zone only, suggesting that slab dynamics were confined to upper mantle convection (e.g. Royden and Faccenna 2018).

4.2.3. *Material types: ultramafic, mafic and sedimentary*

The Alpine Ocean lithosphere comprised an extensive, partly re-fertilized, exhumed mantle that was directly overlain by deep-sea sediments and ophicarbonates (Figure 4.2(e); plate 4.1; Tricart et al. 1982; Lemoine et al. 1987), plus discontinuous patches of gabbroic or basaltic bodies and limited ridge-type magmatism (Manatschal and Müntener 2009).

Peridotites are most commonly serpentinized, re-fertilized abyssal type lherzolites (Barnes et al. 2014). They are compositionally intermediate between oceanic/asthenospheric and sub-continental mantle (e.g. Malenco, Central Alps; Lanzo, Western Alps; McCarthy and Müntener 2015; Picazo et al. 2016).

Systematic lithostratigraphic studies, mostly in the 1970s and 1980s, recognized pre-rift (Triassic), syn-rift (early to mid-Jurassic) and post-rift series in the Liguro-Piemont and Valais oceanic domains and their adjacent continental margins (Lemoine et al. 1986). Post-rift sediments are found almost across the entire domain, mainly as calcschists and flysch deposits. It is often uncertain which of these post-rift sediments were deposited on a truly oceanic substratum and which on the adjacent continental margin.

Sediments truly deposited on ophiolitic seafloor (typically ~200–400 m thick) make up the Schistes Lustrés and Bündnerschiefer complexes (Elter 1971) and comprise radiolarian cherts, shales, pelagic limestones and calcschists (plate 4.2). They range from the Callovo-Oxfordian to the early Paleocene but are dominantly Cretaceous in age (Figure 4.2(e); Caron 1977; De Wever and Caby 1981; Lemoine et al. 1984). These pelagic sediments exhibit syntectonic centimeter-to-meter scale ophiolitic or continental detritus (Tricart et al. 1982; plate 4.1).

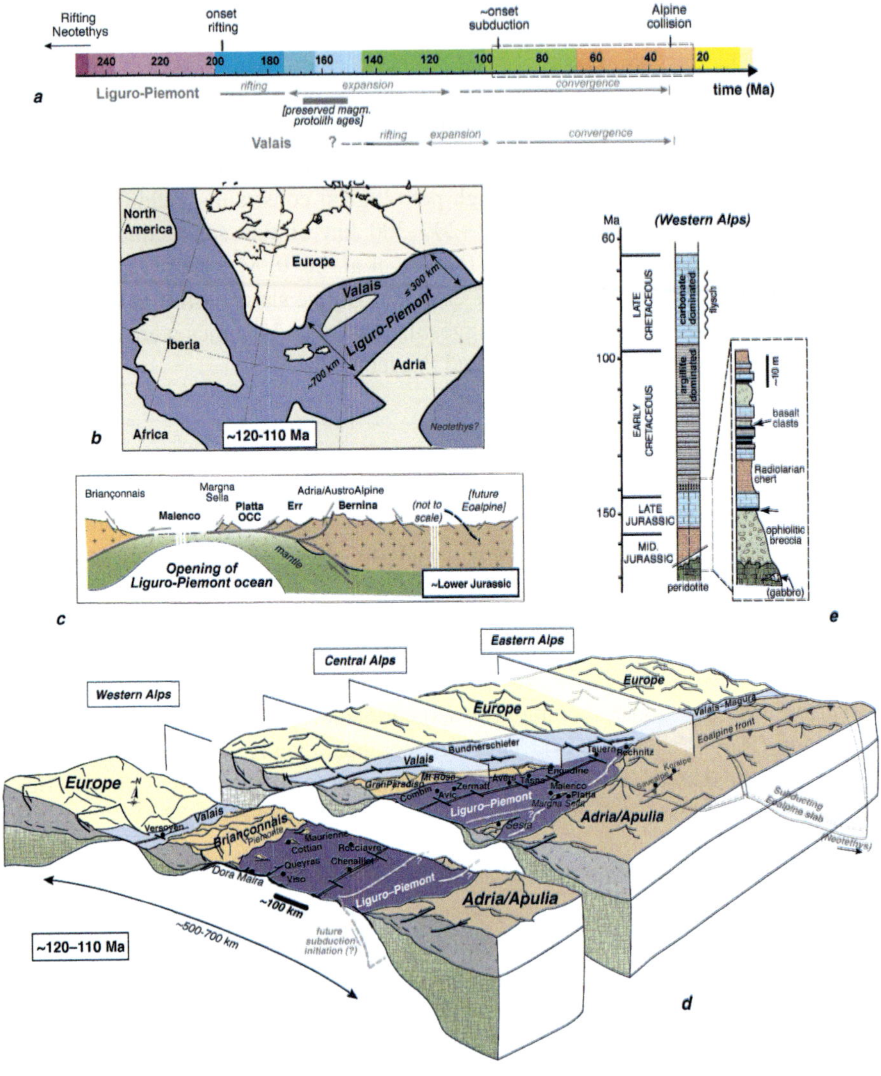

Figure 4.2. *Main Alpine geodynamic stages*

COMMENTS ON FIGURE 4.2.– *a) Chronological overview of the main Alpine geodynamic stages. See text for details. b) Paleogeography of the Alpine realm at ~120 Ma (Aptian) after Vissers et al. (2013), at its maximum extent (see section*

4.2.2). Uncertainties are on the order of 100-200 km at least. c) Paleogeographic restoration of the Central Alps region shortly after the rifting of the Liguro-Piemont domain (after Froitzheim et al. 1996; Mohn et al. 2011) showing key locations of the ocean-continent transitional domain, e.g. the Malenco exhumed sub-continental mantle, Margna-Sella extensional allochthons and Platta oceanic core-complex (see section 4.3.3 and Figure 4.6(c)). d) Schematic 3D view of the Alpine domain at ~120 Ma (compare with Figure 7.1(a)) emphasizing the highly stretched nature of the continental margins, the presence of micro-continental blocks (e.g. Briançonnais, Sesia Zone) or extensional allochthons. The location of the main units discussed in the text is modified after Berger and Bousquet (2008), Manatschal and Müntener (2009) and Picazo et al. (2016). The Eoalpine subduction, whose orientation is still debated, is shown to be the NE of the Alpine domain considered here (Froitzheim et al. 1996; Kästle et al. 2019). The exact location of subduction initiation within the Alpine domain is unknown, but probably to be looked for to the east, as shown here. Abbreviations (same as in the following figures, unless specififed; Continental fragments are capitalized): A: Avers; Av: Avic; Bu: Bündnerschiefer; C: Cottian; Ch: Chenaillet; Co: Combin; DM: Dora Maira; E: Engadine; GP: Gran Paradiso; K: Koralpe; M: Maurienne; Ma: Malenco; MR: Monte Rosa; MS: Margna-Sella; Pl: Platta; Pm: Piemonte units; Q: Queyras; R: Rechnitz; Ro: Rocciavre; Sa: Saualpe; SE: Sesia; T: Tauern; Ta: Tasna; Ve: Versoyen; Vi: Monviso; Z: Zermat-Saas. e) Representative lithostratigraphic column for the metasedimentary domains of the Western Alps (after Lemoine et al. 1986; Deville et al. 1992). Inset: detailed section at the base of one of the Queyras units (Tricart et al. 1982).

Kilometer-sized carbonate (commonly Triassic in age; e.g. "Pre-piedmont" domain; Elter 1971; Tricart et al. 1982; Tricart and Lemoine 1986) or continental fragments are interpreted as extensional allochthons (e.g. Err-Plata; Figure 4.2(c); Manatschal and Nievergelt 1997; Mohn et al. 2011). These allochthons, together with the larger subducted continental fragments (e.g. from largest to smallest: Briançonnais, Sesia Zone, Emilius, Pilonnet; Dal Piaz et al. 2001; Manzotti et al. 2014), underline the considerable width of the former ocean-continent transitional domain.

In contrast, Late Cretaceous to Paleocene flysch sequences is interpreted as trench sediments (e.g. Helminthoid or Prättigau flysch; Merle and Brun 1984; Mueller et al. 2020) mark the former accretionary margin during oceanic subduction.

Overall, these Tethyan sediments dominated by carbonates and organic-rich pelites (the future "calc-schists") contrast with the more mafic graywackes of the circum-Pacific trench fills. The Alpine Tethyan seafloor, thus, did not resemble the thick, stratified Pacific-type crust (Kusky et al. 2014) but rather, in terms of size and constitution, parts of the North Atlantic, such as along an NW–SE transect between Greenland and Norway.

4.3. Subducted oceanic fragments scattered throughout the Alps

Since subducted HP-LT oceanic fragments are better preserved in the Western Alps (Goffé et al. 2004; Figure 4.1(f)) than in parts of the Central Alps having experienced strong collision-related metamorphic overprint (Engi et al. 2004; Berger et al. 2005; Keller et al. 2005; Figure 4.2(b)). The record of the Western Alps is thus examined first (Figure 4.3) and then compared to that of the Central and Eastern Alps (section 4.3.3; Figure 4.4). Plates 1–6 provide some illustrations of classical exposures in the Western Alps.

An observation that pertains to all parts of the Alps is that tectonic units tend to be dominated by one of the two distinct rock types (Agard et al. 2009; Figure 4.3(a); plate 4.3). Firstly, the metamorphosed pelagic sediments that were to become the calcschists of the Schistes Lustrés and Bündnerschiefer herein termed the S (sedimentary) units. Secondly, the MUM (mafic and ultramafic) units, contain only minor calcschists. A crude estimate for the relative proportions of sediment to mafic/ultramafic rocks, with large lateral variations, would be around ~20:1 in S units and ~1:5 in MUM units.

In the Western Alps, the S units are dominantly blueschist facies and are located in the west, whereas the large, mainly eclogitic mafic and ultramafic MUM bodies crop out structurally below and to the east (Figures 4.1(c), (d), 4.3(a) and (b)). The consistency of structural and metamorphic patterns within the various units (either S or MUM) suggests that they were exhumed as individual tectonic slices and nappes along with the subduction interface, and not as a kilometer-scale mélange. Both types were largely exhumed by 35 Ma (Agard et al. 2002), before the collision.

4.3.1. *Metasedimentary-dominated (S) units from the Western Alps*

The S units comprise intensely folded calcschists, rare marbles and minor fragments of other oceanic rocks (plates 1–3). These can be distinguished using lithostratigraphy and metamorphic grade (Plunder et al. 2012; Lagabrielle et al. 2015) and divided into three broad units (Figures 4.3(c) and (d)), whose grade decreases structurally upwards (plate 4.4). Firstly, chloritoid ± garnet ± epidote–bearing lower S units formed at 2.0–2.3 GPa and 470–550°C. Secondly, more calcareous, chloritoid + lawsonite or epidote–bearing middle S units formed at 1.5–2.0 GPa and 370–450°C. Thirdly, lawsonite- and/or Fe–Mg-carpholite–bearing upper S units formed at 1.1 GPa and 330°C and 1.4 GPa and 350°C. In the upper S units, fragments of embedded mafic material show typical lawsonite- and glaucophane-bearing blueschists (plate 4.6).

Continuous evolution of P–T conditions is observed in the Southwestern Alps, from 1.2 to 1.3 GPa and 350°C to ~2.0 GPa and 500°C (Figure 4.3(d); Agard et al. 2001; Ghignone et al. 2020). These metasediments were scraped off the underlying oceanic crust/mantle during subduction, between ~30 and 60 km depth, then underplated from ~62 to ~50–45 Ma (in the Cottian Alps) and at 40–36 Ma (in the Combin Zone). Exhumation rates were on the order of 1–2 mm/year (Figure 4.5(a); Agard et al. 2002).

4.3.2. *Mafic/ultramafic dominated (MUM units from the Western Alps)*

Large, 1–10 km long eclogitized MUM fragments with subordinate metasediments are found in the Liguro–Piemont Ocean (plate 4.3), but not in the Valais Ocean (Figure 4.1(a)). Although strongly deformed and crosscut by shear zones, they are continuous over kilometer lengths (e.g. Zermatt-Saas Antrona, Avic; Tartarotti et al. 2011) and hectometre- to kilometer thick domains (in Monviso, Rocciavre, Lanzo; Figure 4.1(d)). These eclogitized MUM fragments preserve coherent "slab" structures, including sedimentary and/or magmatic features partly reworked during subduction and exhumation (Balestro et al. 2015).

Two representative examples are briefly described here (for more details on these or other Alpine localities, see Agard 2021):

1) In the Zermatt–Saas area, tectonic slices are variably retrogressed to garnet–clinopyroxene ± glaucophane–talc–chloritoid–lawsonite metabasalts (plates 4.5 and 7.6). The P–T conditions are relatively constant at ~2.5–2.8 GPa and 550°C (Angiboust et al. 2009), with the exception of one small, hectometre-thick, 2 km^2 locality hosting coesite-bearing ultrahigh-pressure metasediments (Cignana, ~2.8–3.2 GPa, 600°C). The age constraints for peak burial cluster at 43 ± 4 Ma, with indications for prograde growth at 49–48 Ma (Figure 4.5(a)).

2) In the Monviso area (Lombardo 1978), the ~15 km long, 2–3 km thick Lago Superiore Unit is likely the most intact eclogitized slab fragment recognized worldwide so far (Figure 4.3(f), plate 4.6(g); Angiboust et al. 2011; Locatelli et al. 2019). It comprises serpentinized ultramafics, Mg-rich metagabbros, Fe–Ti metagabbros capped by metabasalts and meter-thick calcschist horizons. Peak burial at ~2.6–2.7 GPa and 550–580°C was attained at 46 ± 3 Ma, following prograde mineral growth at 49 Ma (Figure 4.5(a)). Its detachment from the slab at ~80 km depth was promoted by extensive fracturing, as recorded by eclogite breccia preserved in shear zones (Angiboust et al. 2012a, 2012b; Locatelli et al. 2018; plate 4.6(h–i)).

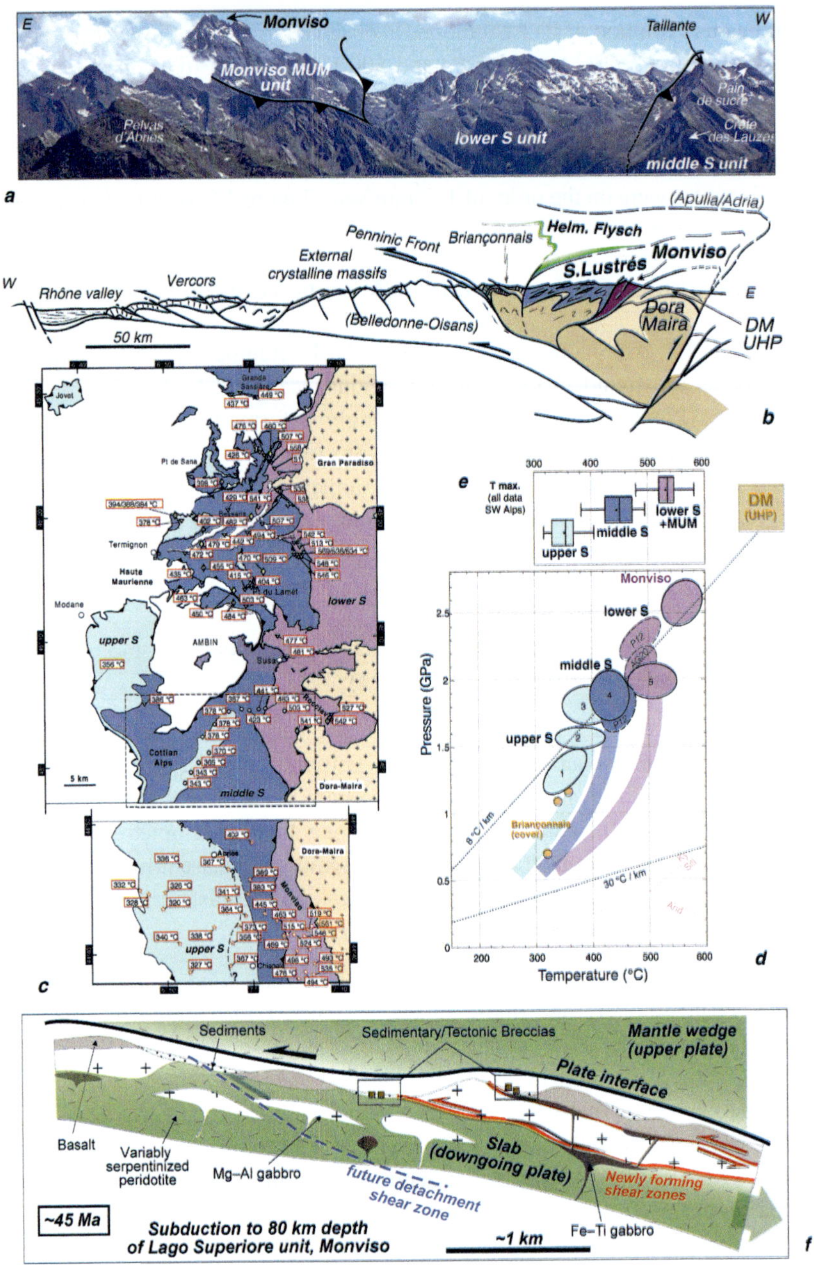

Figure 4.3. *Focus on the subducted fragments of the Western Alps. Abbreviations as in Figure 4.2(d) (except Amb: Ambin; Br: Briançonnais; HF: Helminthoid flysch; PF: Penninic front)*

COMMENTS ON FIGURE 4.3.– *a) View across the main units recognized in the Queyras (following Lagabrielle et al. 2015; view from Bric Froid), the middle, lower and Monviso units. Note the presence of isolated mafic gabbroic masses embedded in the S units (Pelvas d'Abries, Crête des Lauzes). b) Schematic crustal-scale section across the Western Alps along with profile B–B', approximately at the latitude of Briançon (see location on Figure 4.1(a); after Agard et al. 2009). c) Structural map from the Maurienne to Northern Queyras, showing the sub-division into three major units (lower, middle and upper S units). In the Queyras-Cottian Alps, three or four units are recognized depending on latitude and authors (here the two upper S units are after Lagabrielle et al. 2015). Temperatures correspond to the maximum temperature estimated using the irreversible transformation of organic matter in the rocks (data and calibration from Beyssac et al. 2002; data from Gabalda et al. 2009; Plunder et al. 2012; Schwartz et al. 2013). Figure originally modified after Vitale-Brovarone et al. 2014. d) Pressure-temperature estimates for the area shown in Figure 4.3(c) (plain ellipses; updated from Agard et al. 2001, 2002). Dotted ellipses refer to studies further north in the Maurienne (Pl12: Plunder et al. 2012) or further south in the Monviso area (Locatelli et al. 2018 and references therein; AG20: Angiboust and Glodny 2020). e) Boxplots capturing the range of maximum temperatures for the lower, middle and upper units across the area shown in Figure 4.3(c). f) Inferred restoration of the eclogitic Lago Superiore tectonic unit, Monviso massif, at peak burial conditions. This restoration highlights the respective influences of the discontinuous nature of the Alpine oceanic lithosphere (see section 4.2.3) and of later deformation. Some initial sedimentary and/or magmatic contacts were reworked or may have guided deformation during subduction, but new shear zones also formed at ~80 km depth (as testified by the presence of eclogite brecciation; plates 5 and 6; Locatelli et al. 2018 and references therein).*

Although all the MUM units show variations in internal structure and crustal components that can be attributed to the complex ocean-continent transitional domains and/or magmatic processes, all record strikingly similar P–T–t conditions, with peak burial near ~2.6 ± 0.2 GPa, at 550°C, and taking place at 45 ± 5 Ma (Figures 4.5(a) and (b)). The MUM units were, therefore, metamorphosed and detached from the top of the slab at a similar ~80 km depth late in the oceanic subduction history (Figure 4.5(a)). Subsequent deformation and slicing along the plate interface occurred during syn-convergent exhumation at ~42–36 Ma at a rate of ~1 cm/year. Unlike the S units, the MUM units are closely associated with the Internal Basement Complexes (Figure 4.1(d)): they are always tectonically underlain by them, share similarly oriented high-pressure stretching lineations, and have only slightly older peak burial ages (Figure 4.5(a); Gran Paradiso and Monte Rosa: 42–43 Ma; Dora Maira: 40–35 Ma, but probably closer to the lower bound).

4.3.3. *Subducted fragments from the Central and Eastern Alps*

The S-type Valais Units are largely exposed as a ~10 km thick tectonic pile in the eastern part of Switzerland, in the Grisons (Figures 4.4(a), (b) and 4.5(a); Frey and Ferreiro Mählmann 1999). Two large-scale units are recognized, the Grava and Töimul nappes (Weh and Froitzheim 2001 and references therein). These metasediments contain blueschist facies, lawsonite-, Fe–Mg-carpholite- or chloritoid-bearing assemblages. Similar P–T estimates (Figure 4.4(c); 0.9–1.4 GPa, 350–400°C; Wiederkehr et al. 2011) are obtained in the upper S units of the Western Alps (Figure 4.3(d)). They provide evidence for tectonic underplating at ~30–40 km depth (Figure 4.5(a) and (c)), which is close to the Moho depth of the adjacent upper Adriatic Plate, at 42–40 Ma (Wiederkehr et al. 2009). The more deeply buried slices of oceanic metasediments of the lower Glockner Unit of the Tauern Window (~2 GPa, 500°C; Groß et al. 2020) represent fragments entrained by continental subduction and then exhumed as part of the distal European margin.

The Liguro–Piemont units of the Central Alps comprise three units (Figure 4.4(b); Froitzheim et al. 1994, 1996). (1) The lower unit is the metasedimentary Avers S unit (1.1–1.3 GPa, 350–400°C; Ring 1992). (2) The middle unit comprises the MUM units of Malenco, an exhumed sub-continental mantle fragment later subducted to temperatures around 400°C between 63 and 55 Ma (Figure 4.2(c); Trommsdorff et al. 1993; Münterner and Hermann, 1996; Picazo et al. 2019) and Platta, a high-pressure greenschist-facies metamorphosed ocean-continent transitional domain (Handy et al. 1996; Mohn et al. 2011). These units lie structurally below and above, respectively, the high-pressure, greenschist- to blueschist facies Margna–Sella extensional allochthon, a paleogeographic equivalent of the Sesia Zone (Figures 4.2(c) and (d)). Below the Malenco ultramafic massif, the km^2-scale Lanzada tectonic window exposes a fragment of eclogitized metagabbro (~2 GPa, 525°C). Its palaeogeographic origin is uncertain but thought to represent the southernmost extension of the Avers unit (Droop and Chavrit 2014). (3) The upper unit is made up of the ~1–2 km thick Arosa Zone, which was interpreted as the fossil subduction plate interface between ~58 and 47 Ma (Bachmann et al. 2009). Syn-subduction deformation immediately above the subduction thrust, together with upper plate extension recorded by the deposition of the Gosau basins between ~80 and 55 Ma, may reflect a switch from subduction erosion to accretion ~10–15 Ma before ocean closure (Wagreich 1995). Further east, subducted fragments record low-grade blueschist facies conditions (Figure 4.5(a), inset map) in the Engadine Window (1–1.3 GPa, 350–375°C; 42–40 Ma; Bousquet et al. 1998; Wiederkehr et al. 2009), the Tauern Window (~1 GPa; 350–400°C; ~50 Ma; Dingeldey et al. 1997) and the Rechnitz Window (0.9–1.0 GPa; 370°C; 57 ± 3 Ma; Koller 2003; Ratschbacher et al. 2004).

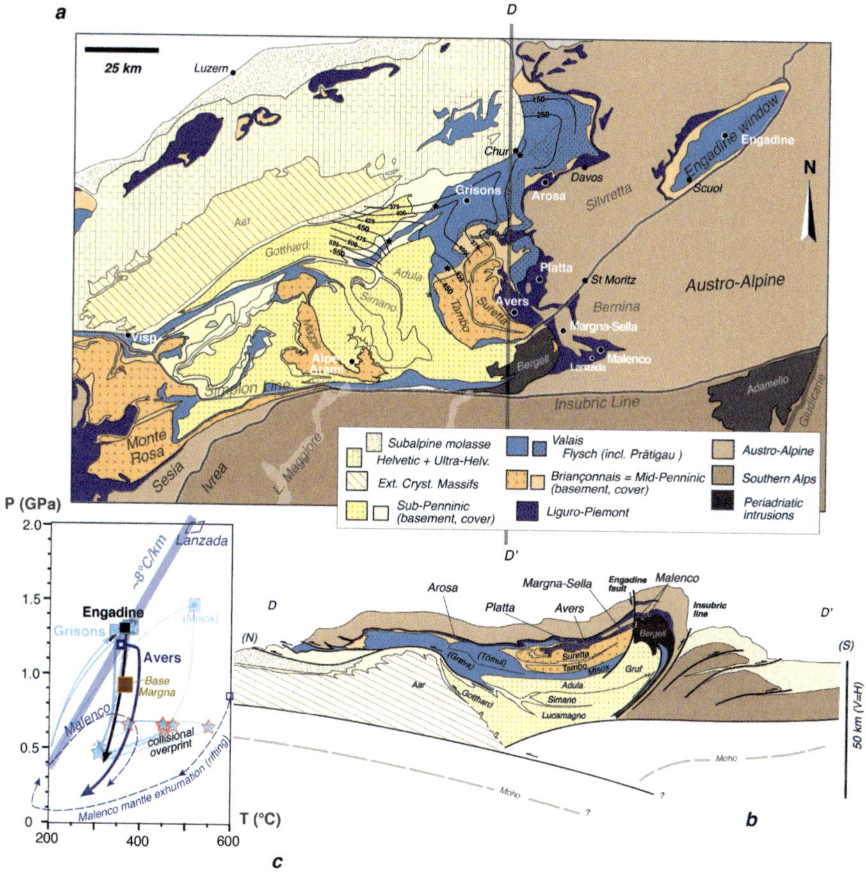

Figure 4.4. *a) Geological map of the Central Alps (after Schmid et al. 1996, 2004; Wei and Froitzhzeim 2001). Same colors as in Figure 4.1(a). Maximum temperatures outline the collisional overprint across the Valais domain (after Wierderkehr et al. 2011). b) Simplified section across D–D'. c) Compilation of P-T paths for the Central Alps (see text for references and Agard 2021). Light blue color paths and corresponding stars correspond to localities of the Valais domain shown in Figure 4.12(a) (see Wiederkehr et al. 2011)*

4.3.4. Comparing restorations at ~60 Ma

For the sake of comparison, two schematic restorations of the subduction configuration at ~60 Ma are shown for the Western and (eastern) Central Alps in Figure 4.6.

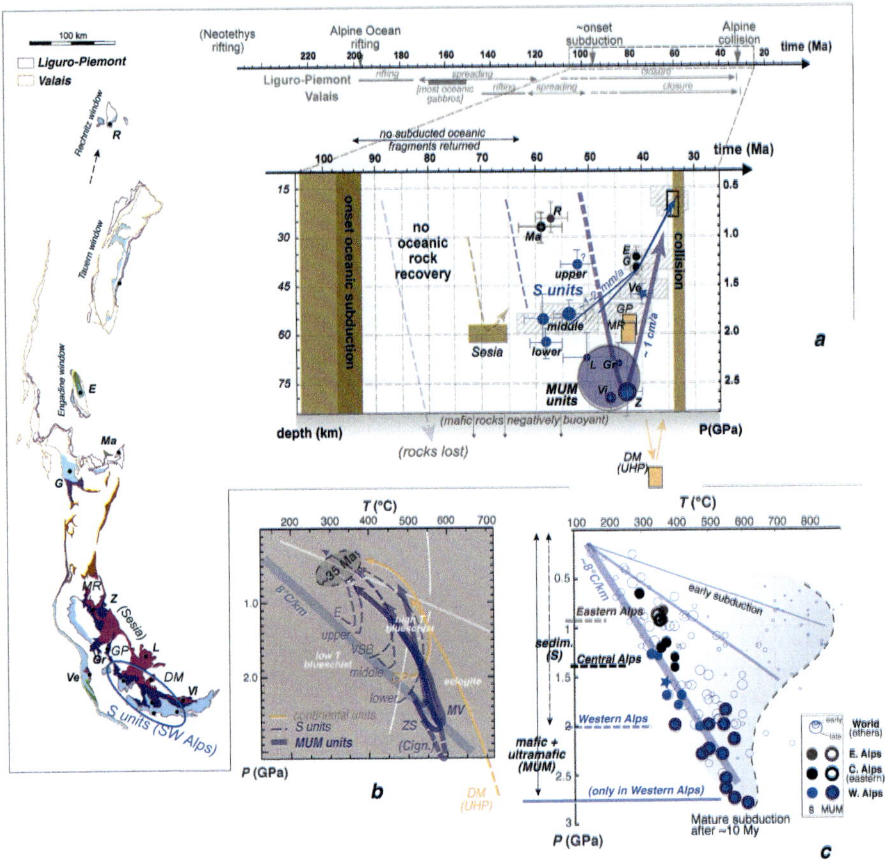

Figure 4.5. *Subducted Alpine fragments*

COMMENTS ON FIGURE 4.5.– *a) Synthetic pressure-time paths of subducted Alpine fragments. Note the contrast between the early period, during which no fragments were recovered, and the late subduction period. Inset map after Bousquet et al. 2008 (see also Figure 4.7). Pressure estimates compiled from the literature (see also Berger and Bousquet 2008; Agard et al. 2009, 2018; Rebay et al. 2018). Since only rare studies provide closely tied P-T and age, average P–T and time estimates for peak burial had to be averaged independently. Circle size is indicative of the number of available age constraints: there are for example at least fourteen instrumentally well-constrained estimates for the peak burial of the Zermatt-Saas area, north of the Aosta valley (references in Agard 2021). Constraints on the exhumation path are comprised of the thicker segment. Fewer studies and ages are available for Monviso. Hatched gray boxes refer to exhumation constraints for metasediments (Agard et al. 2002). Abbreviations as given in the caption of Figure 4.2(d). b) Characteristic shape of P–T paths for subducted alpine units. For oceanic fragments: Engadine (E; after Bousquet et al. 1998, 2002); Valaisan Petit Saint Bernard (VSB; Bousquet et al. 2002); Cottian Alps (lower, middle and upper units; Agard et al. 2001; Plunder et al. 2012; see Figure 4.3(d)); Monviso (MV; Angiboust et al. 2012b; Locatelli et al. 2018); Zermatt-Saas (Groppo et al. 2009; Angiboust and Agard 2010). P–T paths for the continental Internal Basement Complexes are shown for comparison (Dora Maira UHP: after Chopin and Schertl 1999; Groppo et al. 2007; Gran Paradiso: Manzotti et al. 2015). c) Comparison of P–T estimates for all Alpine subducted oceanic fragments, grouped by locality, with diagnostic data worldwide (Agard et al. 2018). For the sake of comparison, the P–T estimates shown here are those listed in Agard et al. (2018). Values for the Central and Eastern Alps, not considered then, are those cited in the text. All localities align along or close to the subduction P–T gradient for mature subduction. This observation rules out significant overpressure in the Alps, which would engender variable P/T ratios in the rock record (see section 4.4.3). Maximum pressures largely differ along strike and in particular, no deeply buried rocks are recovered east of the Simplon/Lepontine Central Alps (save possibly for the very small Lanzada fragment: Figure 4.4(c).*

In the Western Alps (Figure 4.6(a)), this period shortly follows the subduction and partial exhumation of the Sesia micro-continental sliver. Note that the idealized section shown here for the Western Alps is in fact composite since the Sesia Zone is projected onto the southern Western Alps traverse. Assuming convergence velocities around 1–2cm/year, the burial of the MUM eclogitic units to 80 km depth required ≥4 Ma. If the average estimate for their peak burial is 45 Ma (Figure 4.5(a)), then these units started to subduct near 50 Ma. The Lanzo massif, regarded by many as closer to Sesia, may have reached its peak slightly before if an older age is confirmed (52–46 Ma; Rubatto et al. 2008). Given the peak burial depth (from 60–70 to 100 km) and age (between 43 and 35 Ma) of the Internal Basement Complexes (Figure 4.5(a)), 45 Ma also likely coincides with the first impingement of the Briançonnais domain into the subduction zone (Figure 4.6(b)).

In the Central Alps (Figure 4.6(c)), this period coincides with a switch to short-lived subduction accretion, following ~20 Ma of subduction erosion reflected by the reworking of the base of the AustroAlpine upper plate, the relatively small amounts of flysch-type sediments or the mode of deposition of the Coniacian-early Eocene Gosau basins (Wagreich 1995; Wagreich and Decker 2001). Above the Arosa zone, a long-lived record of syn-subduction deformation is indeed documented in the upper plate (e.g. Bernina AustroAlpine units; Bachmann et al. 2009), mostly between ~80 and 60 Ma. In contrast, (de)formation of the Arosa zone is thought to date back to the very last stages of oceanic subduction, between 58 and 47 Ma (Bachmann et al. 2009), immediately prior to continental subduction of the Briançonnais domain (around 50 Ma). Closure of the Liguro-Piemont domain could thus be slightly diachronous, from ~50–45 Ma in the Grisons to ~40 Ma in the Western Alps (see section 4.3.3 and 4.4.1). Age constraints are lacking, however, to assess when the underplating of the Avers Bündnerschiefer took place.

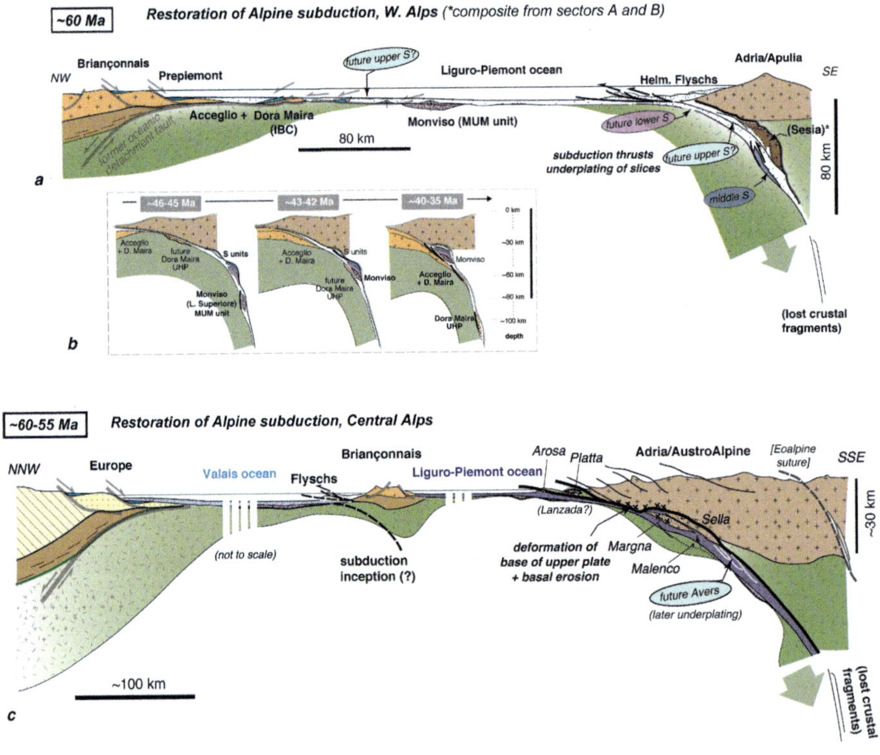

Figure 4.6. *Restoration of the Alpine subduction in the Western and Central Alps*

COMMENTS ON FIGURE 4.6.– *a) Idealized restoration of the subduction configuration in the Western Alps during the Paleocene, at ~60 Ma, after subduction and partial exhumation of the Sesia micro-continental sliver. The section is a composite one since the Sesia Zone is projected here laterally onto the Helminthoid Flysch-Monviso traverse (section B–B'; Figures 4.1(a) and 4.3(b)). The scales of the units and their distances from one another are approximately respected. No vertical exaggeration. Tectonic contacts formed during oceanic stages are featured in grey whereas subduction-related ones are shown in black. b) Tectonic evolution towards the end of oceanic subduction, specifically for the South-western Alps (i.e. Sector A; Maurienne to Queyras; section 4.4.2; Figure 4.7). Uncertainties concern the age of underplating of the upper S units, or the exact age of peak burial for Dora Maira, between 40 and 35 Ma (probably towards the younger bound; Gebauer et al. 1997; section 4.3.3). The evolution considers peak burial at 46–45 Ma for Monviso (Figure 10(b); see text) and ~38 Ma for Dora Maira. Given convergence velocities on the order of 1 cm/a (section 4.2.2), the detachment of the Lago Superiore slice of the Monviso massif (Figures 4.6(a) and 4.8) and part of its exhumation likely relate to the entrance of the Briançonnais micro-continent into subduction. This influence is clearer for sector B, i.e. for the broad Zermatt-Saas area, as shown by Figure 4.5(a), with the entrance of the Gran Paradiso and Monte Rosa. c) Idealized restoration of the subduction configuration in the Central Alps during the Paleocene, at ~60–55 Ma, after subduction and partial exhumation of the Margna-Sella micro-continental sliver. The exact timing of subduction/exhumation of Malenco and the Avers (and Lanzada) units is not precisely known (see text for details) but lies in the range of 60–40 Ma.*

4.4. Discussion: subduction dynamics through space and time

4.4.1. *Contrasting recovery through space and time*

1) Most strikingly, no subducted slab fragments were returned and preserved between 100–95 and 65 Ma throughout the Alps (Figure 4.5(a)). All subducted fragments of the Liguro–Piemont Ocean and the Valais Ocean are younger than ~60–65 Ma, i.e. these fragments post-date the subduction and exhumation of the Sesia Zone.

2) Data constraining the P–T–t paths and lithostratigraphy of Alpine metamorphic rocks reveal a marked contrast between the S and the MUM units (Figure 4.5(a)), particularly in the Western Alps. The MUM units there exhibit larger coherent fragments and/or features attributable to ocean-continent transitional domains: they reached higher peak pressures and temperatures, underwent faster exhumation rates, and were buried (and exhumed) during the final stages of oceanic subduction.

3) While the record of subduction in the Central and Eastern Alps often lacks the spectacular eclogites of the Western Alps, all S and MUM units share the same metamorphic gradient of 8–10°C/km, which is diagnostic of mature subduction (Figure 4.5(c); Agard et al. 2018) and consistent with rocks formed in the same subduction zone. Maximum pressures differ along strike: in general, no deeply buried oceanic rocks (i.e. >1.5 GPa; Figure 4.5(c)) are recovered east of the Simplon and Lepontine Units of the Central Alps, with the exception of the lower Glockner Unit and small Lanzada units (Figure 4.4(c)). The fact that metamorphic pressures vary with temperatures consistently along a common metamorphic gradient, irrespective of location in the Alps, rules out significant overpressure, which would lead to a large scatter in pressure at a given temperature. The same cut-off in maximum pressure is observed in the Alps as in other comparable orogens worldwide (~2.8–3.0 GPa; Agard et al. 2009, 2018), beyond which mafic rocks ± serpentinites become negatively buoyant (Figure 4.5(a)).

4) Offscraping of the S units from the underlying MUM portion of the slab, and partial exhumation along with the plate interface, is documented in the Valais and Liguro-Piemont Oceans between ~60 and 35 Ma (Figure 4.6). Most S units experienced peak metamorphic pressures and temperatures of ~1.0–1.3 GPa and ~350–400°C (~30–40 km depth) which is consistent with most HP-LT rocks worldwide (Agard et al. 2018). The amount of underplated rock varies along strike, from 100 m to several kilometers thick, in both the Valais (Grisons, Valaisan) and Liguro-Piemont Oceans (Avers, Combin, Maurienne to Queyras). But underplating at distinct depths only occurred in the Southwestern Alps.

5) The MUM units are volumetrically insignificant in the Central and Eastern Alps. They experienced peak P–T conditions from unmetamorphosed to blueschist facies <~1 GPa and were subducted between 60 and 50 Ma. In the Western Alps, all MUM units show comparable eclogitic conditions and reached peak burial late in the oceanic subduction history at ~45 Ma, at most a few million years before continental subduction (Figure 4.5(a)). Assuming convergence velocities around 1–2 cm/year, the burial of these MUM eclogitic units to 80 km depth must have started near 50 Ma (Figure 4.5(a)). The ~5 My difference in peak burial between the MUM and the Internal Basement Complexes (45 Ma vs. 35–42 Ma) shows that they were located 50–100 km apart, i.e. that MUM units were located close to the continent (Figure 4.6(a)). It also suggests that the entrance of the continental margin into the subduction zone at ~45 Ma triggered the detachment of MUM units from the slab, possibly facilitated by lateral discontinuities in the heterogeneous oceanic lithosphere, as in Monviso (Figure 4.3(f)). The continental Internal Basement

Complexes subsequently accompanied the MUM units along with the plate interface during part of their return path (only its very late ascent for Monviso; Angiboust and Glodny 2020), as shown by their spatial association, their commonly fast exhumation rates and their tight P–T paths.

6) The present location of the MUM units coincides with the areal extent of the subducted Briançonnais, Sesia and Margna–Sella continental fragments (Figure 4.7). More MUM units are found near the Gran Paradiso, Monte Rosa and Sesia massifs (Figures 4.1(c) and (d)) than near Dora Maira. This suggests a strong influence of both initial margin architecture and continental subduction, as discussed below.

4.4.2. *Four distinct sectors: paleogeographic contrasts in subduction dynamics*

Based on these first-order contrasts in subducted-related history, and with some reservations due to P-T-t-d-cartographic-paleogeographic uncertainties, it is possible to identify several distinct sectors along the Alpine belt (Figure 4.7):

– Sector A, south of Gran Paradiso (along CIFALPS and south of the ECORS geophysical profiles; Figure 4.1(a), encompasses the Queyras, Cottian, Maurienne, Rocciavre, Monviso and Dora Maira sectors of the internal Alps. This sector is characterized by S units subducted across a range of depths and recovered across a relatively long period (~60–40 Ma), by the presence of pluri-km large gabbroic bodies (Rocciavre, Monviso) and, compared to the other sectors, a lesser influence of collision. It shows, at least from present exposures, the longest (mid-Penninic) Briançonnais margin subducted between high-grade blueschists and eclogites, with Ambin, Acceglio and Dora Maira (see the detailed reconstruction by Michard et al. 2004 and Figure 4.6(b)). The Dora Maira Internal Basement Complex (IBC) is in fact the only Briançonnais-derived continental UHP massif and the most composite one, with five or six hectometre- to kilometer thick slices, including one with coesite recording higher pressures than its adjacent MUM (i.e. Monviso). The extensively thinned continental crust near Dora Maira may have been more readily subducted to a great depth, yet not buoyant enough to help exhume dense MUM units. It is also worth noting, in passing, that maximum continental subduction was to about the same depths in the Western Alps (~3.5 GPa for Dora Maira) as for the Central Alps (4–5 GPa for the Alpe Arami distal European margin) and was approximately coeval at ~40–35 Ma. There is finally almost no Valais domain along with this sector A (Figure 4.7).

– Sector B, Gran Paradiso and NE Monte Rosa (and possibly into the western Central Alps; along with the NFP20-W profile) encompass the most voluminous MUM massifs (Zermatt-Saas, Antrona, Avic, Lanzo), which also show the largest individual mantle-dominated bodies (Lanzo, Avic). The S units are less abundant than in sector A and maximum pressures are far less variable. No ages older than ~45 Ma were found for these metasedimentary units. The Gran Paradiso and Monte Rosa Internal Basement Complexes record P-T conditions similar to or slightly lower than the MUM units. Neither of them are UHP nor show a complex stack of units (two only in the Gran Paradiso; Manzotti et al. 2015). This sector is characterized by the presence of the subducted continental Sesia Zone (±Ivrea-Canavese), which is located in a more internal position (hinterland) than ocean-derived units. This sector is also where very low-grade nappes stacked in the Préalpes preserve a record of accretionary wedge formation; and where the Ivrea body, interpreted as a shallow piece of upper plate mantle, is also found (Figure 4.7(c)).

– Sector C, (eastern) the Central Alps and Engadine window (along with the NFP20-E profile), is overall metasedimentary-dominated and oceanic fragments reached maximum pressures around 1.2–1.3 GPa. The S units can be several kilometers thick but are not subdivided as many times (yet?) as in sectors A or B. The MUM remnants are made of recognizable and well-characterized OCT and subcontinental mantle which was not deeply subducted. Subducted oceanic remnants are younger than 65 Ma, although subduction initiated at ~100 Ma (Bachmann et al. 2009). A continental sliver, Margna-Sella, smaller than the Sesia Zone, is present in the same structural position. Contrary to sectors A and B the Briançonnais massifs are fairly restricted (Tambo, Suretta; absence of Internal Basement Complex). On the other hand, the very distal European margin experienced locally UHP conditions (e.g. 4–5 GPa in Alpe Arami).

– Sector D, east of the Engadine window (i.e. along with the TRANSALP profile), only preserves volumetrically small subducted fragments of the Alpine ocean (e.g. Matrei, Rechnitz) with maximum pressures around 1.2–1.3 GPa. There is no Briançonnais domain. The European margin was subducted along with sector C but far less deep (e.g. 1.8 GPa in the Eclogite zone). This sector also coincides with the area affected by the former Eoalpine (Cretaceous) metamorphism in the Austro-Alpine nappes.

For sectors A and B, the eclogitic facies MUM units are closely related to the Internal Basement Complexes. UHP conditions in Penninic domains (oceanic: Cignana; continental: Dora Maira) are only found in these two sectors. There is evidence for subduction erosion of the upper plate basement in sectors B–D, marked by prolonged deformation and recovery of subducted micro-continental slivers at 55–50 Ma at the base of the Austro-Alpine domain (B: Dent Blanche, Angiboust et al. 2014; C: Bernina, Bachmann et al. 2009; D: Tauern, Dingeldey et al. 1997; see also Polino et al. 1990).

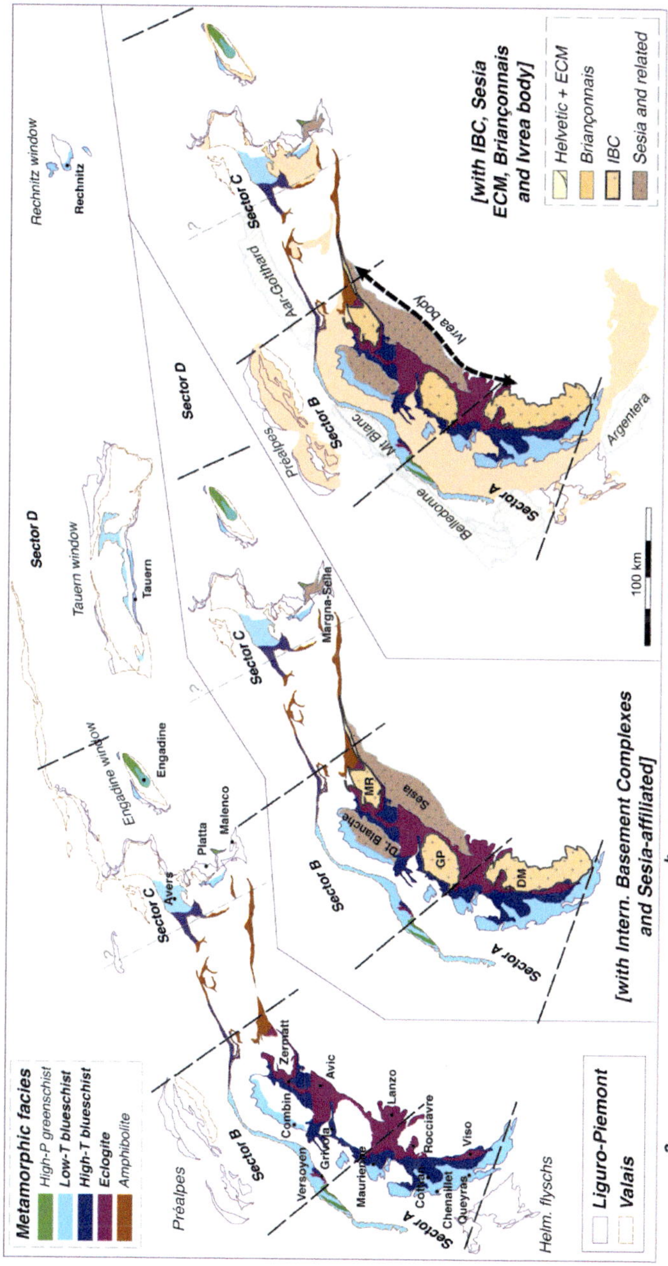

Figure 4.7. *Distribution of the metamorphic facies of all subducted Alpine oceanic domains*

COMMENTS ON FIGURE 4.7.– *a) Distribution of the metamorphic facies of all subducted Alpine oceanic domains (after Bousquet et al. 2008). Sectors A–D are defined solely on the basis of marked contrasts in the history of subducted fragments (see section 4.4.2 for details). Sector boundaries are indicative (and therefore drawn as straight lines) since they were certainly deformed by later collision. b) Same as a) for sectors A–C featuring the Internal Basement Complexes and the Sesia Zone: an obvious correlation exists (i) between the spatial extent of the mid-Penninic internal Briançonnais Internal Basement Complex (IBC) and the location of deeply buried – yet recovered – subducted oceanic fragments, and (ii) between the Sesia Zone and sector B (see text). c) Similar map with Helminthoid flyschs, Préalpes, External crystalline massifs and the entire Briançonnais domain for comparison. Note the match between sector B and the Préalpes or Ivrea body (after Zhao et al. 2015, 2020). Sector C was strongly affected by continental collision in its western part, which may suggest further subdivision of this domain.*

These lateral contrasts in subduction dynamics along the Alpine belt are depicted in a snapshot of the Alpine oceanic subduction at ~40 Ma, towards the end of oceanic subduction (Figure 4.8(a), to be compared with Figure 4.2(d)). Initial teleseismic P-wave tomography from the AlpArray seismic experiment shows that most of the subducted Alpine Tethyan lithosphere is now detached and lies partly in the Mantle Transition Zone (Figure 4.8(b)). However, some of the first-order contrasts in the structure of the lithosphere imaged by geophysics spatially overlap with the segmentation outlined by sectors A–D.

Further work could help determine whether this segmentation of the former Alpine subduction zone chiefly results from (i) margin segmentation inherited from the Variscan orogeny (von Raumer et al. 1999, 2002; Manzotti et al. 2016; Ballevre et al. 2018), (ii) rifting processes and/ or magmatic production (Mohn et al. 2011), (iii) differential response during material subduction, i.e. mechanical (rheology) or chemical (via fluid liberation), (iv) differential kinematics and/or (v) sedimentary inputs.

4.4.3. *A fragmentary yet faithful (rock) record of subduction*

The present study confirms that HP-LT oceanic fragments from the Alps provide a reliable P–T record diagnostic of cold subduction zones worldwide, one of the most extensive fossil examples so far:

i) Several reasons allow us to rule out significant tectonic overpressure in subducted Alpine fragments (see also Froitzheim 2020). Subducted remnants consistently align along the same P/T gradient, typical of mature subduction worldwide (Figure 4.5(c)). Any significant overpressure would engender variable P/T ratios in the oceanic rock record. Maximum recorded pressures for mafic eclogites (~2.8–3 GPa) match the threshold beyond which slab crust becomes negatively buoyant with respect to the surrounding mantle (Figure 4.5(a)) and fully coupled to the upper plate (Wada et al. 2008; Agard et al. 2020). Mineral parageneses found either in rheologically weak (sediments) or strong material (mafics and ultramafics) yield the same pressure estimates. This does not preclude minor, local overpressure. The hm-thick Cignana UHP sub-unit could for example reflect ~10% overpressure, at most (3.2 vs. 2.8 GPa; Figure 4.5(b)), or represent a deeper, thinner strip of the slab.

ii) The structural relationships and P–T–time-lithostratigraphic-petrological data preserve good access to the subduction history: the collision-related metamorphic overprint is moderate or insignificant in many places (e.g. Berger and Bousquet 2008); the distribution of tectonic units and deformation shows that subduction structures are not completely messed up and that Alpine fragments correspond to tectonic slices/imbrications formed along with the plate interface which was later variably disrupted by deformation (Angiboust and Agard 2010; Tartarotti et al. 2019; Locatelli et al. 2018). These Alpine features allow us to study the former geometry of the plate interface, in some places at the hm-km scale (e.g. down-dip geometries of metasedimentary-dominated units south of Gran Paradiso; geometry of OCT domains and/or specific horizons; shallow accretion of the Préalpes) or dynamic processes such as underplating, strain localization, subduction rheology or fluid release (Herviou et al. 2021). Rheological contrasts between specific sedimentary horizons, for example, may explain preferential localization of décollements in metapelite- or carbonate-dominated sequences, hence the successive accretion of the upper (more pelitic) and middle (more carbonate-rich) S units.

iii) The along-strike contrasts in subduction dynamics of sectors A–D help refine the initial setting/paleogeography of the Alpine subduction (Figure 4.7): the Briançonnais continental margin was likely more stretched along with sector A during rifting (i.e. Dora Maira); the presence of the Sesia Zone resulted in specific subduction dynamics along sector B; its subduction triggered a change in boundary conditions which initiated the recovery of oceanic rocks (~60–40 Ma). Importantly, these lateral contrasts appear to have largely persisted during continental subduction and collision, as shown, for example, by the Préalpes and Ivrea body located along with sector B (Figure 4.7(c)). The divides between the different sectors do in fact coincide with some key features detected by teleseismic P-wave tomography (Figure 4.8b). From ~65–60 to ~50 Ma, subduction appears to have been erosive along with sectors B–D (particularly in the eastern Central Alps) but accretionary along with sector A. During the first period (pre-Sesia subduction), the Alpine subduction was either (entirely?) in erosive mode, sediment starved and/or efficiently lubricated so that underplating was impeded at depth.

4.4.4. *What the Alpine example typifies*

While subducted fragments are fairly abundant in the Western Alps, no oceanic rocks were recovered between the onset of subduction (~100 Ma) and ~60 Ma. This process only started after the subduction of the Sesia Zone (Figure 4.5(a)). Subduction may have been more "classical" before 70–65 Ma, i.e. not returning anything, as many large-scale subduction zones today (e.g. Chile or Japan). The 165-150 Ma cluster of Liguro-Piemont magmatic ages (Figure 4.2(a)) shows that the oceanic lithosphere formed between ~150 and 120 Ma (possibly even with a thicker crust) was entirely subducted, over roughly the same amount of time (~30 Ma, from ~100 to 70 Ma).

The relative abundance of subducted oceanic fragments most likely results from the slow-spreading character of the ocean and the presence of OCT domains. This configuration provides many primary discontinuities/weaknesses facilitating later strain localization and offscraping along the subduction interface (Ruh et al. 2015), together with relatively buoyant material. The Alpine record is indeed characterized by (i) a marked contrast between units that are metasedimentary-dominated (S units) and others representing km-thick portions of slab lithosphere (MUM units); (ii) a contrast in their relative timing of burial and exhumation; (iii) the role of continental margin subduction on initiating detachment from the slab (and partly exhumation; Figure 4.6(b)); (iv) the preferential recovery of OCT domains.

The Alpine situation stands in contrast with large-scale subduction zones from fast-spreading oceans characterized by a more regular "ocean-plate stratigraphy" (Kusky et al. 2014; Wakabayashi 2015, 2017) and scarcier rock return. It supports the view that there is a recovery/exhumation filter in the rock record, which preferentially returns eclogitic fragments from slow-spreading oceans or thinned margins (Agard et al. 2018). The intermittent rock recovery accompanying the subduction of oceanic lithosphere is therefore controlled by both plate interface rheology and kinematics as well as by structures inherited from the former (passive) margin, seafloor structure and/or sediment type.

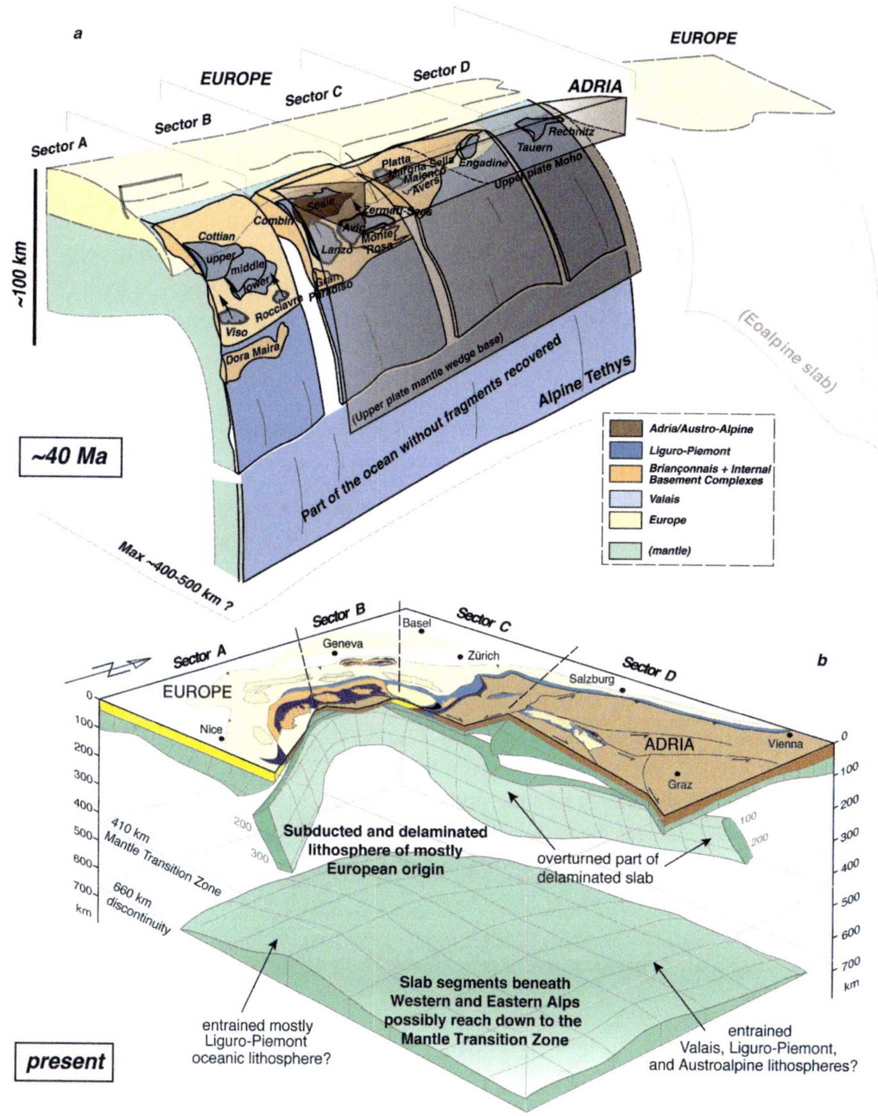

Figure 4.8. *a) Idealized view of the Alpine subduction zone at ~40 Ma, showing the relative location of subducted fragments discussed in the text (Figure 4.2(d)). b) Present-day slab configuration, after Agard and Handy 2021. This drawing is based on geophysical profiles (the grey transect line in Figure 4.1(a)) and preliminary interpretations of teleseismic tomography of the AlpArray working group by Paffrath et al.*

The lack of a magmatic arc atop the subduction zone was also recently associated with the slow-spreading nature of the ocean (McCarthy et al. 2018, 2020). These authors suggested that serpentinization of the upper portion of the sinking lithospheric mantle and efficient decoupling from the dry mantle below had caused its shallow accretion, thereby preventing enough water transfer at depth. Evidence for significant underplating of serpentinized mantle or shallow serpentinite diapirs (such as in the Izu-Bonin subduction zone; Tamblyn et al. 2019) is lacking in the Alpine record. Either these have been eroded or it may be, alternatively, that subduction of ~300 km long oceanic lithosphere may not be sufficient to hydrate the mantle wedge at sub-arc depths and/or to trigger sufficient counter-flow and mantle upwelling (Agard et al. 2020).

4.5. Conclusion

The rock record of the Alpine belt enables restoring a very significant part of the history and dynamics of oceanic subduction. Realizing the importance of seafloor (and margin) inheritance has also enabled us to access even further details of subduction processes. Several major conclusions can be drawn:

1) *The Alpine P-T record is representative of subduction processes.*

The consistent P/T gradient of oceanic fragments, similar to mature subduction, their recovery below a maximum pressure threshold (2.8–3 GPa) and their coherent distribution along spatial gradients and within units indicate that the oceanic subduction record can be trusted and used as a reliable proxy of the pressure and temperature regime of subduction, and that tectonic overpressure is limited (<~10%).

2) *Subduction dynamics can be reconstructed and influenced by later contraction.*

During the first half of the subduction period, between the onset of subduction (~100–95 Ma) and ~70–65 Ma, no subducted oceanic fragment was recovered; this only started after the subduction of the Sesia Zone. Subduction to ~80 km depth, a few Ma before continental subduction, of the MUM units of the Western Alps, which correspond to fragments of oceanic crust formed early at ~165–150 Ma, shows that only oceanic domains next to the continental margin and/or OCT domains are recovered. All other fragments formed between ~150 and 120 Ma have disappeared over a similar amount of time, during the first 30 Ma of subduction.

Along-strike contrasts allow recognition of four main sectors, respectively south of Gran Paradiso (A), between the Gran Paradiso and NE Monte Rosa (B), between the (eastern) Central Alps and the Engadine window (C), and east of the Engadine window (D). These differences obviously persisted long after the end of oceanic subduction, since some are still visible today (Figure 4.8(b)). Finally, between ~65 and 50 Ma, subduction erosion dominated along with sectors B–D (particularly in the eastern Central Alps), whereas subduction accretion prevailed along sector A.

3) *The Alpine subduction archive is remarkable but not atypical.*

The Alpine belt exposes many subducted fragments, particularly in the Western Alps, and documents the closure of a short-lived (60 Ma), slow-spreading, slowly closing ocean. The fate of the Alpine ocean, whose past seafloor essentially comprised variably re-fertilized exhumed mantle, irregularly distributed magmatic rocks and pelagic sediments together with OCT domains and extensional allochthons, exemplifies the preferential recovery of eclogitic fragments from slow-spreading oceans. Nothing such as the incoming ocean plate stratigraphy observed off Chile or Japan today. In addition to plate kinematics, syn-subduction sedimentation and plate interface rheology inherited features (i.e., former Variscan- or rifting-related structures, distribution of magmatic bodies/provinces and/or sediments) therefore largely impacted the subduction dynamics of the oceanic lithosphere. This conclusion is strengthened by the spectacular subduction-related rock record of the Cyclades/Pindos "ocean", recently reappraised as a former stretched continental margin (Schmid et al. 2020). In contrast with the Cyclades, however, at least ~200–300 km of real Atlantic-type seafloor disappeared in the Alps. With such a short slab the Alpine subduction likely never reached beyond the Mantle Transition Zone, another significant difference with large-scale subduction zones associated with wholesale mantle convection.

4.6. Acknowledgments

I would like to express my gratitude to C.L. Rosenberg and N. Bellahsen for their support and for pushing me to write this in the first place (this is in fact, what triggered the spark for the Agard 2021 paper cited here). Needless to say, this contribution would not have been possible without my former students who hopped the alpine trails with me, as well as many colleagues over the years.

4.7. References

Agard, P. (2021). Subduction of oceanic lithosphere in the Alps: Selective and archetypal from (slow-spreading) oceans. *Earth-Science Reviews*, 214, 103517.

Agard, P. and Handy, M.R. (2021). Ocean subduction dynamics in the Alps. *Elements*, 17, 9–16.

Agard, P., Jolivet, L., Goffé, B. (2001). Tectonometamorphic evolution of the Schistes Lustre's complex: Implications for the exhumation of HP and UHP rocks in the western Alps. *Bull. Soc. Geol. Fr.*, 172, 617–636.

Agard, P., Monie, P., Jolivet, L., Goffé, B. (2002). Exhumation of the Schistes Lustres complex: In situ laser probe Ar-40/Ar-39 constraints and implications for the Western Alps. *J. Metamorph. Geol.*, 20, 599–618.

Agard, P., Yamato, P., Jolivet, L., Burov, E. (2009). Exhumation of oceanic blueschists and eclogites in subduction zones: Timing and mechanisms. *Earth-Science Reviews*, 92, 53–79.

Agard, P., Plunder, A., Angiboust, S., Bonnet, G., Ruh, J. (2018). The subduction plate interface: Rock record and mechanical coupling (from long to short timescales). *Lithos*, 320–321, 537–566.

Agard, P., Prigent, C., Soret, M., Dubacq, B., Guillot, S., Deldicque D. (2020). Slabitization: Mechanisms controlling subduction development and viscous coupling. *Earth-Science Reviews*, 208, 103259.

Ampferer, O. and Hammer, W. (1911). Geologischer Querschnitt durch die Ostalpen vom Allgäu zum Gardasee. *Jahrbuch der Kaiserlich-Königlichen Geologischen Reichsanstalt*, 61, 532–710.

Amstutz, A. (1951). Sur l'évolution des structures alpines. *Arch. Sci.*, 4, 323–329.

Angiboust, S. and Agard, P. (2010). Initial water budget: The key to detaching large volumes of eclogitized oceanic crust along the subduction channel? *Lithos*, 120, 453–474.

Angiboust, S. and Glodny, J. (2020). Exhumation of eclogitic ophiolitic nappes in the W. Alps: New age data and implications for crustal wedge dynamics. *Lithos*, 356–357, 105374.

Angiboust, S., Agard, P., Jolivet, L., Beyssac, O. (2009). The Zermatt-Saas ophiolite: The largest (60-km wide) and deepest (c. 70–80 km) continuous slice of oceanic lithosphere detached from a subduction zone? *Terr. Nova*, 21, 171–180.

Angiboust, S., Agard, P., Raimbourg, H., Yamato, P., Huet, B. (2011). Subduction interface processes recorded by eclogite-facies shear zones (Monviso, W. Alps). *Lithos*, 127, 222–238.

Angiboust, S., Agard, P., Yamato, P., Raimbourg, H. (2012a). Eclogite breccias in a subducted ophiolite: A record of intermediate-depth earthquakes? *Geology*, 40, 707–710.

Angiboust, S., Langdon, R., Agard, P., Waters, D., Chopin, C. (2012b). Eclogitization of the Monviso ophiolite (W. Alps) and implications on subduction dynamics. *J. Metamorph. Geol.*, 30, 37–61.

Angiboust, S., Glodny, J., Oncken, O., Chopin, C. (2014). In search of transient subduction interfaces in the Dent Blanche-Sesia Tectonic System (W. Alps). *Lithos*, 205, 298–321.

Bachmann, R., Glodny, J., Oncken, O., Seifert, W. (2009). Abandonment of the South Penninic–Austroalpine palaeosubduction zone, Central Alps, and shift from subduction erosion to accretion: Constraints from Rb/Sr geochronology. *Journal of the Geological Society*, 166, 217–231.

Balestro, G., Festa, A., Dilek, Y., Tartarotti, P. (2015). Pre-Alpine extensional tectonics of a peridotite-localized oceanic core complex in the late Jurassic, high-pressure Monviso ophiolite (Western Alps). *Episodes*, 38, 266–282.

Ballèvre, M., Manzotti, P., Dal Piaz, G.V. (2018). Pre-Alpine (Variscan) inheritance: A key for the location of the future Valaisan Basin (Western Alps). *Tectonics*, 37, 786–817.

Barnes, J.D., Beltrando, M., Lee, C.-T.A., Cisneros, M., Loewy, S., Chin, E., (2014). Geochemistry of Alpine serpentinites from rifting to subduction: A view across paleogeographic domains and metamorphic grade. *Chemical Geology*, 389, 29–47.

Bebout, G.E. (2007). Metamorphic chemical geodynamics of subduction zones. *Earth and Planetary Science Letters*, 260, 373–393. DOI: j.epsl.(2007).05.050

Beltrando, M., Lister, G.S., Forster, M., Dunlap, W.J., Fraser, G., Hermann, J. (2009). Dating microstructures by the 40Ar/39Ar step-heating technique: Deformation–pressure–temperature–time history of the Penninic Units of the Western Alps. *Lithos*, 113, 801–819.

Beltrando, M., Compagnoni, R., Lombardo, B. (2010). (Ultra-)high-pressure metamorphism and orogenesis: An Alpine perspective. *Gondwana Research*, 18, 147–166.

Berger, A. and Bousquet, R. (2008). Subduction-related metamorphism in the Alps: Review of isotopic ages based on petrology and their geodynamic consequences. *Geological Society, London, Special Publications*, 298, 117–144.

Berger, A., Mercolli, I., Engi, M. (2005). The central Lepontine Alps: Notes accompanying the tectonic and petrographic map sheet Sopra Ceneri (1: 100,000). *Schweizerische Mineralogische und Petrographische Mitteilungen*, 85, 109–146.

Beyssac, O., Goffé, B., Chopin, C., Rouzaud, J. (2002). Raman spectra of carbonaceous material in metasediments: A new geothermometer. *Journal of Metamorphic Geology*, 20, 859–871.

Bousquet, R., Oberhansli, R., Goffé, B., Jolivet, L., Vidal, O. (1998). High-pressure-low-temperature metamorphism and deformation in the Bundnerschiefer of the Engadine window: Implications for the regional evolution of the eastern Central Alps. *J. Metamorph. Geol.*, 16, 657–674.

Bousquet, R., Goffé, B., Vidal, O., Oberhansli, R., Patriat, M. (2002). The tectono-metamorphic history of the Valaisan domain from the Western to the Central Alps: New constraints on the evolution of the Alps. *Geol. Soc. Am. Bull.*, 114, 207–225.

Bousquet, R., Engi, M., Gosso, G., Berger, A., Spalla, M.I., Zucali, M., Goffé, B. (2004). Explanatory notes to the map: Metamorphic structure of the Alps. Project, CGMW – Metamorphic and Magmatic Maps.

Bousquet, R., Oberhänsli, R., Goffé, B., Wiederkehr, M., Koller, F., Schmid, S.M., Schuster, R., Engi, M., Berger, A., Martinotti, G. (2008). Metamorphism of metasediments at the scale of an orogen: A key to the Tertiary geodynamic evolution of the Alps. *Geological Society, London, Special Publications*, 298, 393–411.

Bousquet, R., Oberhänsli, R., Schmid, S.M., Berger, A., Wiederkehr, M., Robert, C., Möller, A., Rosenberg, C.L., Zeilinger, G., Molli, G., Koller, F. (2012). Metamorphic framework of the Alps. Map 1: 1,000,000. CCGM/CGMW (Commission for the Geological Map of the World), Paris [Online]. Available at: https://ccgm.org/en/product/carte-metamorphique-des-alpes/.

Calahorrano, A., Sallarès, V., Collot, J.-Y., Sage, F., Ranero, C.R. (2008). Nonlinear variations of the physical properties along the southern Ecuador subduction channel: Results from depth-migrated seismic data. *Earth and Planetary Science Letters*, 267, 453–467.

Chopin, C. (1984). Coesite and pure pyrope in high-grade blueschists of the Western Alps: A first record and some consequences. *Contributions to Mineralogy and Petrology*, 86, 107–118.

Chopin, C. and Schertl, H.-P. (1999). The UHP unit in the Dora-Maira massif, western Alps. *International Geology Review*, 41, 765–780.

Coltice, N., Husson, L., Faccenna, C., Arnould, M. (2019). What drives tectonic plates? *Science advances*, 5, eaax4295.

Dal Piaz, G.V. (1971). Alcune considerazioni sulla genesi delle ofioliti piemontesi e dei giacimenti ad esse associati. *Boll. Associazione Mineraria Subalpina, Torino*, 8, 365–388.

Dal Piaz, G.V. and Ernst, W. (1978). Areal geology and petrology of eclogites and associated metabasites of the Piemonte ophiolite nappe, Breuil–St. Jacques area, Italian Western Alps. *Tectonophysics*, 51, 99–126.

Dal Piaz G.V., Hunziker J.C., Martinotti G. (1972). La zona Sesia-Lanzo e l'evoluzione tettonico-metamorfica delle Alpi nord-occidentali interne. *Memorie della Società Geologica Italiana*, 11, 433–466.

Dal Piaz, G.V., Cortiana, G., Del Moro, A., Martin, S., Pennacchioni, G., Tartarotti, P., (2001). Tertiary age and paleostructural inferences of the eclogitic imprint in the Austroalpine outliers and Zermatt–Saas ophiolite, western Alps. *International Journal of Earth Sciences*, 90, 668–684.

De Wever, P. and Caby, R. (1981). Datation de la base des schistes lustrés postophiolitiques par des radiolaires (Oxfordien-Kimmeridgien moyen) dans les Alpes Cottiennes (Saint Véran, France). *Compte rendu de l'Académie des Sciences de Paris*, 292, 467–472.

Debelmas, J. and Lemoine, M. (1970). The western Alps: Palaeogeography and structure. *Earth-Science Reviews*, 6, 221–256.

Dercourt, J., Zonenshain, L., Ricou, L.-E., Kazmin, V., Le Pichon, X., Knipper, A., Grandjacquet, C., Sbortshikov, I., Geyssant, J., Lepvrier, C. (1986). Geological evolution of the Tethys belt from the Atlantic to the Pamirs since the Lias. *Tectonophysics*, 123, 241–315.

Deville, E., Fudral, S., Lagabrielle, Y., Marthaler, M., Sartori, M. (1992). From oceanic closure to continental collision: A synthesis of the "Schistes lustrés" metamorphic complex of the Western Alps. *Geological Society of America Bulletin*, 104, 127–139.

Dingeldey, C., Dallmeyer, R.D., Koller, F., Massonne, H.-J. (1997). P-T-t history of the Lower Austroalpine Nappe Complex in the "Tarntaler Berge" NW of the Tauern Window: Implications for the geotectonic evolution of the central Eastern Alps. *Contributions to Mineralogy and Petrology*, 129, 1–19.

Droop, G.T. and Chavrit, D. (2014). Eclogitic metagabbro from the Lanzada Window, eastern Central Alps: Confirmation of subduction beneath the Malenco Unit. *Swiss Journal of Geosciences*, 107, 113–128.

Elter, G. (1971). Schistes lustrés et ophiolites de la zone piémontaise entre Orco et Doire Baltée (Alpes Graies). Hypothèses sur l'origine des ophiolites. *Géologie Alpine*, 47, 147–169.

Engi, M., Bousquet, R., Berger, A. (2004). Explanatory notes to the map: Metamorphic structure of the Alps. Project, CGMW – Metamorphic and Magmatic Maps.

Ernst, W. (1971). Metamorphic zonations on presumably subducted lithospheric plates from Japan, California and the Alps. *Contributions to Mineralogy and Petrology*, 34, 43–59.

Forsyth, D. and Uyeda, S. (1975). On the relative importance of the driving forces of plate motion. *Geophys. J. R. Astr. Soc.*, 43, 163–200.

Frey, M. and Ferreiro Mählmann, R. (1999). Alpine metamorphism of the Central Alps. *Schweizerische Mineralogische und Petrographische Mitteilungen*, 79, 135–154.

Froitzheim, N. (2020). Deep subduction and exhumation of continental crust in the Alps. *EGU General Assembly 2020*, EGU2020-8378. DOI: 10.5194/egusphere-egu2020-8378.

Froitzheim, N., Schmid, S.M., Conti, P. (1994). Repeated change from crustal shortening to orogen-parallel extension in the Austroalpine units of Graubünden. *Eclogae Geol. Helv.*, 87, 559-612.

Froitzheim, N., Schmid, S.M., Frey, M. (1996). Mesozoic paleogeography and the timing of eclogite-facies metamorphism in the Alps: A working hypothesis. *Eclogae Geologicae Helvetiae*, 89, 81.

Gebauer, D., Schertl, H.-P., Brix, M., Schreyer, W. (1997). 35 Ma old ultrahigh-pressure metamorphism and evidence for very rapid exhumation in the Dora Maira Massif, Western Alps. *Lithos*, 41, 5–24.

Ghignone, S., Balestro, G., Gattiglio, M., Borghi, A. (2020). Structural evolution along the Susa Shear Zone: The role of a first-order shear zone in the exhumation of meta-ophiolite units (Western Alps). *Swiss Journal of Geosciences*, 113, 1–16.

Goffé, B. and Chopin, C. (1986). High-pressure metamorphism in the Western Alps: Zoneography of metapelites, chronology and consequences. *Schweizerische mineralogische und petrographische Mitteilungen*, 66, 41–52.

Goffé, B., Goffé-Urbano, G., Saliot, P. (1973). Sur la présence d'une variété magnésienne de la ferrocarpholite en Vanoise (Alpes françaises) : sa signification probable dans le métamorphisme alpin. *Comptes rendus de l'Académie des Sciences Paris*, 277, 1965–1968.

Goffé, B., Schwartz, S., Lardeaux, J.-M., Bousquet, R. (2004). Metamorphic structure of the Western and Ligurian Alps. *Mitt. Osterr. Miner. Ges.*, 149.

Groppo, C.T., Castelli, D.C.C., Rolfo, F. (2007). HT, Pre-Alpine relics in a spinel-bearing dolomite marble from the UHP Brossasco-Isasca Unit (Dora-Maira Massif, western Alps). *Periodico di Mineralogia*, 76(2–3), 155–168.

Groppo, C.T., Beltrando, M., Compagnoni, R. (2009). The P–T path of the ultra-high pressure Lago di Cignana and adjoining high-pressure meta-ophiolitic units: Insights into the evolution of the subducting Tethyan slab. *Journal of Metamorphic Geology*, 27, 207–231.

Groß, P., Handy, M.R., John, T., Pestal, G., Pleuger, J. (2020). Crustal-scale sheath folding at HP conditions in an exhumed alpine subduction zone (Tauern window, Eastern Alps). *Tectonics*, 39, e2019TC005942.

Handy, M.R. and Oberhänsli, R. (2004). Explanatory notes to the map: Metamorphic structure of the Alps – Age map of metamorphic structure of the Alps – Tectonic interpretation and outstanding problems. *Mitt. Österr. Miner. Ges.*, 149, 201–225.

Handy, M.R., Herwegh, M., Kamber, B.S., Tietz, R., Villa, I.M. (1996). Geochronologic, petrologic and kinematic constraints on the evolution of the Err-Platta boundary, part of a fossil continent-ocean suture in the Alps (eastern Switzerland). *Schweizerische Mineralogische und Petrographische Mitteilungen*, 76, 453–474.

Handy, M.R., Schmid, S.M., Bousquet, R., Kissling, E., Bernoulli, D. (2010). Reconciling plate-tectonic reconstructions of Alpine Tethys with the geological–geophysical record of spreading and subduction in the Alps. *Earth-Science Reviews*, 102, 121–158.

Herviou C., Verlaguet A., Agard P., Locatelli M., Raimbourg H., Lefeuvre B., Dubacq B. (2021). Along dip variations of subduction fluids: The 30-80 km depth traverse of the Schistes Lustrés complex (Queyras-Monviso, W. Alps). *Lithos*, 394–395, 106168

Jolivet, L., Faccenna, C., Goffé, B., Burov, E., Agard, P. (2003). Subduction tectonics and exhumation of high-pressure metamorphic rocks in the Mediterranean orogens. *Am. J. Sci.*, 303, 353–409.

Kästle, E.D., Rosenberg, C.L., Boschi, L., Bellahsen, N., Meier, T., El-Sharkawy, A. (2019). Slab break-offs in the Alpine subduction zone. *International Journal of Earth Sciences*, 109, 587–603.

Keller, L.M., Hess, M., Fügenschuh, B., Schmid, S.M. (2005). Structural and metamorphic evolution of the Camughera–Moncucco, Antrona and Monte Rosa units southwest of the Simplon line, Western Alps. *Eclogae Geologicae Helvetiae*, 98, 19–49.

Kienast, J. and Velde, B. (1970). Le métamorphisme alpin dans les Alpes franco-italiennes : mise en évidence d'un gradient de température et de pression. *CR Acad. Sci.*, 271, 637–640.

Koller, F. (2003). 5th workshop of Alpine geological studies. Field trip guide E5. Low T–High P metamorphism in the Tarntal Mountains (Lower Austroalpine Unit). *Geologisch-Paläontologische Mitteilungen Innsbruck*, 26, 47–59.

Kusky, T.M., Windley, B.F., Safonova, I., Wakita, K., Wakabayashi, J., Polat, A., Santosh, M. (2013). Recognition of ocean plate stratigraphy in accretionary orogens through Earth history: A record of 3.8 billion years of sea floor spreading, subduction, and accretion. *Gondwana Research*, 24, 501–547.

Lagabrielle, Y. and Cannat, M. (1990). Alpine Jurassic ophiolites resemble the modern central Atlantic basement. *Geology*, 18, 319–322.

Lagabrielle, Y., Brovarone, A.V., Ildefonse, B. (2015). Fossil oceanic core complexes recognized in the blueschist metaophiolites of Western Alps and Corsica. *Earth-Sci. Rev.*, 141, 1–26.

Lapen, T.J., Johnson, C.M., Baumgartner, L.P., Mahlen, N.J., Beard, B.L., Amato, J.M. (2003). Burial rates during prograde metamorphism of an ultra-high-pressure terrane: An example from Lago di Cignana, western Alps, Italy. *Earth and Planetary Science Letters*, 215, 57–72.

Lefeuvre, B., Agard, P., Verlaguet, A., Dubacq, B., Plunder, A. (2020). Massive lawsonite formation in carbonate-rich subducted metasediments: Implications for mass transfer, decarbonation and fluid-mediated processes (Schistes Lustrés, W. Alps), *Lithos*, 370, 105629.

Lemoine, M. and Tricart, P. (1986). Les Schistes lustrés piémontais des Alpes occidentales : approche stratigraphique, structurale et sédimentologique. *Eclogae Geologicae Helvetiae*, 79, 271–294.

Lemoine, M., Steen, D., Vuagnat, M. (1970). Sur le probleme stratigraphique des ophiolites piemontaises et de roches sédimentaires associees : observations dans le massif de Chabriire en Haute-Ubaye (Basses-Alpes, France), *CR Sot. Sci. Nat. GenGve*, 5, 44–59.

Lemoine, M., Marthaler, M., Caron, M., Sartori, M., Amaudric du Chaffaut, S. (1984). Découverte de foraminifères planctoniques du Crétacé supérieur dans les schistes lustrés du Queyras (Alpes occidentales). Conséquences paléogéographiques et tectoniques. *Comptes-rendus des séances de l'Académie des Sciences. Série 2 : Mécanique-physique, chimie, sciences de l'univers, sciences de la terre*, 299, 727–732.

Lemoine, M., Bas, T., Arnaud-Vanneau, A., Arnaud, H., Dumont, T., Gidon, M., Bourbon, M., de Graciansky, P.-C., Rudkiewicz, J.-L., Megard-Galli, J. (1986). The continental margin of the Mesozoic Tethys in the Western Alps. *Marine and Petroleum Geology*, 3, 179–199.

Lemoine, M., Tricart, P., Boillot, G. (1987). Ultramafic and gabbroic ocean floor of the Ligurian Tethys (Alps, Corsica, Apennines): In search of a genetic imodel. *Geology*, 15, 622–625.

Lippitsch, R., Kissling, E., Ansorge, J. (2003). Upper mantle structure beneath the Alpine orogen from high-resolution teleseismic tomography. *Journal of Geophysical Research: Solid Earth*, 108.

Locatelli, M., Verlaguet, A., Agard, P., Federico, L., Angiboust, S. (2018). Intermediate-depth brecciation along the subduction plate interface (Monviso eclogite, W. Alps). *Lithos*, 320, 378–402.

Lombardo, B. (1978). Osservazioni preliminari sulle ofioliti metamorfiche del Monviso (Alpi Occidentali). *Rendiconti Della Societa Italiana di Mineralogia e Petrologia*, 34, 235–305.

Manatschal, G. and Müntener, O. (2009). A type sequence across an ancient magma-poor ocean-continent transition: The example of the western Alpine Tethys ophiolites. *Tectonophysics*, 473, 4–19.

Manzotti, P., Ballevre, M., Zucali, M., Robyr, M., Engi, M. (2014). The tectonometamorphic evolution of the Sesia–Dent Blanche nappes (internal Western Alps): Review and synthesis. *Swiss Journal of Geosciences*, 107, 309–336.

Manzotti, P., Poujol, M., Ballèvre, M. (2015). Detrital zircon geochronology in blueschist-facies meta-conglomerates from the Western Alps: Implications for the late Carboniferous to early Permian palaeogeography. *International Journal of Earth Sciences*, 104, 703–731.

Manzotti, P., Ballèvre, M., Poujol, M., (2016). Detrital zircon geochronology in the Dora-Maira and Zone Houillère: A record of sediment travel paths in the Carboniferous. *Terra Nova*, 28, 279–288.

McCarthy, A., Chelle-Michou, C., Müntener, O., Arculus, R., Blundy, J. (2018). Subduction initiation without magmatism: The case of the missing Alpine magmatic arc. *Geology*, 46, 1059–1062.

Merle, O. and Brun, J. (1984). The curved translation path of the Parpaillon Nappe (French Alps). *Journal of Structural Geology*, 6, 711–719.

Michard, A. (1977). Charriages et métamorphisme haute pression dans les Alpes cottiennes méridionales : à propos des schistes à jadéite de la bande d'Acceglio. *Bulletin de la Société Géologique de France*, 7, 883–892.

Michard, A., Avigad, D., Goffé, B., Chopin, C. (2004). The high-pressure metamorphic front of the south Western Alps (Ubaye-Maira transect, France, Italy). *Schweiz. Mineral. Petrogr. Mitt.*, 84, 215–235.

Mohn, G., Manatschal, G., Masini, E., Müntener, O., (2011). Rift-related inheritance in orogens: A case study from the Austroalpine nappes in Central Alps (SE-Switzerland and N-Italy). *International Journal of Earth Sciences*, 100, 937–961.

Mueller, P., Maino, M., Seno, S., (2020). Progressive deformation patterns from an accretionary prism (Helminthoid Flysch, Ligurian Alps, Italy). *Geosciences*, 10, 26.

Müntener, O. and Hermann, J. (1996). The Val Malenco lower crust–upper mantle complex and its field relations (Italian Alps). *Schweizerische Mineralogische und Petrographische Mitteilungen*, 76, 475–500.

Nagel, T.J. (2008). Tertiary subduction, collision and exhumation recorded in the Adula nappe, central Alps. *Geological Society, London, Special Publications*, 298, 365–392.

Oberhansli, R. and Goffé, B. (2004). Explonatory notes to the map: Metamorphic structure of the Alps introduction. *Mitteilungen der Österreichischen Mineralogischen Gesellschaft*, 149, 115–123.

Picazo, S., Müntener, O., Manatschal, G., Bauville, A., Karner, G., Johnson, C. (2016). Mapping the nature of mantle domains in Western and Central Europe based on clinopyroxene and spinel chemistry: Evidence for mantle modification during an extensional cycle. *Lithos*, 266, 233–263.

Picazo, S.M., Ewing, T.A., Müntener, O. (2019). Paleocene metamorphism along the Pennine–Austroalpine suture constrained by U–Pb dating of titanite and rutile (Malenco, Alps). *Swiss Journal of Geosciences*, 112, 517–542.

Plunder, A., Agard, P., Dubacq, B., Chopin, C., Bellanger, M. (2012). How continuous and precise is the record of P-T paths? Insights from combined thermobarometry and thermodynamic modelling into subduction dynamics (Schistes Lustres, W. Alps). *J. Metamorph. Geol.*, 30, 323–346.

Pognante, U. and Kienast, J.-R. (1987). Blueschist and eclogite transformations in Fe-Ti gabbros: A case from the Western Alps ophiolites. *Journal of Petrology*, 28, 271–292.

Polino, R., Dal Piaz, G.V., Gosso, G. (1990). Tectonic erosion at the Adria Margin and accretionary processes for the Cretaceous orogeny of the Alps. In *Deep Structure of the Alps*, Roure, F., Heitzman, P., Polino, R. (eds). Volume Speciale Società Geologica Italiana, Rome.

Polino, R., Dela Pierre, F., Borghi, A., Carraro, F., Fioraso, G., Giardino, M., with contributions from Bellardone, G., Conti, A., Gattiglio, M., Malusà, M., Mosca, P. (2002). Note illustrative della Carta Geologica d'Italia alla scala 1:50.000, Foglio 132-152-153 Bardonecchia. Servizio Geologico d'Italia – Regione Piemonte, Direzione Regionale Servizi Tecnici di Prevenzione, Turin.

Rebay, G., Zanoni, D., Langone, A., Luoni, P., Tiepolo, M., Spalla, M.I. (2018). Dating of ultramafic rocks from the Western Alps ophiolites discloses Late Cretaceous subduction ages in the Zermatt-Saas Zone. *Geological Magazine*, 155, 298–315.

Reinecke, T. (1998). Prograde high-to ultrahigh-pressure metamorphism and exhumation of oceanic sediments at Lago di Cignana, Zermatt-Saas Zone, western Alps. *Lithos*, 42, 147–189.

Ring, U. (1992). The Alpine geodynamic evolution of Penninic nappes in the eastern Central Alps: Geothermobarometric and kinematic data. *Journal of Metamorphic Geology*, 10, 33–53.

Ring, U., Ratschbacher, L., Frisch, W., Durr, S., Borchert, S. (1991). Late-stage east-directed deformation in the Platta, Arosa, and lowermost Austroalpine nappes (eastern Swiss Alps). *Neues Jahrbuch für Geologie und Paläontologie-Monatshefte*, 357–363.

Rosenbaum, G. and Lister, G.S. (2005). The Western Alps from the Jurassic to Oligocene: Spatio-temporal constraints and evolutionary reconstructions. *Earth-Science Reviews*, 69, 281–306.

Royden, L. and Faccenna, C. (2018). Subduction orogeny and the Late Cenozoic evolution of the Mediterranean arcs. *Annual Review of Earth and Planetary Sciences*, 46, 261–289.

Rubatto, D. and Angiboust, S. (2015). Oxygen isotope record of oceanic and high-pressure metasomatism: A P–T–time–fluid path for the Monviso eclogites (Italy). *Contributions to Mineralogy and Petrology*, 170, 44.

Rubatto, D. and Hermann, J. (2003). Zircon formation during fluid circulation in eclogites (Monviso, Western Alps): Implications for Zr and Hf budget in subduction zones. *Geochimica et Cosmochimica Acta*, 67, 2173–2187.

Rubatto, D., Müntener, O., Barnhoorn, A., Gregory, C. (2008). Dissolution-reprecipitation of zircon at low-temperature, high-pressure conditions (Lanzo Massif, Italy). *American Mineralogist*, 93, 1519–1529.

Ruh, J.B., Le Pourhiet, L., Agard, P., Burov, E., Gerya, T. (2015). Tectonic slicing of subducting oceanic crust along plate interfaces: Numerical modeling. *Geochemistry, Geophysics, Geosystems*, 16, 3505–3531.

Schmid, S.M. and Kissling, E. (2000). The arc of the western Alps in the light of geophysical data on deep crustal structure. *Tectonics*, 19, 62–85.

Schmid, S.M., Fügenschuh, B., Kissling, E., Schuster, R. (2004). Tectonic map and overall architecture of the Alpine orogen. *Eclogae Geologicae Helvetiae*, 97, 93–117.

Schmid, S.M., Fügenschuh, B., Kounov, A., Matenco, L., Nievergelt, P., Oberhänsli, R., Pleuger, J., Schefer, S., Schuster, R., Tomljenovic, B., Ustaszewski, K., van Hinsbergen, D.J.J. (2020). Tectonic units of the Alpine collision zone between Eastern Alps and western Turkey. *Gondwana Research*, 78, 308–374.

Schwartz, S., Guillot, S., Reynard, B., Lafay, R., Debret, B., Nicollet, C., Lanari, P., Auzende, A.L. (2013). Pressure–temperature estimates of the lizardite/antigorite transition in high-pressure serpentinites. *Lithos*, 178, 197–210.

Simon-Labric, T., Rolland, Y., Dumont, T., Heymes, T., Authemayou, C., Corsini, M., Fornari, M. (2009). 40Ar/39Ar dating of Penninic Front tectonic displacement (W. Alps) during the Lower Oligocene (31–34 Ma). *Terra Nova*, 21, 127–136.

Stampfli, G. and Marchant, R. (1997). Geodynamic evolution of the Tethyan margins of the Western Alps. In *Deep Structure of the Swiss Alps-Results from NRP 20*, Plomerová, J. (ed.). Birkhaüser Verlag AG, Basel.

Stampfli, G., Mosar, J., Marquer, D., Marchant, R., Baudin, T., Borel, G. (1998). Subduction and obduction processes in the Swiss Alps. *Tectonophysics*, 296, 159–204.

Tamblyn, R., Zack, T., Schmitt, A., Hand, M., Kelsey, D., Morrissey, L., Pabst, S., Savov, I. (2019). Blueschist from the Mariana forearc records long-lived residence of material in the subduction channel. *Earth and Planetary Science Letters*, 519, 171–181.

Tartarotti, P., Zucali, M., Panseri, M., Lissandrelli, S., Capelli, S., Ouladdiaf, B. (2011). Mantle origin of the Antrona serpentinites (Antrona ophiolite, Pennine Alps) as inferred from microstructural, microchemical, and neutron diffraction quantitative texture analysis. *Ofioliti*, 36, 167–189.

Tartarotti, P., Guerini, S., Rotondo, F., Festa, A., Balestro, G., Bebout, G.E., Cannaò, E., Epstein, G.S., Scambelluri, M. (2019). Superposed sedimentary and tectonic block-in-matrix fabrics in a subducted serpentinite mélange (High-pressure Zermatt saas ophiolite, western Alps). *Geosciences*, 9, 358.

Thöni, M. (1999). A review of geochronological data from the Eastern Alps. *Schweizerische Mineralogische und Petrographische Mitteilungen*, 79, 209–230.

Tricart, P. and Lemoine, M. (1991). The Queyras ophiolite West of Monte Viso (Western Alps): Indicator of a peculiar ocean floor in the Mesozoic Tethys. *Journal of Geodynamics*, 13, 163–181.

Tricart, P., Bourbon, M., Lagabrielle, Y. (1982). Révision de la coupe Péouvou-Roche Noire (zone piémontaise, Alpes franco-italiennes) : bréchification synsédimentaire d'un fond océanique ultrabasique. *Géol. Alp. Grenoble*, 58, 105–113.

Trommsdorff, V., Piccardo, G., Montrasio, A. (1993). From magmatism through metamorphism to sea floor emplacement of subcontinental Adria lithosphere during pre-Alpine rifting (Malenco, Italy). *Schweizerische Mineralogische und Petrographische Mitteilungen*, 73, 191–203.

Van Hinsbergen, D.J., Torsvik, T.H., Schmid, S.M., Maţenco, L.C., Maffione, M., Vissers, R.L., Gürer, D., Spakman, W. (2020). Orogenic architecture of the Mediterranean region and kinematic reconstruction of its tectonic evolution since the Triassic. *Gondwana Research*, 81, 79–229.

Vissers, R.L., van Hinsbergen, D.J., Meijer, P.T., Piccardo, G.B. (2013). Kinematics of Jurassic ultra-slow spreading in the Piemonte Ligurian ocean. *Earth and Planetary Science Letters*, 380, 138–150.

Vitale Brovarone, A. and Agard, P. (2013). True metamorphic isograds or tectonically sliced metamorphic sequence? New high-spatial-resolution petrological data for the New Caledonia case study. *Contrib. Mineral. Petrol.*, 166, 451–469.

Vitale Brovarone, A., Picatto, M., Beyssac, O., Lagabrielle, Y., Castelli, D. (2014). The blueschist-eclogite transition in the Alpine chain: P-T paths and the role of slow-spreading extensional structures in the evolution of HP-LT mountain belts. *Tectonophysics*, 615, 96–121.

Von Raumer, J., Abrecht, J., Bussy, F., Lombardo, B., Menot, R.-P., Schaltegger, U. (1999). The Paleozoic metamorphic evolution of the Alpine external massifs. *Schweizerische Mineralogische und Petrographische Mitteilungen*, 79, 5–22.

Wada, I., Wang, K., He, J., Hyndman, R.D. (2008). Weakening of the subduction interface and its effects on surface heat flow, slab dehydration, and mantle wedge serpentinization. *Journal of Geophysical Research: Solid Earth*, 113.

Wagreich, M. (1995). Subduction tectonic erosion and Late Cretaceous subsidence along the northern Austroalpine margin (Eastern Alps, Austria). *Tectonophysics*, 242, 63–78.

Wagreich, M. and Decker, K. (2001). Sedimentary tectonics and subsidence modelling of the type Upper Cretaceous Gosau basin (Northern Calcareous Alps, Austria). *International Journal of Earth Sciences*, 90, 714–726.

Wakabayashi, J. (2015). Anatomy of a subduction complex: Architecture of the Franciscan Complex, California, at multiple length and time scales. *International Geology Review*, 57, 669–746.

Wakabayashi, J. (2017). Structural context and variation of ocean plate stratigraphy, Franciscan Complex, California: Insight into mélange origins and subduction-accretion processes. *Progress in Earth and Planetary Science*, 4, 18.

Weh, M. and Froitzheim, N. (2001). Penninic cover nappes in the Prattigau half-window (Eastern Switzerland): Structure and tectonic evolution. *Eclogae Geologicae Helvetiae*, 94, 237–252.

White, D.A., Roeder, D.H., Nelson, T.H., Crowell, J.C. (1970). Subduction. *Geological Society of America Bulletin*, 81, 3431–3432.

Wiederkehr, M., Sudo, M., Bousquet, R., Berger, A., Schmid, S.M. (2009). Alpine orogenic evolution from subduction to collisional thermal overprint: The 40Ar/39Ar age constraints from the Valaisan Ocean, central Alps. *Tectonics*, 28.

Wiederkehr, M., Bousquet, R., Ziemann, M.A., Berger, A., Schmid, S.M. (2011). 3-D assessment of peak-metamorphic conditions by Raman spectroscopy of carbonaceous material: An example from the margin of the Lepontine dome (Swiss Central Alps). *International Journal of Earth Sciences*, 100, 1029–1063.

Zhao, L., Paul, A., Guillot, S., Solarino, S., Malusà, M.G., Zheng, T., Aubert, C., Salimbeni, S., Dumont, T., Schwartz, S. (2015). First seismic evidence for continental subduction beneath the Western Alps. *Geology*, 43, 815–818.

Plate 4.1. *Typical oceanic sediments deposited on ophiolitic seafloor in the Western Alps (i.e. calcschists, pelagic limestones and shales, radiolarian cherts. See Figure 4.2(e), which also shows detrital intercalations (ophiolitic arenites, breccias and olistoliths))*

COMMENTS ON PLATE 4.1.– *a) Direct sedimentary contact between serpentinized and brecciated peridotite criss-crossed by calcite veins ("ophicalcite") and metasediments (Col d'Urine, Pelvas d'Abries; see Lagabrielle et al. 2015). The first horizon is a fine-grained mixture of ultramafic debris and pelagic sediments. The sequence is progressively richer in sedimentary fraction, then changes to carbonate-dominated. These rocks were affected by blueschist facies metamorphism. b) Ophicalcite (Queyras). c) Calcschist with ultramafic and mafic pebbles (Zermatt-Saas area). d) Calcschist with gabbroic clasts (Avic area). e) Calcschist hosting a former debris flow made of mafic and sedimentary rocks (near Cignana, Zermatt-Saas area). f) Boudinaged mafic horizon in impure marble (Zermatt-Saas area). g and h) Typical, more marble-looking calcschists found in mafic and ultramafic units, particularly in eclogitized ones. i) Coesite-bearing quartzites (meta-cherts) from Lago di Cignana (Reinecke et al. 1991, 1998).*

Plate 4.2. *Typical lithologies encountered in the Schistes Lustrés, emphasizing the variable pelitic fractions in the calcschists. a–d) Strongly pelitic and organic-rich calcschists, with numerous quartz ± calcite veins, some carpholite-bearing (in d). Shearing and boudinage of the veins in b) corresponds to the early exhumation deformation stage (~45 Ma for the upper S unit; D2: Agard et al. 2001, 2002). e) Alternation of pelitic- and carbonate-rich horizons. The pelitic fraction is littered with black lawsonite crystals. f and g) Regular alternations of pelitic- (white arrows) and carbonate-rich horizons (black arrows), overall more carbonate-rich (particularly in f). h–j) Carbonate-rich sections, with alternating carbonates and cherts (grey arrow) in i). Arrows in h) indicate the larger fold hinges. Brittle-ductile shear bands indicating top to the west shear senses are coeval with the late exhumation stage in the area (~38–37 Ma; D3: Agard et al. 2001, 2002; see Figure 4.6(a), step 3)*

Plate 4.3. *Panoramic views across the subducted oceanic units of the Western Alps*

COMMENTS ON PLATE 4.3.– *a) Typical landscape of the metasedimentary Schistes Lustrés (view to the NW from the path to Col d'Urine, near Abries; Queyras). Note the regular west-dipping schistosity and variations in erodability associated with subtle changes in the sedimentary protolith and/or presence of pre-rift extensional allochthons or mafic/ultramafic bodies. b) Recognizable Late Jurassic marble horizon of Crête de la Taillante, north of Col Agnel (Queyras). They tend to be more abundant in the eastern (lower) SL units. c) Large km-scale extensional allochthon in the Haute Ubaye valley, in the western part of the Queyras Schistes Lustrés. The well-stratified dolomitic Triassic limestone from the Peouvou, which dominates the scenery to the right/east of the picture, is wrapped up in organic-rich calcschists (left part of the picture). Dm-scale, very fresh blueschist facies carpholite crystals can be found in the screen from landslides reworking the calschists. d) Carbonates directly overlapping onto a large mafic body (north of Rocca Bianca, Queyras; see also Lagabrielle et al. 2015). e) View to the east from Pain de Sucre (courtesy Baxter) looking onto Monviso in the clouds. The carbonate-dominated calcschists (and very lawsonite-rich) form abrupt ridges highlighting the schistosity. Note the contact with softer metasediments in the lower right of the picture. f) The iconic Matterhorn/Cervinio massif (view to the west from ~Brittania Hut, Zermatt). The relatively soft slopes of the oceanic units are overlooked by the continental Dent Blanche unit at the summit. g and h) Large, several km-thick metaperidotite bodies are found near Avic, as also shown by the sparse vegetation. Some hm-scale gabbroic horizons are found here and there (as in g). i) View onto the west-dipping contact separating the Monviso and the blueschist to eclogite facies Monviso unit (left) from the eclogitic Lago Superiore unit (Viso Mozzo). j) View onto the west-dipping contact separating the Lago Superiore unit from the thin sedimentary cover of one of the continental Dora Maira units. Note the distinctive color of the basal metaperidotites from the oceanic unit.*

Plate 4.4. *Typical mineralogy of S units across the Western Alps. a) Polished section of a folded calcschist (shown in plate 4.2(d)) literally riddled with lawsonite, here distinctively black from the many inclusions of organic matter in the crystals. Note the fuzzy ankerite horizon separating the carbonate and pelitic horizons reflecting the prograde exchange reaction transforming calcite and chlorite into an assemblage of lawsonite, ankerite and quartz (after Lefeuvre et al. 2020). b) Large cm-scale lawsonite pseudomorphs in mafic-rich horizons from calschists (middle S unit; Figure 4.3(c)). c) Lawsonite-ankerite mode-1 vein adjacent to lawsonite rosettes in the schist (upper S unit, Cottian Alps). d) Lawsonite-bearing quartz vein (with minor ankerite). e) Almost pure and very fresh carpholite-bearing veins in pelitic schist (Ubaye valley, Queyras; plate 4.3(c)). f) Typical chloritoid-bearing calcschist (Val Clavalité, south part of the Aosta valley; see Angiboust and Agard 2010). g) Fresh mm- to cm-scale lawsonite crystals hosting a large diffuse fraction of organic matter (together with rutile) and showing evidence for growth zoning. h) Fresh carpholite crystals from quartz-carpholite veins (Haute-Ubaye valley; courtesy Herviou). Such needles are commonly well-preserved in quartz, i.e. as long as sub-grain migration/rotation processes in quartz are insignificant, allowing detection of a former subduction history in even retrogressed calcschists (e.g. Bousquet et al. 1998, 2002; Agard et al. 2001, 2002). i) Strongly deformed chloritoid-chlorite-bearing mica(calc)schist, with phengite composition attesting to high-pressure conditions. Chloritoid is partly retrogressed here*

Plate 4.5. *Typical mineralogy of MUM units across the Western Alps (a–f) and peculiar features of the Lago Superiore unit, Monviso massif (g–j)*

COMMENTS ON PLATE 4.5.– *a–c) Lawsonite pseudomorphs in Zermatt (a), Avic (b) and Monviso (c). d) Metapillows, pillow-breccias and surrounding hyaloclastites metamorphosed under eclogite facies conditions. Note the contrast between the lawsonite-rich rim of the elliptic former pillow, hydrated during seafloor hydrothermal alteration, and the dry garnet-omphacite-bearing core. Lawsonite pseudomorphs are ubiquitous across km-scale exposures between Zermatt and Täsch. e) Metagabbros boudins, with composition intermediate between magnesian and Fe-Ti rich, in strongly deformed basals (Zermatt-Saas region). f) Garnet-chloritoid-glaucophane ± omphacite-bearing metabasalt (Val Clavalité, Aosta valley). g) View onto one of the best-preserved "slab" fragments in the world (Figure 4.3): the 2–2.5 km-thick Lago Superiore unit (Monviso massif, Italy). Note the tectonic contact with the overlying blueschist facies Monviso unit, whose lithostratigraphy is completely overturned. In the Lago Superiore unit, the pristine lithological succession of the downgoing lithosphere is preserved to a first approximation with, from bottom to top: ~500m thick serpentinized basal peridotites, eclogitized Mg-rich metagabbros, sills, dykes and a discontinuous layer of Fe-Ti metagabbros hosting garnet, omphacite, abundant rutile and lawsonite pseudomorphs, partly retrogressed metabasalts and a few meters of metamorphosed calcschists at the very top, below the Monviso unit. h–j) Eclogite breccia (Lago superiore unit, Monviso massif; Colle di Luca): fragments torn apart (h); set of eclogite fragments boudinaged and later embedded in serpentinites (i); talc-ankerite-rich metasomatic rind formed at the contact of the mafic and ultramafic rocks, largely due to fluid ingression (j). For thin-section views see plate 4.6(c–f).*

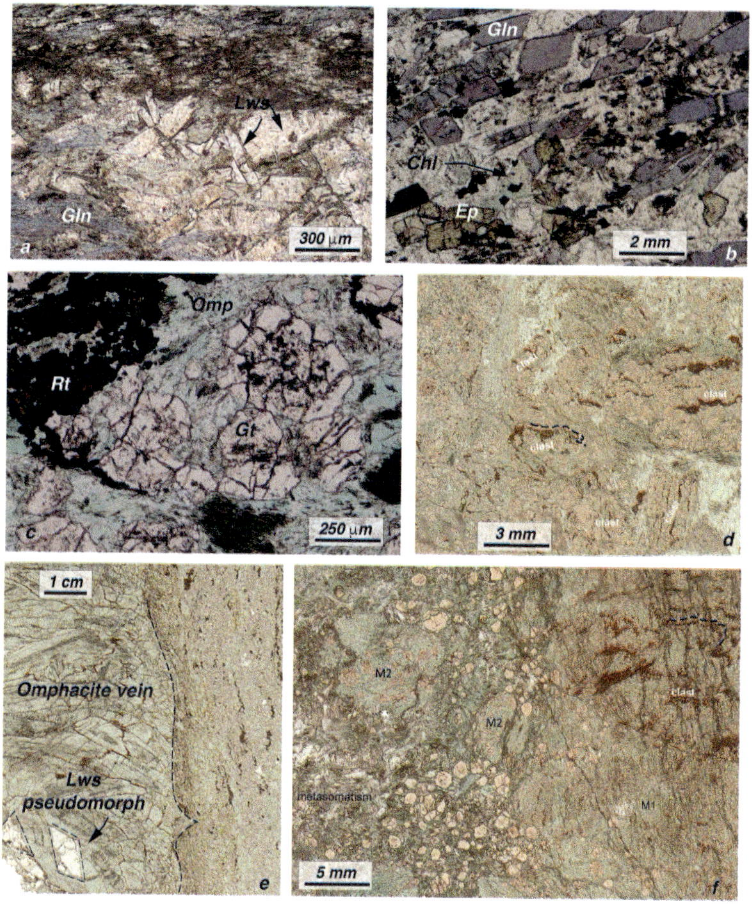

Plate 4.6. *Mineralogy of mafic protoliths in the Western Alps. a) Lawsonite blueschist facies metabasalt from the Queyras, preserving fresh lawsonite and glaucophane (upper S unit; courtesy Ballèvre). b) Epidote-glaucophane-bearing metabasalt from the Cottian Alps (middle S unit). c) Typical garnet-omphacite-rutile-bearing Fe-Ti-rich metagabbro from the Monviso massif (Lago Superiore unit). d) Thin-section view of brecciated eclogite. Clasts are easily identified by their trails of rutile folded before brecciation. e) Omphacite vein forming in a mylonite of Fe-Ti-rich metagabbro. f) Successive generations of the matrix (M1–M3) forming within eclogitic breccias from the Lago superiore unit (Lower Shear zone, Monviso massif; Locatelli et al. 2019b). Fluid in equilibrium with the M1 cement is "internally" derived from adjacent (<~1 m away) Mg-rich metagabbros. In contrast, M2 and particularly M3 show the influence of externally-derived fluids dominated by an ultramafic component. Progressive disaggregation of the rock during M2 preludes to full-scale connectivity (i.e. permeability, fluid ingression and strain localization) across the shear zone*

5

Continental Subduction in the Alps: From Field Data to Kinematic Models

Paola MANZOTTI[1] and Michel BALLÈVRE[2]

[1] *Department of Geological Sciences, Stockholm University, Sweden*

[2] *Géosciences Rennes-UMR 6118, Université de Rennes 1, France*

5.1. Historical background and main questions

With the development of plate tectonics in the 1960s, a major difference was put forward between the denser oceanic lithosphere, which was disappearing into subduction zones, and the lighter continental lithosphere, which was not able to subduct. Plate convergence first resulted in oceanic subduction, which was eventually followed by continental collision to build mountain belts (e.g. Dewey and Bird 1970). High-pressure/low-temperature rocks (i.e. blueschists and eclogites) were thus associated with the subduction of the oceanic lithosphere (e.g. Ernst 1971). In this model, the continental crust was not able to subduct because of its lower density, hence greater buoyancy. However, since the early 1970s, the case for continental subduction was put forward, because of observations in Alpine rocks, with the two most significant being recalled below.

First, early mapping by the Italian geologists of the internal part of the Western Alps led to the identification of a zone displaying garnet-glaucophane-omphacite

Geodynamics of the Alps 2,
coordinated by Claudio L. ROSENBERG and Nicolas BELLAHSEN.

micaschists ("eclogitic micaschists") (Stella 1894). Later on, in this zone, jadeite-quartz assemblages, typifying eclogite-facies conditions, were found in an undeformed pod of metagranite from the Sesia Zone (Monte Mucrone, Figure 5.1(a)) (Compagnoni and Maffeo 1973). This was a hallmark because typical eclogite-facies minerals were found in a magmatic rock iconic for the continental crust (see for example Turner (1981)). Because eclogite-facies assemblages were known in all other lithologies from the Sesia Zone (i.e. not only in the granites, but also in the country-rocks of the granitic intrusions), it was understood that the entire Sesia Zone behaved as a coherent continental body involved in the subduction process, the only location on Earth where the low geothermal gradients may have been responsible for the development of these high-pressure (HP) – low-temperature mineral assemblages. It is thus not surprising that the earliest plate-tectonic models for the Western Alps involved continental subduction, of the Sesia Zone in this case (Dal Piaz et al. 1972)

Second, the report of coesite – a polymorph of SiO_2 stable at very HP – in the whiteschists from the Dora-Maira massif (Chopin 1984), together with specific mineral assemblages in associated rocks (Figure 5.1(b)), implies pressure higher than about 3 GPa. This has given rise to the concept of ultra-HP (UHP) metamorphism (with key indicators coesite, diamond). The Dora-Maira massif is considered a piece of strongly deformed rocks of undisputable continental derivation because it is made of large volumes of orthogneisses deriving from porphyritic granitoids. Assuming that pressure was lithostatic, this indicated that pieces of continental crust may have been subducted to depths of the order of 100 km (or even more) and then returned back to the surface. Because coesite has been found in many other continent-derived (and oceanic) rocks, continental subduction is therefore becoming an integral part of the early history of the subduction-collision belts.

Continental subduction is hereafter defined as continental, dominantly crustal, material recording HP metamorphism (in most cases at eclogite-facies P-T conditions), without any prejudice upon the pre-orogenic position and syn-orogenic burial mechanism. Once the fact of continental subduction was largely accepted (by the end of the 1980s), it immediately opened up an array of major questions, amongst which the most important are summarized below:

– Does continental subduction always succeed oceanic subduction, with the buoyant continental crust being dragged down by the subducting slab of oceanic lithosphere?

– Which parts of the continental crust are subducted? Is it the entire palaeomargin, or just parts of it, for example, extensional allochthons? What is the role of the inherited structures associated with the rifting history of the palaeomargin?

– What is the age of the HP/UHP metamorphism? Is it the same along as well as across the belt, or is this metamorphism diachronous? In the latter case, does this reflect the progressive burial of a single continental plate, or are there several continental domains separated by oceanic domains?

– What are the mechanisms for the exhumation of the HP-UHP rocks? Is erosion the driving force for exhumation? What is the record of erosion in the nearby sedimentary basins? Erosion may have been combined with other tectonic processes, like the buoyant uprise of the HP-UHP bodies, associated or not with plate divergence, slab roll-back, etc. Such questions involve a discussion of the dynamics of the mountain belts, i.e. the forces responsible for the building of the orogenic prism, and their evolution trough time.

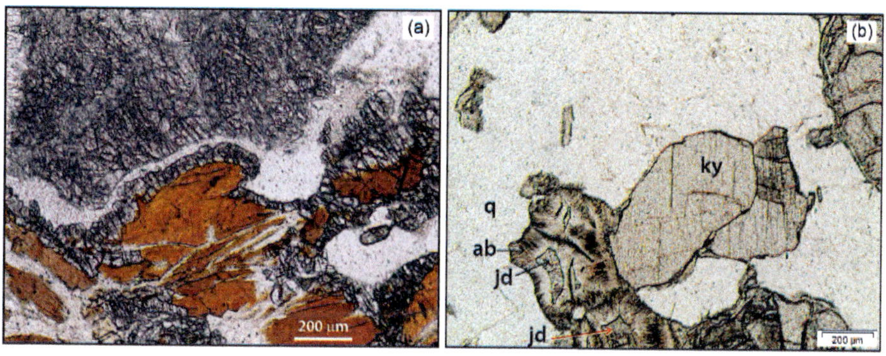

Figure 5.1. *a) Pseudomorphic-coronitic transformation in an eclogite-facies metagranite (Monte Mucrone, Sesia Zone). Magmatic biotite is rimmed by a corona of garnet and high-Si muscovite (phengite) at the contact with the former magmatic plagioclase, now transformed into a very fine-grained aggregate consisting of jadeite, zoisite, kyanite and quartz. Image courtesy of Alessandro Da Mommio. b) A typical UHP paragenesis from the Brossasco-Isasca Unit (Dora-Maira nappe), consisting of garnet, jadeite (partially replaced by fibrous albite), kyanite, quartz, indicating complete recrystallization of the rock above the stability field of paragonite*

To answer these questions in the specific case of the European Alpine belt, this chapter will review the available data – a huge amount of data indeed, the Alps being one of the most intensively studied belts in the world. We will first characterize the type of crustal material that was involved in the subduction process, in order to understand its previous history (lower crust, upper crust or both?) and its location before the subduction (tilted blocks along the distal part of a palaeomargin, extensional allochthons?). We will then review the basic knowledge about the HP-UHP rocks in terms of P-T paths and timing. Finally, a discussion of the present geometry of the subducted continental units will lead to some considerations about the kinematic evolution of the mountain belt.

5.2. A reference P-T space

Many discussions in the following will involve considerations based on the evolving position of the studied rocks in the P-T space. In order to give a consistent overview, all data will be plotted on the same P-T space (Figure 5.2), as follows. Pressure will be labeled in GPa (for clarity, remember that 1 GPa = 10 kbar), and temperature in degrees Celsius.

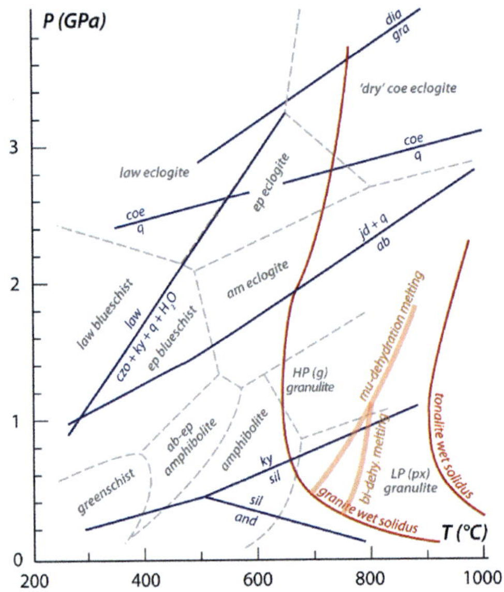

Figure 5.2. *Distribution in the pressure (P) - temperature (T) space of the main metamorphic facies (in italics, after Ernst and Liou (2008)) with respect to some useful reaction curves in different chemical systems. See text for more details. The high-grade amphibolite facies corresponds to the field to the right of the granite wet solidus. Mineral abbreviations from Holland and Powell (1998). LP = low pressure; HP = high pressure*

A few reference lines are plotted on this P-T space, namely:

– three polymorphic transitions (among the aluminosilicates andalusite, sillimanite and kyanite; among the two polymorphs of SiO_2, namely quartz and coesite; and between the two forms of C, namely graphite and diamond);

– the univariant reactions albite = jadeite + quartz and lawsonite = clinozoisite + kyanite + quartz + H_2O; melting reactions at high temperatures. The minimum melting conditions for most crustal material may be approximated by the "granite wet solidus" (i.e. the reaction albite + orthoclase + quartz + H_2O = melt). This reaction is only valid for water-saturated rocks, a condition that is rarely achieved. Therefore, water is extracted from OH-bearing minerals, during reactions involving the breakdown of such minerals. Muscovite-dehydration melting proceeds at lower T with respect to biotite-dehydration melting (Le Breton and Thompson 1988; Spear et al. 1999).

Metamorphic facies	Metabasites	Metapelites
greenschist	ab-chl-act-ep	q-mu-chl q-mu-chl-bi
law blueschist	gl-law-ttn	q-mu-ctd-car-chl
ep blueschist	g-gln-ep-ru	q-mu-g-ctd-chl
law eclogite	law-cpx-g-ru	q-mu-g-ctd
ep eclogite am eclogite	g-cpx-ru-q-am-ep	q-mu-g-ky-ru
"dry" coe eclogite	g-cpx-ru-coe	coe-mu-g-ky-ru
ab-ep amphibolite	ab-chl-hb-ep	q-mu-g-bi-chl
amphibolite	pl-hb-ilm	q-mu-g-st-bi- q-mu-g-bi-ky/sil
high-grade amphibolite	pl-hb-ilm	q-g-bi-sil-melt
LP granulite	pl-cpx-opx-(hb)	q-kfs-g-crd/opx-(bi)-sil-melt
HP granulite	pl-cpx-g-(hb)	q-kfs-g-opx-ky-(bi)-melt

Table 5.1. *Key mineral assemblages in metabasites and metapelites defining the main metamorphic facies. Mineral abbreviations from Holland and Powell (1998)*

Most geologists are familiar with the concept of metamorphic facies (Eskola 1920), defined with respect to the mineral assemblages occurring in rocks of basaltic composition. Other rocks metamorphosed at the same P-T conditions will display different assemblages (Table 5.1). The exact location in the P-T space of the different metamorphic facies has evolved with the progress in experimental petrology. Note that the facies boundaries are, by definition, dependent upon the exact bulk rock composition, and are therefore shown as dashed lines in the P-T space.

The blueschist and eclogite facies occupy a very large range in the P-T space. They are therefore divided for convenience into different domains. Specifically, the lawsonite blueschist facies is located at lower *T* with respect to the epidote blueschist facies. Lawsonite eclogite occurs at lower *T* with respect to the other eclogite types (i.e. epidote eclogite, amphibole eclogite and "dry" coesite-bearing eclogite).

5.3. A reference map and cross-section of the Alps

5.3.1. *Large-scale overview of the Alpine belt*

The Alps is an arcuate mountain belt extending from Nice (France) to Vienna (Austria) (Figure 5.3), for a length of about 1000 km (e.g. Laubscher 1971; Schmid et al. 2004, 2017; Pfiffner 2014). It is made of a stack of nappes deriving from both continental and oceanic domains, some of them having been involved in the subduction zone before the final collision. The former European palaeomargin consists of (i) Mesozoic sequences outcropping in the Jura Mountains and the external fold-and-thrust belt and (ii) reworked Paleozoic basements making up the bulk of the External Crystalline Massifs (e.g. Bellahsen et al. 2014). The distal part of the European palaeomargin (i.e. the sub-Penninic nappes) has been severely deformed and metamorphosed, and it includes some eclogite-facies units (Adula nappe, Eclogite Zone in the Tauern window) that will be discussed below. The Penninic nappes include units derived from the Valais basin (Lower Penninic) and the Piemonte-Liguria Ocean (Upper Penninic), and a complex stack of units (Middle Penninic) derived from the intervening continental Briançonnais Domain. The Adriatic palaeomargin is represented by the Sesia-Dent Blanche nappes, all involved in the subduction zone, and the Austro-Alpine nappes, some of them having been dragged down to eclogite-facies conditions (Ötztal, Texel, Schober, Saualpe, Koralpe and Porhorje) (Figure 5.4).

The different sections across the Alps display differences in geometry, kinematics and timing (e.g. Schmid et al. 2004, 2017; Pfiffner 2014). The *Western Alps* (i.e. from the Mediterranean coast to the Simplon Fault) display the largest amount of crustal material recording HP to UHP metamorphism (from ca. 80–35 Ma), and therefore our discussion will focus on this area. In part of the *Central Alps* (i.e. from the Simplon Fault to the Engadine window, Figure 5.4), a significant overprint in amphibolite facies has taken place at a late stage (the Oligocene "Lepontine dome"). The *Eastern Alps* (i.e. from the Engadine window to Vienna) are dominated by the Austroalpine nappes (Figures 5.3 and 5.4), some of them having also been involved in the subduction process much earlier (Cretaceous ca. 100–90 Ma) than in the Western Alps.

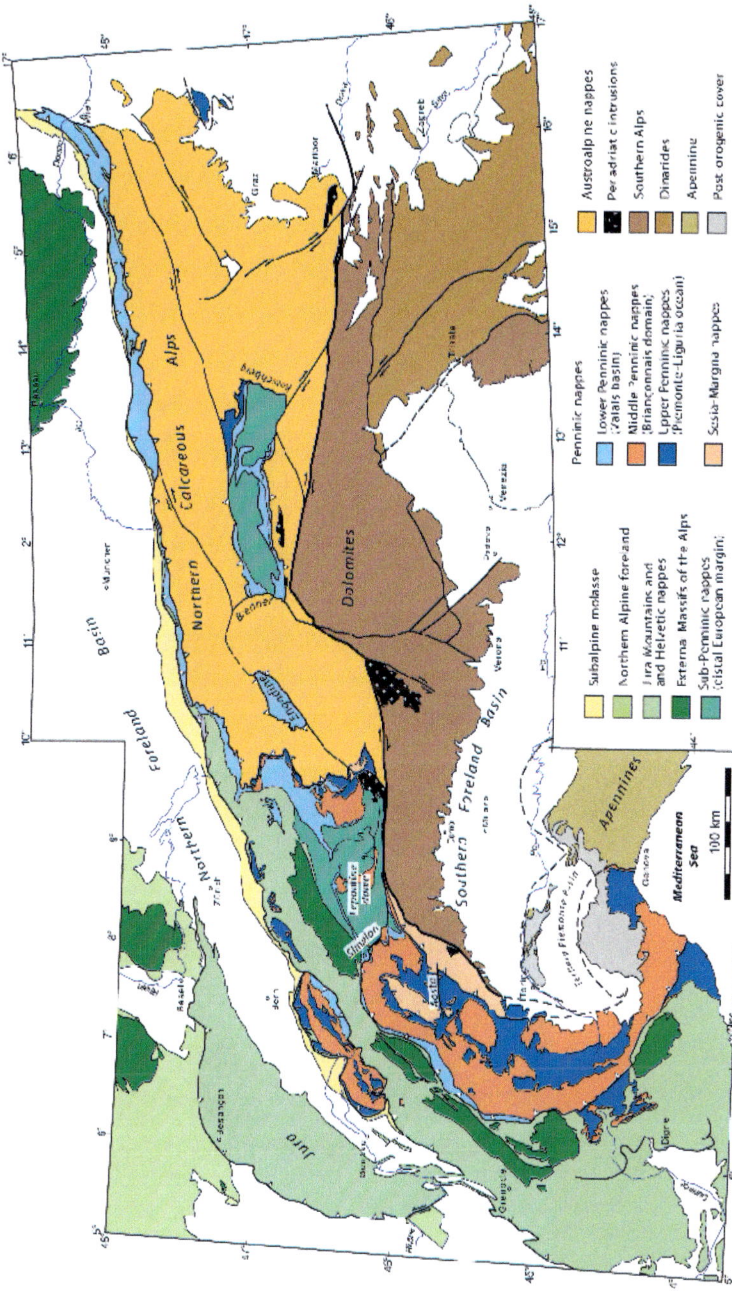

Figure 5.3. *Tectonic map of the Alps displaying the main localities where the eclogite-facies continental crust has been found (modified from Schmid et al. (2004))*

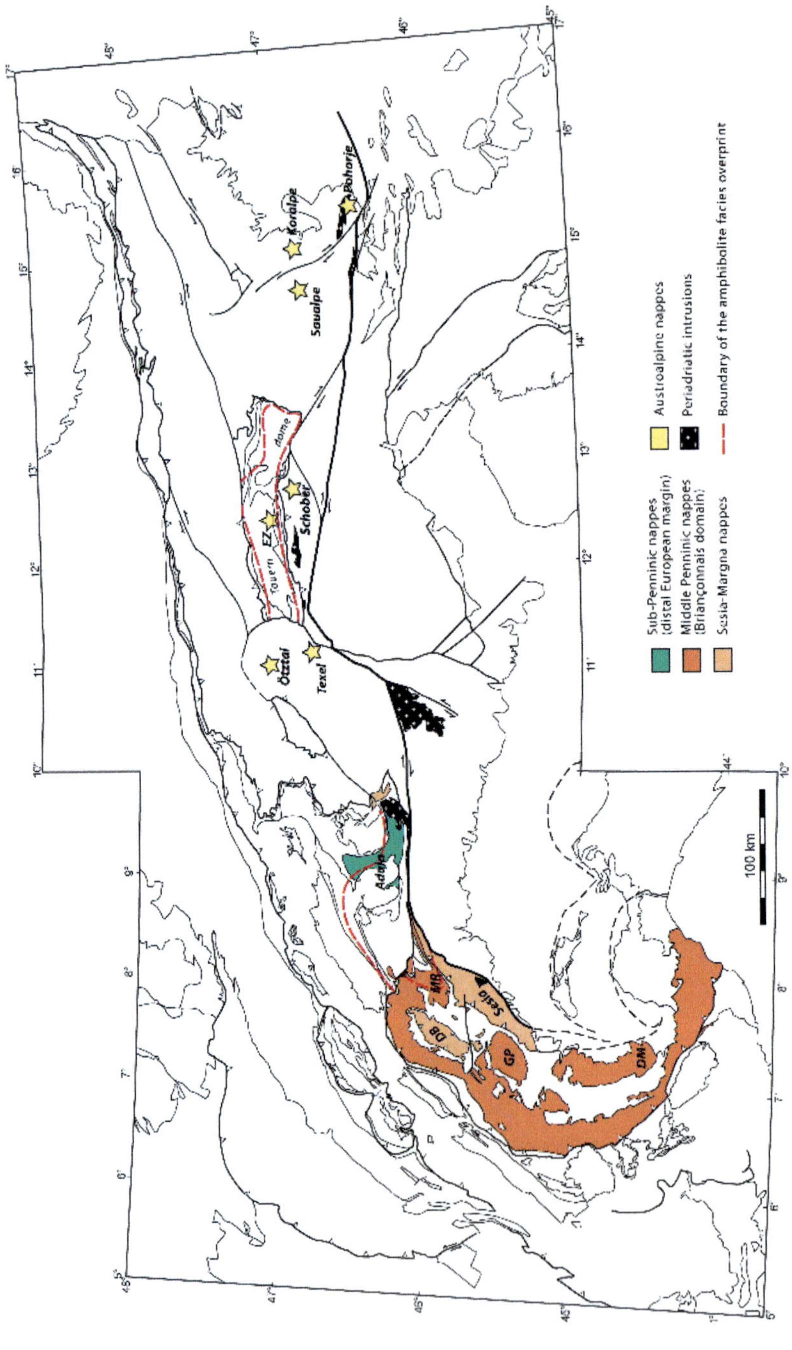

Figure 5.4. *Tectonic map of the Alps displaying the location of the allochthonous units essentially made of subducted continental crust*

5.3.2. *Large-scale structure of the Western Alps*

The Western Alps may be subdivided into two major structural domains, separated by the Insubric Line (e.g. Schmid et al. 2004; Steck et al. 2015). East or south-east of the Insubric Line, the Southern Alps are made by pre-Alpine high- to medium-grade metamorphic rocks and Mesozoic sediments that escaped significant ductile deformation and metamorphism during the Alpine orogeny. Therefore, the Southern Alps belonged to the upper plate during the Alpine orogeny. To the west or north-west of the Insubric Line and east of the Penninic Thrust, the nappe stack involves units that have been all involved in the subduction zone and are therefore part of the lower plate. In detail, the nappe stack is made, from top to bottom, of the following main units (Figure 5.5).

On top of the stack, the Sesia-Dent Blanche nappes (Compagnoni et al. 1977; Manzotti et al. 2014a) consist of granitic orthogneisses, metagabbros, and a moderately to strongly reworked polycyclic basement, displaying incipient blueschist to eclogite-facies metamorphism. The Mesozoic cover is only locally preserved (Roisan Zone) and is used to subdivide the stack into subnappes. Erosion has separated frontal, north-western outliers, namely the Dent Blanche and Pillonet klippen, from their root zone, the Sesia Zone.

The Piemonte-Liguria ocean-derived ophiolites and the associated sediments (see Chapter 4 for further details on these units) are classically divided into an upper Combin Zone, with scarce relicts of blueschist-facies metamorphism (Manzotti et al. 2021), and a lower Zermatt Zone, with well-preserved eclogite-facies rocks (e.g. Bucher et al. 2005).

Intensely deformed basement rocks crop out in windows below the oceanic units of the Zermatt Zone. These are, from the Swiss-Italian border to the Ligurian Alps, the Monte Rosa, Gran Paradiso, and Dora-Maira nappes (sometimes referred to as Internal Crystalline Massifs). Their Mesozoic cover is discontinuously exposed, and sometimes lacking, and they are essentially made of strongly deformed porphyritic metagranites (now augengneisses) and their pre-Alpine country-rocks. Monocyclic, graphite- and conglomerate-rich metasediments were deposited during the Upper Carboniferous crop out in windows below the polycyclic basement. These monocyclic units, namely the Pinerolo Unit in the Dora-Maira (Novarese 1898; Manzotti et al. 2016) and the Money Unit in the Gran Paradiso (Compagnoni et al. 1974; Manzotti et al. 2014c; Manzotti et al. 2015a), represent the deepest structural level observed in the nappe stack.

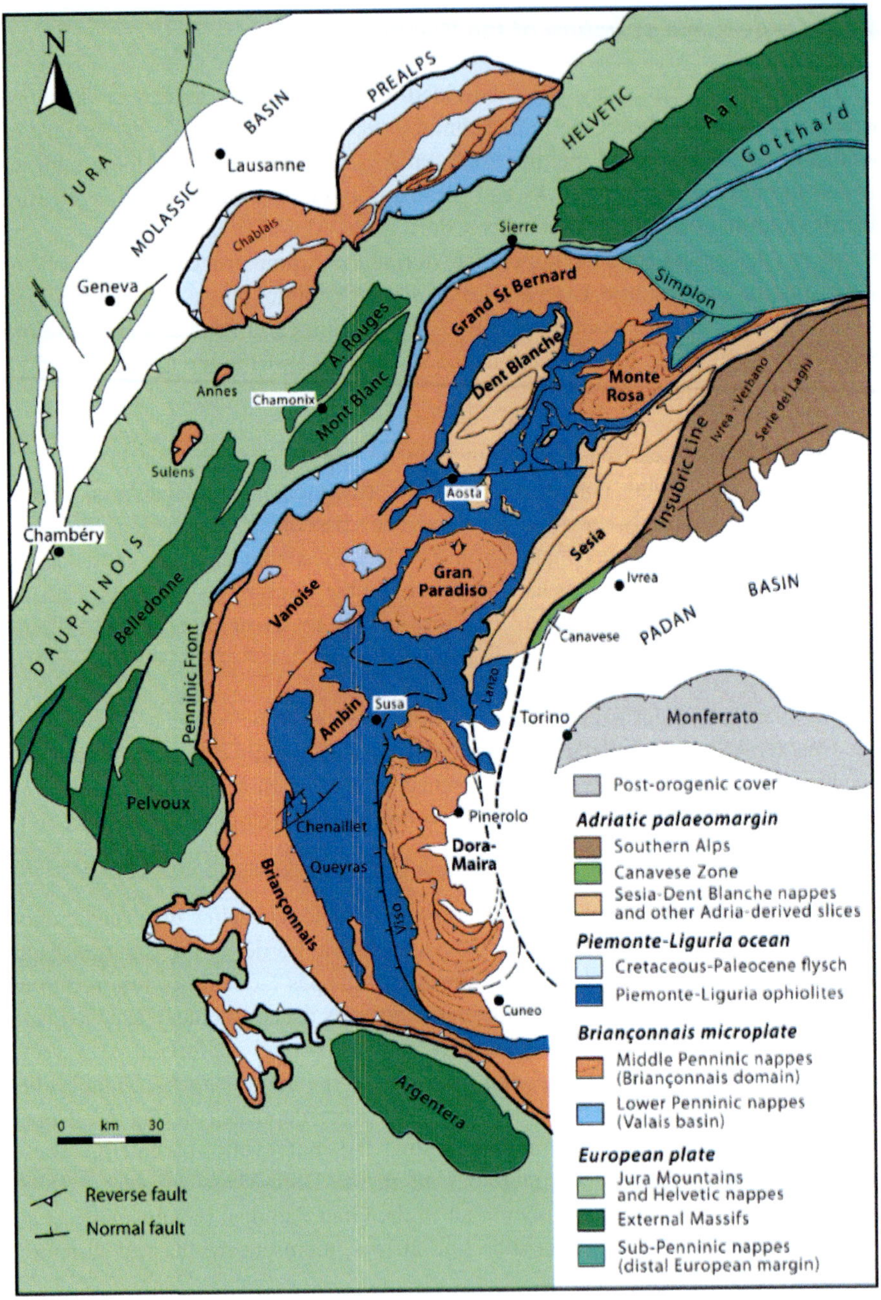

Figure 5.5. *Tectonic map of the Alps (modified after Schmid et al. (2004))*

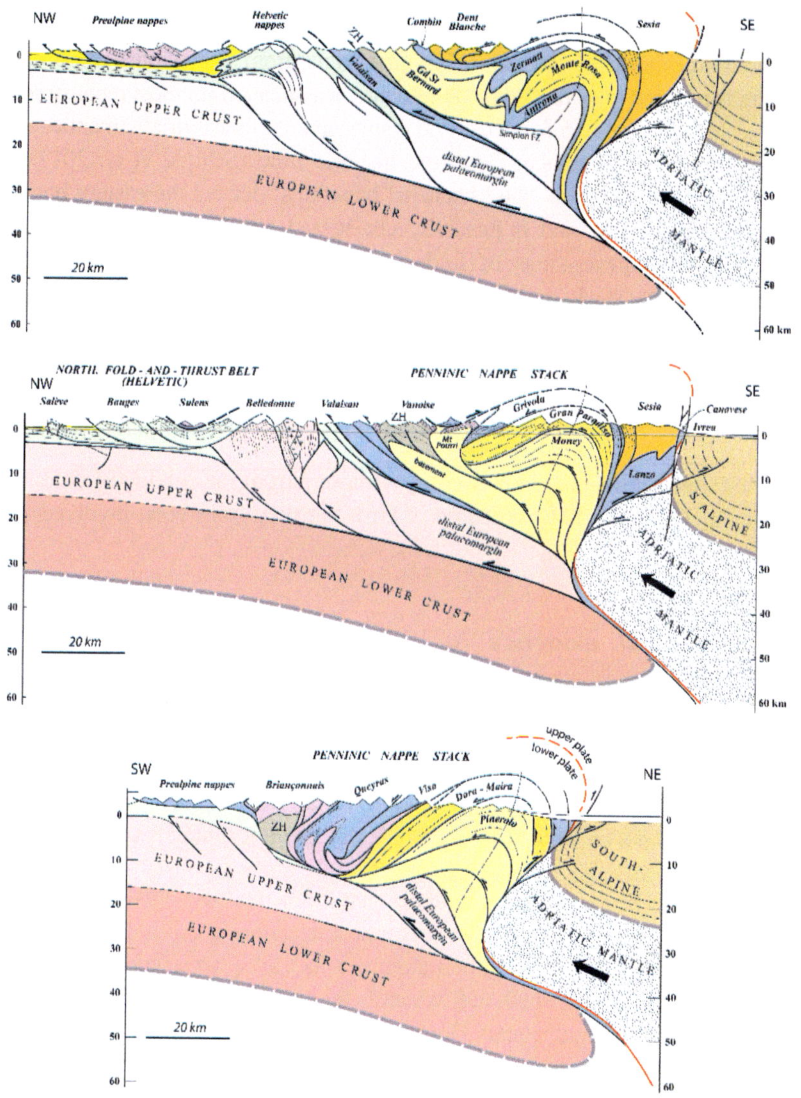

Figure 5.6. *Schematic structural cross-sections of the Western Alps, illustrating the main points relevant to continental subduction (modified from Ballèvre et al. (2018) and Ballèvre et al. (2020))*

Finally, the external part of the Briançonnais Zone is made of Carboniferous sediments overlain by their Mesozoic cover, frequently detached and piled up, and of more internal units essentially made of a Paleozoic basement.

As displayed by the cross-sections (Figure 5.6), the nappe stack is deformed by late folds. An antiformal stack whose axial trace runs from the Monte Rosa to the Dora-Maira is referred to as the Vanzone fold and is associated (in most reconstructions at crustal-scale, as in Figure 5.6) to the rigid displacement of the Adriatic mantle, coeval with top-to the SE displacement along the Insubric Line (Schmid et al. 2017). The second set of post-nappe folds (the Mischabel backfolds) defines the eastern boundary of the Briançonnais Zone and is generally considered older than the Vanzone fold. Although these two generations of folds have a key role in shaping the present geometry of the nappe stack, their role in exhuming HP-UHP rocks is minor, with most of the exhumation having taken place before their development.

5.4. Subduction of the Mesozoic cover

The Mesozoic sediments represent only a minor part of the subducted continental crust (their thickness did not exceed ca. 1 km), but they have been involved in the subduction process.

5.4.1. *Petrological evidence for subduction metamorphism in the Mesozoic covers*

Indeed, a great effort has been made to unravel their progressive metamorphism across the Alpine belt. Because most Mesozoic metasediments derive from platform carbonates (limestones and dolostones) or basinal marls (now transformed into calcschists), it was not an easy task to identify key metamorphic minerals.

Aragonite, the HP polymorph of calcite, has been reported only in the Western Vanoise, probably because it readily inverts into calcite during exhumation (Gillet and Goffé 1988; Hacker et al. 2005). When there is a lack of quartz, dolomite is a stable phase up to relatively high T (600–650°C), and therefore the Triassic dolomites, which are so frequent in the Briançonnais Domain, are an "unreactive" lithology. If Al_2O_3 and SiO_2 are present in the system (e.g. in the metamarls), lawsonite may develop, as reported in Mesozoic metasediments from Vanoise (Ellenberger 1960; Goffé 1979), Acceglio and Grana and Maira Valleys (Michard 1967). Fe-Mg carpholite has been recorded in a large number of occurrences, mainly as fibers in quartz veins developed in Ca-poor metasediments (e.g. black shales of Cretaceous age) (Goffé 1984; Goffé and Chopin 1986; Goffé et al. 2004; Michard et al. 2004).

Two other lithologies displaying HP minerals are found in the Mesozoic covers. First, residual sediments developed during the Liassic uplift and erosion of the Briançonnais Domain (i.e. meta-bauxites) record specific mineral assemblages (Goffé and Saliot 1977; Goffé 1980, 1984; Chopin et al. 2003). Secondly, some condensed layers, a few cm thick, and enriched in Mn or Fe^{3+}, have developed blue amphiboles ("glaucophane" sensu lato), sometimes overgrowing preserved microfossils (Raguin 1925; Lemoine 1960; Le Guernic 1967; Lefèvre and Michard 1976). The latter is especially important because they constrain the age of the HP metamorphism in the Briançonnais Domain, which cannot be older than the Paleocene (ca. 56 Ma).

5.4.2. *Structural considerations*

The stratigraphy of the Mesozoic covers is characterized by a lower part dominantly made of conglomerates and quartzites (Early Triassic), and an upper part essentially consisting of limestones, dolostones, and marls (Middle Triassic to Palaeocene). At the base of the upper part, one or more layers of evaporites are present. The latter constitute the rheologically weakest material and act as decollement levels during the Alpine shortening (Figure 5.7). It follows that a large amount of these covers have been detached, and stacked close to the surface, as displayed in the Prealpine Nappes from Switzerland. There, unmetamorphosed sequences similar to those occurring in the internal zones of the Alpine belt have been found. For example, the Mesozoic series of the Préalpes médianes rigides show strong similarities with those from the Briançonnais area and with the more metamorphosed Vanoise and Barrhorn areas (e.g. Ellenberger 1952; Sartori 1990). Another striking example is given by the similarities between the Nappe de la Brèche in the Préalpes and the Evolène Series of the Mont Fort Nappe (Pantet et al. 2020).

There are a few exceptions to this general rule (Figure 5.7), when the Triassic evaporites have been eroded during the pre-orogenic history (e.g. Acceglio Zone), or have not been deposited, like in the Dent Blanche Nappe (Roisan-Cignana Shear Zone: Manzotti et al. 2014b) and in the Sesia Zone (Scalaro Unit: Venturini et al. 1994; Giuntoli et al. 2016; Regis et al. 2016). In such cases, the Mesozoic covers are still adherent to their Paleozoic basement, and both record the same P-T history.

To sum up, the fate of the Mesozoic covers during the Alpine subduction varies considerably and is primarily controlled by the tectonostratigraphic inherited from the pre-orogenic history. When the Mesozoic cover has been detached at an early stage of the orogenic history, the ophiolitic units derived from the Piemonte-Liguria Ocean may be in close or even direct contact with the subducted Paleozoic basements, like in the eastern part of the Gran Paradiso nappe (Figure 5.8) and most of the Monte Rosa nappe.

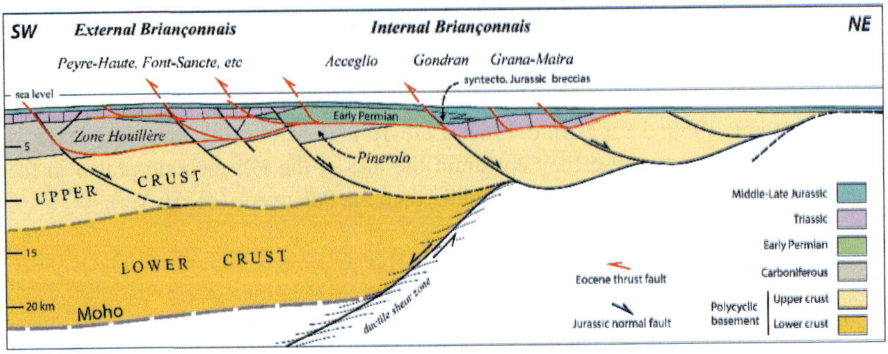

Figure 5.7. *Proposed correlations between detached cover nappes (External and Internal Briançonnais) and subducted basement slices (Dora-Maira Massif) along a transect in the southern part of the Western Alps (Ballèvre et al. 2020)*

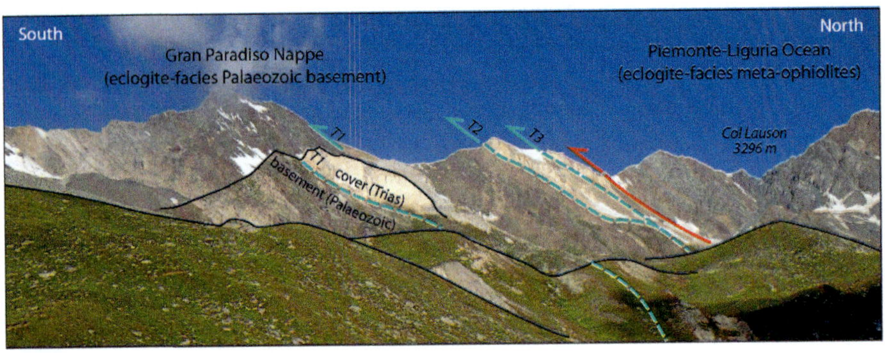

Figure 5.8. *A field section across the contact between the Piemonte-Liguria Ocean (Grivola Zone, to the right of the image) and the underlying window of continentally derived units (northern margin of the Gran Paradiso Nappe, to the left of the image). Note remnants of Mesozoic covers from the Gran Paradiso basement, now cleaved and stacked into three slices (T1–T3), each with a considerably reduced thickness, and disappearing eastwards at map scale. The Mesozoic covers essentially consist of pale yellow dolomitic limestones, and associated cargneules (=Rauhwacke), the latter deriving from the evaporitic detachment layers*

5.5. Subduction of the Paleozoic basement

The main question to be answered in this section concerns the identification of the type of crust that was involved in the subduction process. A straightforward answer would be to distinguish, from their pre-Alpine relicts, the upper crust (dominantly granitic, or made of low-grade metamorphic rocks) from the lower crust (dominantly made of high-grade, partially melted, metamorphic rocks). The answer is not in fact so simple, for two reasons.

First, we cannot simply assume that all high-grade metamorphic rocks (i.e. those displaying garnet-biotite-sillimanite assemblages), of the pre-Alpine age, stayed in the lower crust once formed in the lower crust. There are many examples, outside the Alpine belt, where partially-melted rocks, typically considered representative of the lower crust, are outcropping at the surface since at least the Triassic (in the Armorican Massif, the French Massif Central, the Vosges-Black Forest and the Bohemian Massif). These were exhumed before the end of the Paleozoic, and occupy, with respect to the Alpine orogeny, a position in the upper crust.

Secondly, during the rifting processes preceding the Alpine orogeny (i.e. during the Triassic and the Jurassic), the continental crust has been deformed. This leads to a spatial reorganization of the pre-Alpine material. There may have been a decoupling between the upper crust (more brittle) and the lower crust, which some authors consider as more rigid than the upper crust, while others consider it as more ductile, depending on the rheological properties or materials considered or used for analog or numerical modeling (e.g. Brun and Beslier 1996; Bürgmann and Dresen 2008) and/or on the previous lithological heterogeneities (e.g. Petri et al. 2019). This may lead to the formation of extensional allochthons in the extended margins, separated by exhumed subcontinental mantle, but the material constituting these extensional allochthons may vary substantially.

5.5.1. *The pre-Alpine crust of the Adriatic plate*

5.5.1.1. *A reference cross-section in the Southern Alps*

To begin the discussion, let us examine a reference section across the pre-Alpine crust, in the South-Alpine domain, where a tilted section through the crust has been identified (e.g. Mehnert 1975; Fountain 1976; Handy et al. 1999). The Mesozoic sediments of the South-Alpine domain were deposited in basins controlled by normal faults active during the Triassic and the Jurassic (Bertotti et al. 1993; Bertotti 2001). Felsic explosive volcanic rocks of the Permian age are locally preserved below these Mesozoic sediments. The upper crust (i.e. the Serie dei Laghi) is dominantly made of paragneisses and amphibolites intruded by Permian granitoids.

This amphibolite-facies basement locally retains evidence of a pre-Permian orogenic cycle (Strona-Ceneri Zone: Boriani et al. 1990; Romer and Franz 1998; Franz and Romer 2007; Zurbriggen 2017). The lower crust is exposed in the Ivrea-Verbano Zone and consists of upper amphibolite to granulite facies paragneisses and marbles, intruded by massive gabbroic bodies (termed the Basic Formation). A major discontinuity (the Cossato-Mergozzo-Brissago Line) separates the Ivrea-Verbano Zone from the Serie dei Laghi (Boriani et al. 1990).

The late-Paleozoic evolution of this crustal section may be explained by partial melting in the upper mantle during lithospheric extension and subsequent magmatic underplating at the Moho transition, with high heat flow inducing partial melting of the lower crust. The resulting magmas migrated upward and either collected in large granitoid plutons or led to eruptions, forming caldera-like volcanoes in the Permian (Quick et al. 2009; Karakas et al. 2019). In detail, uncertainties remain regarding (i) the geometry and kinematics of high-temperature shear zones, (ii) their temporal relation to the granulite-facies metamorphism, and (iii) the exact timing of the subsequent cooling.

5.5.1.2. *The Sesia-Dent Blanche nappes*

The interpretation of the continental material subducted during the early stages of the Alpine orogeny is frequently made by reference to the section exposed in the Southern Alps. Indeed, the Sesia-Dent Blanche nappes are considered part of the Adriatic palaeomargin (Manzotti et al. 2014a) and should therefore easily be compared with the Southern Alps.

The Valpelline Series in the Dent Blanche nappe, the Upper Unit of the Mont Mary nappe, and the II DK in the Sesia Zone represent slices of the Permian lower crust (Carraro et al. 1970; Manzotti et al. 2014a) (Figure 5.9). The main argument in favor of this view is that (i) these units consist of high-temperature rocks (paragneisses, amphibolites, marbles) equilibrated at 0.6–0.8 GPa (Figures 5.10 and 5.11), (ii) they display abundant evidence of partial melting (Figure 5.10) and lack granitoid intrusions, and (iii) the age of the high-grade metamorphism is Permian (Kunz et al. 2018). There is only scarce evidence for the exhumation of these rocks before the Alpine orogeny, and they are thought to have slowly cooled at depth, in a similar way to the Ivrea-Verbano Zone (Figures 5.9 and 5.10).

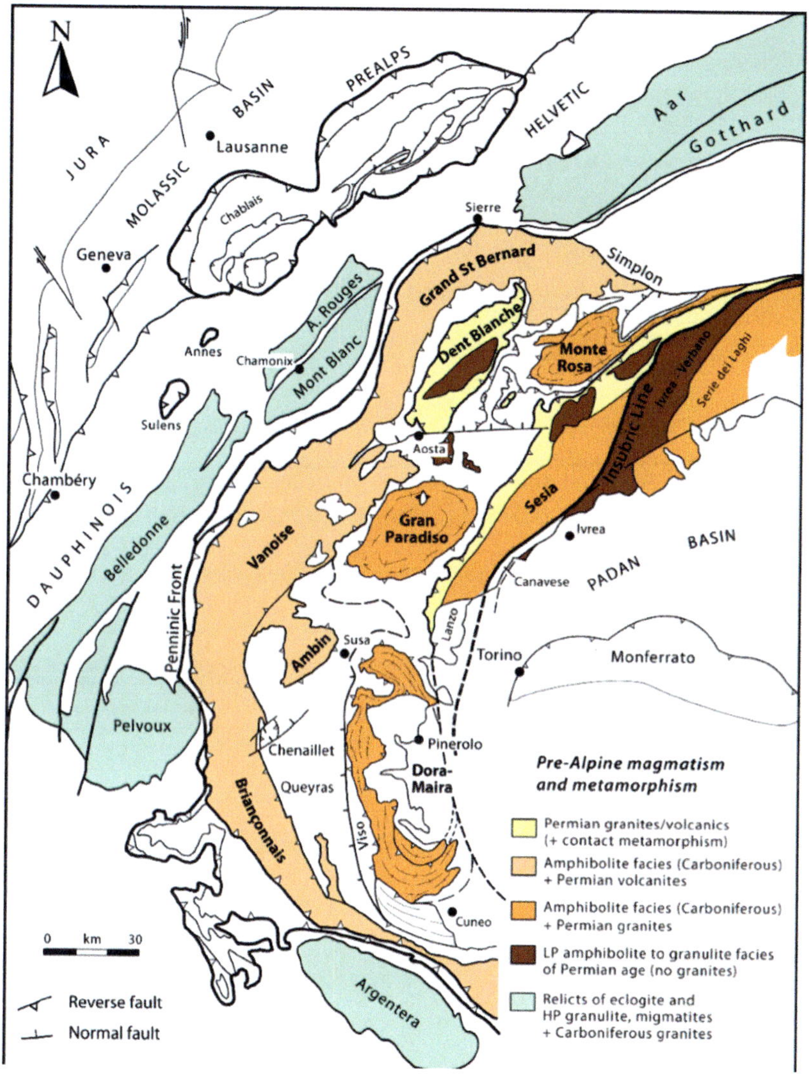

Figure 5.9. *A simplified map of the pre-Alpine metamorphism and magmatism in the Western Alps. In this map, colors are only shown for the pre-Alpine basement involved in the Alpine orogeny, belonging to the Helvetic Zone (External Crystalline Massifs), the Briançonnais nappes (External Briançonnais and Internal Crystalline Massifs), the Sesia-Dent Blanche nappe and the Southern Alps (Ivrea-Verbano and Serie dei Laghi). Mesozoic covers, oceanic units (e.g. Zermatt, Viso, and so forth) and Oligo-Miocene foreland basins are left as blank magmatic underplating with respect to the regional granulite-facies event. The crustal section was modified during Triassic to Jurassic rifting and also during the Alpine orogeny (Handy et al. 1999)*

Figure 5.10. *Some well-preserved examples of pre-Alpine magmatism: a) amphibole-bearing granite from the Arolla Series; b) undeformed, coarse-grained gabbro from the Arolla Series, Monte Cervino) and metamorphism; c) high-grade amphibolite to granulite paragneisses with folded boudins of amphibole, O. Müntener for scale; d) pegmatite veins cutting across a folded layer of amphibolite; e) evidence for incipient partial melting in the amphibolites, with coarse-grained, peritectic pyroxene in the leucosomes) in the Valpelline Series, Dent-Blanche nappe*

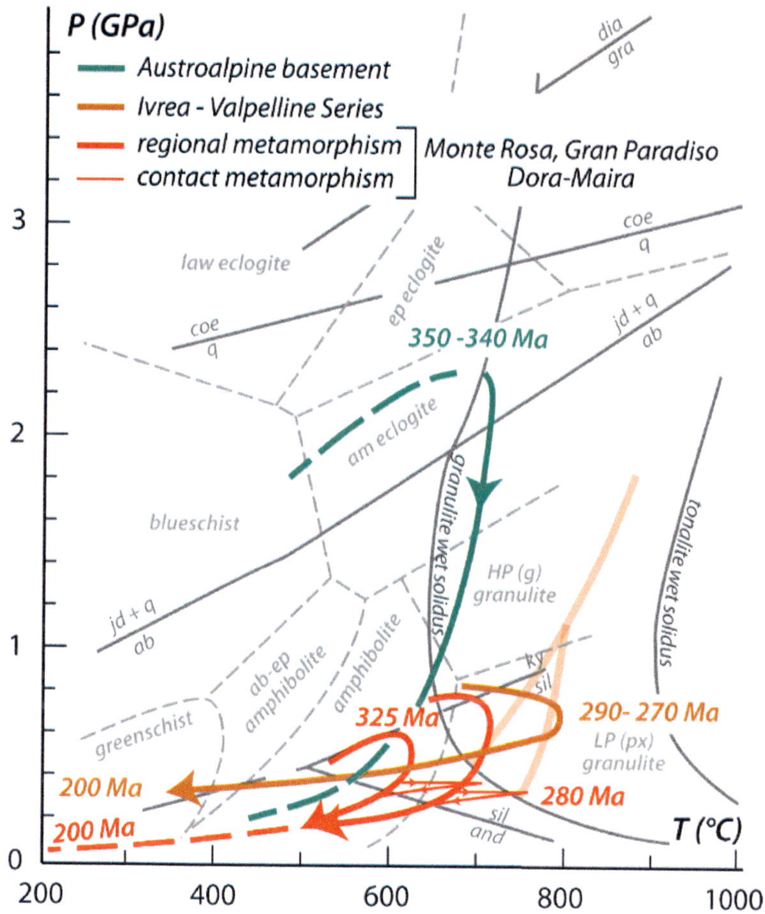

Figure 5.11. *Different types of pre-Alpine P-T paths of the continental crust involved in the Alpine orogeny. Some units (e.g. the Hochgrössen unit in the Austroalpine) record an early Carboniferous eclogite-facies stage, followed by decompression before the end of the Carboniferous (Faryad et al. 2002; Melcher and Meisel 2004). Other units record an amphibolite to high-grade amphibolite regional metamorphism (Dal Piaz 1971; Le Bayon et al. 2006), of probable Late Carboniferous age, followed by decompression and granitic intrusions during the Early Permian. Lastly, some units record high-grade metamorphism (at the transition from amphibolite to granulite facies) of the Permian age, followed by essentially isobaric cooling (Gardien et al. 1994; Manzotti and Zucali 2013)*

By contrast, the Arolla Series and the external part of the Sesia Zone (Figure 5.9) are predominantly formed by a large volume of meta-aluminous granitoids (Figure 5.10), a few layered gabbroic complexes (Cervino, Collon, Sassa, Berrio) and some felsic volcanics (e.g. Manzotti et al. 2017; Manzotti et al. 2018a). All have provided Permian ages. In rare cases, the granitoids are emplaced into partially-melted paragneisses developed at lower P (0.3–0.5 GPa) than in the preceding units. The Arolla Unit is therefore considered representative of the upper crust. The Mesozoic sediments of the Roisan Zone may have been deposited on top of this basement.

In the Sesia Zone (Figure 5.9), the internal zone (i.e. the Eclogitic Micaschist Complex, hereafter referred to as EMC) contains large volumes of pre-Alpine paragneiss (locally with garnet–biotite–sillimanite relicts, Dal Piaz et al. (1972); Compagnoni (1977)), amphibolites and rare granulites, now transformed into eclogites (Gosso et al. 2010) and marbles (also displaying pre-Alpine relicts, Castelli 1991). Very few gabbroic intrusives are found in the EMC. U-Pb data on titanite (Castelli and Rubatto 2002) and zircon (Kunz et al. 2018) suggest that the high-temperature metamorphism is Permian in age. These characteristics are therefore similar to the Ivrea-Verbano Zone (Figure 5.9). However, in the Sesia Zone, these lithologies were intruded by granitoids of the Permian age (e.g. Monte Mucrone), developing contact aureoles in their country rocks, including skarns at the expense of the marbles. These features identify the EMC as a fragment of the Permian middle crust or, better, of a lower crust already partly exhumed prior to Permian magmatism.

5.5.1.3. *The Austroalpine nappes*

The pre-Alpine basement of the Austroalpine nappes displays relicts of eclogite-facies metamorphism, like in the Silvretta, Ötztal and Ulten Units (Tumiati et al. 2003) west of the Tauern window (Thöni 2006), and the Hochgrössen Unit east of the Tauern window (Faryad et al. 2002; Melcher and Meisel 2004). Some of these eclogites may have been developed at the expense of an early Paleozoic oceanic crust, involved in a subduction zone during the Variscan orogeny (Melcher and Meisel 2004). They are characterized by decompression P-T paths at high temperatures, inducing partial melting in the surrounding metasediments (Figure 5.11). Other units in the Austroalpine nappes record amphibolite-facies metamorphism (e.g. Rappold Complex (Gaidies et al. 2008); Campo Unit (Petri et al. 2016)).

	P-T conditions of pre-Alpine metamorphism	Field relations	Isotopic age	Reference
Adula	Eclogite facies		> 333 Ma (Lu-Hf WR-Grt) 324 ± 7 Ma (Lu-Hf Wr-Grt)	Herwartz et al. 2011 Sandmann et al. 2014
Savona	Eclogite facies			Messiga et al. 1992a
Ruitor – Gd St Bernard	Amphibolite facies		330 Ma (U-Pb monazite)	Giorgis et al. 1999
Siviez-Mischabel	Eclogite facies			Thélin et al. 1990; Rahn 1991
Suretta	Eclogite-facies	Pre-granitic (Permian)		Nussbaum et al. 1998
Monte Rosa	High-grade amphibolite facies	Pre-granitic (Permian)		Bearth 1952
Gran Paradiso	Amphibolite facies	Pre-granitic (Permian)		Le Bayon et al. 2006
Dora Maira (UHP)	Amphibolite facies	Pre-granitic (Permian)		Compagnoni and Rolfo 2003
Dora Maira (HP)	Amphibolite facies	Pre-granitic (Permian)	324 ± 6 Ma (U-Pb monazite)	Nosenzo et al. 2022
Emilius	High-grade amphibolite to granulite facies		280–285 Ma (U-Pb zircon)	Pennacchioni 1996; Kunz et al. 2018

	P-T conditions of pre-Alpine metamorphism	Field relations	Isotopic age	Reference
Dent Blanche (Arolla)	High-grade amphibolite facies	Syn-granitic	280–285 Ma (U-Pb zircon)	Manzotti et al. 2018a
Dent Blanche (Valpelline)	High-grade amphibolite facies to granulite facies	Cut by pegmatitic dykes	290 Ma, 270 Ma (U-Pb zircon)	Gardien et al. 1994; Manzotti and Zucali 2013; Kunz et al. 2018
Sesia (II DK)	High-grade amphibolite facies to granulite facies		290 Ma, 270 Ma (U-Pb zircon)	Vuichard 1987; Kunz et al. 2018
Sesia (Internal Zone)	High-grade amphibolite facies to granulite facies	Pre-granitic (Permian)	295–280 Ma (U-Pb zircon)	Giuntoli et al. 2018a; Kunz et al. 2018
Silvretta	Eclogite facies		350 Ma	Ladenhauf et al. 2001
Ötztal	Eclogite facies		347 ± 9 Ma	Miller and Thöni 1995
Ulten	Eclogite facies		336 ± 4 Ma	Tumiati et al. 2003
Schobergruppe	Eclogite facies		Min 313 Ma	Hauke et al. 2019a
Hochgrössen	Eclogite facies		397 ± 8 Ma Ar-Ar on edenite	Faryad et al. 2002; Melcher and Meisel 2004

Table 5.2. *Grade and age of the pre-Alpine metamorphism recorded in continental crust involved in the Alpine orogeny. Units in white belong to the distal European palaeomargin, in green to the external Briançonnais, in yellow to the internal Briançonnais, in pink to the Sesia-Dent Blanche nappes and in orange to the Austro-Alpine nappes*

5.5.2. *The pre-Alpine crust of the Briançonnais domain*

The Paleozoic basement of the Briançonnais domain is known in a large number of units, extending from the Ligurian Alps to the Tambo and Suretta nappes in the Central Alps. This Paleozoic basement consists of two main sequences, namely (i) late Neoproterozoic to early Carboniferous sedimentary and magmatic rocks variably involved in the Variscan orogeny, and (ii) Late Carboniferous and Permian sediments and volcanics unconformably deposited over the Variscan basement. The pre-Alpine metamorphism in the Briançonnais units is of interest for the present purpose (for a recent summary, see Ballèvre et al. (2018)).

Pre-Alpine eclogite-facies rocks are known in some Briançonnais units, notably the Savona (Messiga et al. 1992a), Siviez-Mischabel (Thélin et al. 1990; Rahn 1991) and Suretta (Nussbaum et al. 1998) nappes (Table 5.2 and Figure 5.11). Because the pre-Alpine eclogite-bearing rocks are cut by high-level granitic intrusives of the Permian age, these units were exhumed well before the Alpine orogeny, and therefore occupied an upper crustal position during the Alpine orogeny.

Other Briançonnais units are characterized by amphibolite-facies metamorphism of the Late Carboniferous age (e.g. Ruitor – Grand Saint Bernard (Giorgis et al. 1999); Monte Rosa (Bearth 1952; Dal Piaz 1971); Gran Paradiso (Le Bayon et al. 2006); Dora-Maira (Nosenzo et al. 2022)), culminating with cordierite-bearing partial melts, the latter being intruded by Permian granites.

One important aspect of the Briançonnais units is indeed the widespread Permian magmatism. The latter is recorded by widespread sills, dykes and lava flows in the Zone Houillère (Ballèvre et al. 2020), and by a large amount of plutonic bodies (Monte Rosa: Pawlig 2001; Luisier et al. 2021; Gran Paradiso: Bertrand et al. 2005; Ring et al. 2005; Dora-Maira: Bussy and Cadoppi 1996; Gauthiez-Putallaz et al. 2016). In the Central Alps, the Gruf Complex may represent a piece of the Permian lower crust stacked during the Alpine orogeny (Galli et al. 2012, 2013; Nicollet et al. 2018).

5.5.3. *Tectonic implications for building the Alpine stack*

The continental material subducted in the Western and Central Alps belongs to three main units, whose characteristics are briefly recalled here, from top to bottom.

The Sesia-Dent Blanche nappes contain three types of crustal units. Lower crustal units (Valpelline, II DK) cooled down at depth following the high-temperature Permian event and stayed at the base of the crust up to the onset of the Alpine orogeny (Table 5.2). Upper crustal units are essentially made of granitoids (Arolla), with local relicts of their country rocks, displaying evidence for partial melting at a lower *P*. Granitic intrusions and high-grade metamorphism are Permian in age (Manzotti et al. 2018a). Intermediate units (the internal Sesia Zone) are characterized by a relatively high-grade metamorphic episode, followed by an early decompression during which some granitic intrusions have been emplaced during the Permian. The existence of Mesozoic covers deposited on top of the pre-Alpine basement from these intermediate units (and potentially now defining subunits inside the internal Sesia Zone) has been proposed (Venturini et al. 1994; Giuntoli and Engi 2016).

The Emilius and related slices share similarities with the internal Sesia Zone (i.e. the intermediate units) for their pre-Alpine evolution (Table 5.2). However, they differ from these by a smaller amount of Permian granitoids and by a younger age for the Alpine eclogite-facies metamorphism. No evidence for their Mesozoic cover has been found up until now.

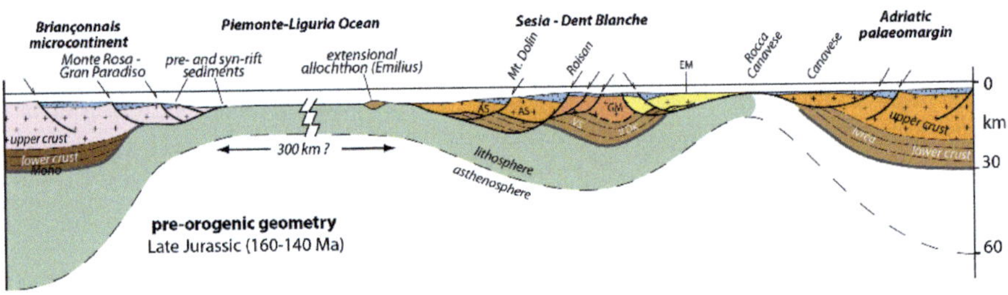

Figure 5.12. *A reconstruction of the tectonic situation before the Alpine orogeny (modified after Manzotti et al. (2014a)). Note significant changes in crustal thickness and composition, due to heterogenous stretching during the Jurassic extension, with implications for rheological behavior during Alpine subduction*

In the Briançonnais nappes, only slices of the upper crustal units are presently recognized, consistent with the observation that most of them were unconformably covered by Triassic to Cretaceous sediments. One remarkable aspect of the Permian magmatism in these nappes is that it involves a significant crustal component (Buzzi and Gaggero 2008). This suggests that a large amount of felsic melts have been extracted from rocks at depth during the Permian, building a granulitic lower crust similar to the one cropping out in the Ivrea-Verbano Zone. Note, however, that this high-grade lower crust is not exposed today in any of the Briançonnais units of the Western and Central Alps.

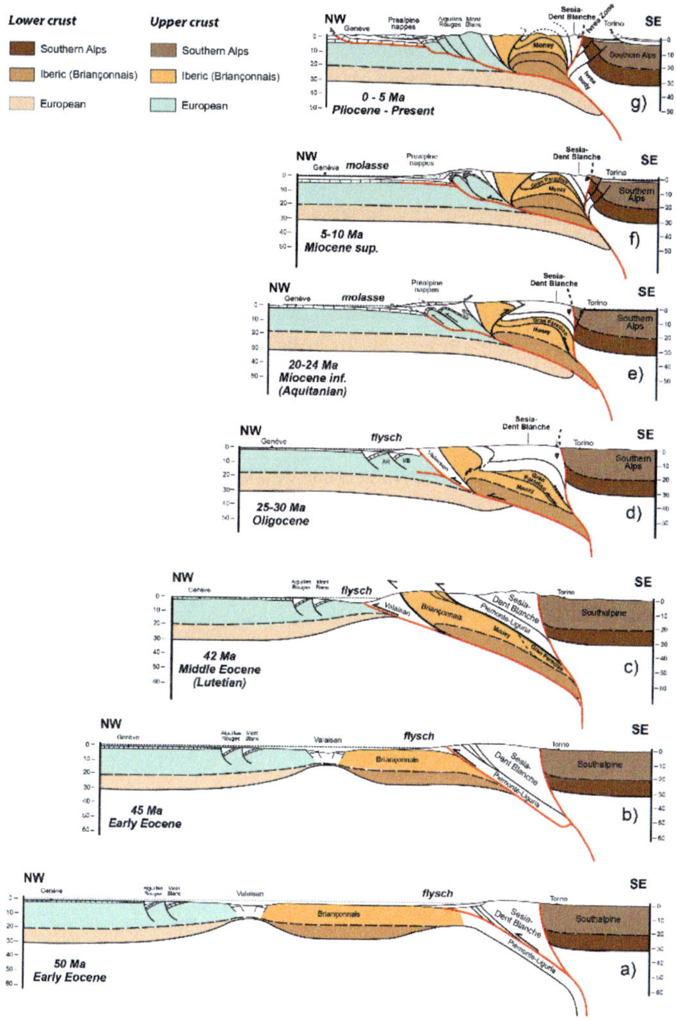

Figure 5.13. *A schematic representation of the fate of the pre-Alpine lower crust during the Alpine subduction-collision. The European lower crust remains at depth, while the Briançonnais and Adriatic lower crust is involved in the subduction process. Please note, however, that the lower crust from the Briançonnais is detached from its upper crustal counterpart and contributes to building the mountain root. In the Sesia-Dent Blanche nappes the distinction between lower and upper crust is not shown (see Manzotti et al. (2014a) for further details)*

During the Triassic and the Jurassic, the continental crust has been extended (e.g. Lemoine et al. 1986; Lemoine and Trumpy 1987; Froitzheim and Manatschal 1996; Manatschal and Bernoulli 1998; Manatschal 2004), leading to the exhumation of the subcontinental mantle and the opening of two or three oceanic basins (e.g. Lemoine et al. 1987; Lagabrielle and Cannat 1990; McCarthy et al. 2020). Stretching of the continental crust is associated with the activity of synsedimentary normal faults, and is characterized by its heterogeneity at all scales, leading to the development of extensional allochthons (Figure 5.12). In addition, the extension was probably largely asymmetrical, with the Briançonnais Domain occupying a position of upper plate with respect to the Adria-derived units (Lemoine et al. 1987). Therefore, in the Briançonnais Domain, the upper crust may have been detached from the corresponding lower crust during the Jurassic extension, in which case these units may derive from extensional allochthons made only of the upper crust (Figure 5.12).

In summary, the pieces of continental crust that will be involved in the subduction zones consist of a wide range of materials and structures. Their rheological behavior during subduction will vary according to the nature of the material involved (e.g. dry granulitic lower crust, dominantly granitic upper crust, upper Paleozoic sediments and volcanics), and the structural inheritance associated with the Jurassic extension.

The schematic kinematic model takes into account these peculiarities (Figure 5.13): the European lower crust is presently subducted, the Briançonnais lower crust is stacked at depth in the orogenic root and slices of the Adriatic lower crust are now outcropping in the Sesia-Dent Blanche nappes.

5.6. Deformation history of the subducted crust in the Alps

In this section, we will now address the question of the internal deformation (i.e. the strain) of the subducted continental rocks. Field observations are key elements for understanding the rheological behavior of the subducted crust, bearing in mind the following questions. Is the subducted crust pervasively deformed or heterogeneously deformed? Is the main ductile deformation acquired during burial (i.e. along the prograde, up-P, path), or at some stage during the decompression history? Do we have evidence for brittle behavior at peak P-T conditions or at any place along the P-T path?

5.6.1. *Evidence for heterogeneous deformation during subduction*

In the Alps, the subducted continental units display evidence of Alpine ductile deformations associated with different metamorphic conditions, overprinting pre-Alpine structures and metamorphism. Deformation is heterogeneous at all scales and varies from almost no signs of strain and metamorphism in the Valpelline Series and II DK to pervasive overprinting under eclogite facies (EMC) and/or greenschist facies (Arolla Series, Gneiss Minuti Complex) conditions. Undeformed volumes with diverse sizes are observed in all units (Table 5.3). These volumes are especially important because they display relicts of pre-Alpine history.

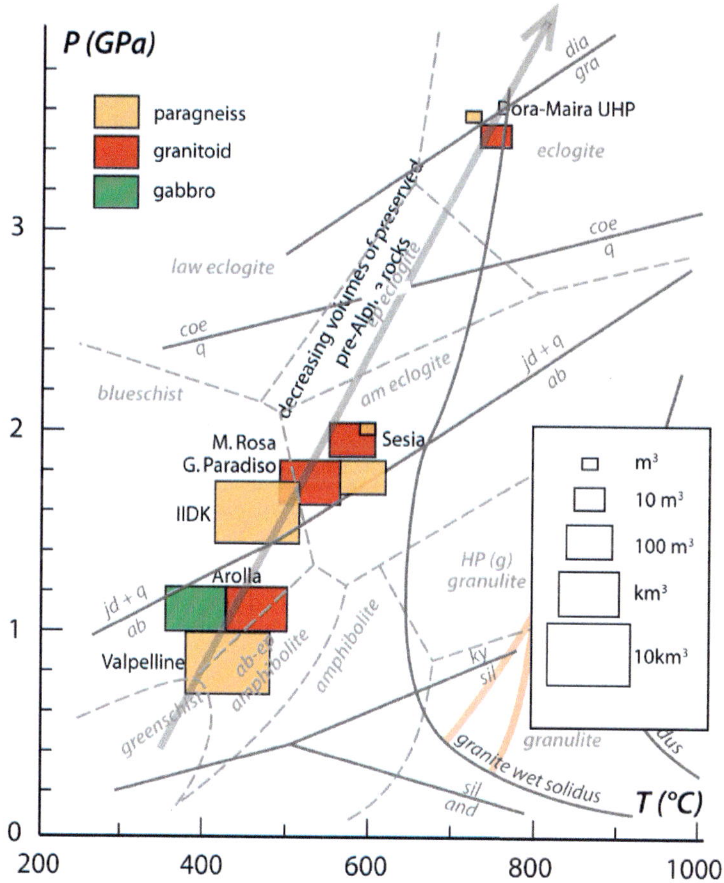

Figure 5.14. *A schematic representation of the size of the low-strain volumes in the subducted continental crust from the Western Alps with respect to their peak P-T Alpine metamorphic conditions. Note that the Mesozoic sequences are absent from this diagram (because they are always pervasively deformed)*

In the Sesia-Dent Blanche nappes, large portions have escaped Alpine deformation, and, for this reason, have acquired special status and received specific names (Figure 5.14). The Valpelline Series displays sections hundreds of meters thick made of paragneisses with interlayered marbles and amphibolites. The paragneisses and the amphibolites show evidence of partial melting, before the intrusion of several sets of pegmatitic dykes. The Alpine overprint is very weak, or even absent, and is restricted to the static development of chlorite on biotite, and chloritoid and kyanite on cordierite (Kienast and Nicot 1971). Greenschist-facies shear zones bound the Valpelline Series at the contact with the surrounding units. However, in the Valpelline Series, the Alpine deformation is also marked by ultracataclasites and pseudotachylites (Masson 1938; Pennacchioni and Guermani 1993; Manzotti and Ballèvre 2017; Menant et al. 2018).

Several examples of undeformed volumes occur in the units attributed to the Briançonnais micro-continent (Figure 5.14). For example, the eastern face of the Monte Rosa (Bearth 1952) displays a section nearly 2 km thick (Figure 5.15). The two main lithologies are Permian granitoids and their derivates, intruding pre-Alpine, high-grade, paragneisses. The intrusive relationships are nicely preserved in the lower part of the face (for more than one km in thickness), and the paragneisses show, in addition to the regional, high-grade, metamorphism, the contact metamorphism associated with the intrusion of the Permian granitoids (Dal Piaz 2001). Similar field observations can be made at a smaller scale in most of the Internal Crystalline Massifs (e.g. Gran Paradiso (Bertrand 1968; Callegari et al. 1969; Le Bayon and Ballèvre 2006; Menegon and Pennachioni 2010); Dora-Maira (Compagnoni and Rolfo 2003)). To the east, significant volumes escaping Alpine deformation have also been reported in the Tambo (Marquer et al. 1996) and Suretta (e.g. Nussbaum et al. 1998) nappes.

In the distal part of the European palaeomargin, the Adula Nappe is a coherent body made of (i) a large amount of paragneisses, some of them being intruded by Ordovician orthogneisses, (ii) orthogneiss of Early Permian age and (iii) scarce relicts of a Mesozoic cover (e.g. Cavargna-Sani et al. 2014). Undeformed volumes during the Alpine orogeny have not been recognized up until now in this unit. However, relicts of the pre-Alpine, eclogite-facies and metamorphism are described.

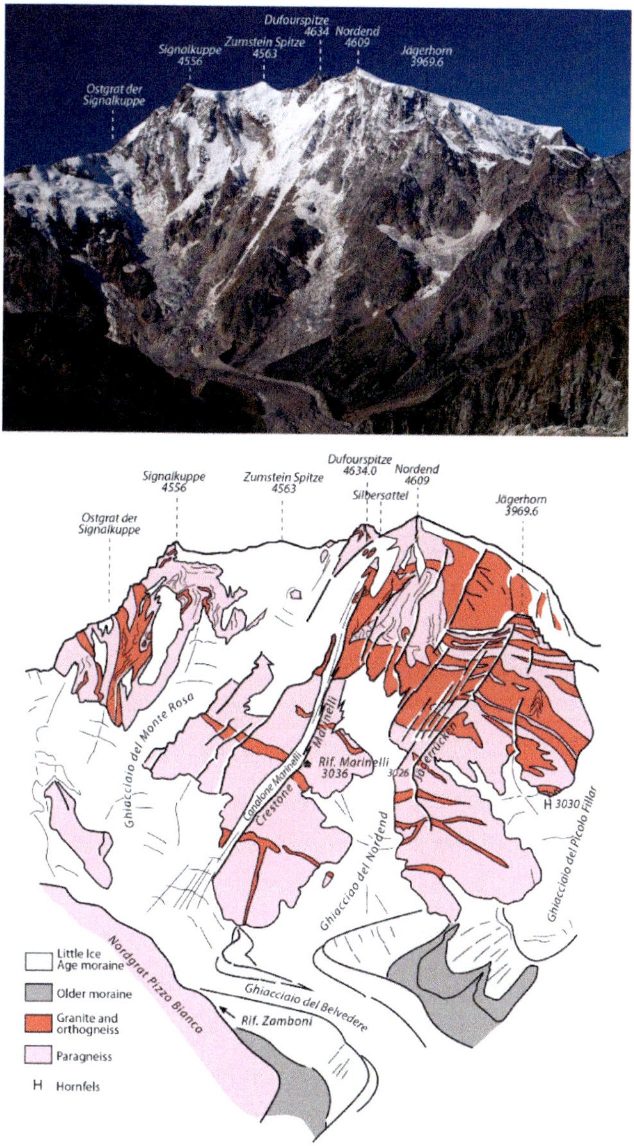

Figure 5.15. *The eastern face of the Monte Rosa, one of the highest cliffs in the Alps (ca. 2 km high), displays pre-Alpine paragneisses cut by a dense set of dykes and stocks of porphyritic Permian granites (redrawn after Bearth (1952)). This low-strain volume, of km^3 scale, occurs in the core of a continental unit subducted to eclogite facies depths*

	Mesozoic cover	Granites and gabbros	Polycyclic basement
Adula	Dolomite carbonates, with fossils! (Jurassic?)		
Suretta	Dolomite boudins in calcite marbles (Triassic)		Minor volumes of mafic rocks, and paragneisses (intruded by granitoids)
Monte Rosa			Large volumes of paragneisses and associated granitic intrusives (of the order of several km^3)
Gran Paradiso	Dolomite boudins in calcite marbles (Triassic)	Large volumes (several km^3) in the Orco Valley	Moderate volumes in paragneisses up to 1 km^3 (Bardoney, Valeille, Valnontey Valleys)
Dora-Maira (HP)	Dolomite boudins (Triassic)		Moderate volumes in orthogneisses and paragneisses up to 1 km^3 (Germanasca Valley)
Dora-Maira (UHP)		Rare and minor volumes (Brossasco, a few 10m^3)	Rare and minor volumes (Case Garneri, a few dm^3)
Dent Blanche (Roisan + Arolla)	Dolomites, with fossils (up to 10-100 m^3) (Triassic)	Granites (km^3) and gabbros (a few 100 m^3 to km^3)	Mont Morion (ca. 1 km^3)
Dent Blanche (Valpelline)			Large volumes (of the order of 10 km^3) of Valpelline paragneisses, marbles, and amphibolites
Sesia (External Zone)	Dolomites (Triassic)	Rare and very small volumes (if a single one?)	
Sesia (II DK)			Large volumes (of the order of 10 km^3) of II DK paragneisses, marbles, and amphibolites
Sesia (Internal Zone)	Dolomites (Triassic)	Monte Mucrone, Lago della Vecchia (at most a few 10 m^3)	Rare and small volumes, a few m^3 in size

Table 5.3. *Selected examples of undeformed volumes of continental crust preserved in tectonic units subducted during the Alpine orogeny. Units in white belong to the distal European palaeomargin, in yellow to the internal Briançonnais, in pink to the Sesia-Dent Blanche nappes and in orange to the Austro-Alpine nappes*

5.6.2. *Rheological behavior of the subducted continental crust*

5.6.2.1. *Pervasive ductile deformation at peak HP-UHP*

The rheological behavior of the continental crust at HP-UHP may only be investigated in places where the HP-UHP metamorphism is well preserved, like in the internal part of the Sesia Zone. There, the following observations relevant to rheology have been made. As a whole, rare and small volumes escaped ductile deformation during the Alpine, HP (ca. 2.0 GPa, 550–600°C), episode (see above). Several superposed ductile deformations have occurred at eclogite facies conditions, or close to peak P-T conditions (e.g. Gosso 1977; Lardeaux et al. 1982; Zucali et al. 2002; Babist et al. 2006). A range of behavior is observed, with the "usual" competence contrasts, from the weakest (marbles, dominated by calcite) to the strongest (eclogites, dominated by garnet and pyroxene) lithologies (Figure 5.16). This range of behavior is also seen at the grain scale, because garnet, jadeite/omphacite and glaucophane porphyroblasts (in decreasing order) display stronger competence than their mica and quartz "matrix" (e.g. Ildefonse et al. 1990; Trepmann and Stöckert 2001, 2002).

Figure 5.16. *Pervasive ductile deformation at HP conditions (Cima di Bonze, internal Sesia Zone). In this area, former amphibolites, transformed into eclogites, are boudinaged within eclogitic micaschists a), indicating the persistence of a significant rheological difference between contrasting lithologies fully recrystallized at the same P-T conditions. In the micaschists b) (Daniele Regis for scale), an early pervasive foliation is deformed by open folds, generally associated with crenulation cleavages at various scales, both schistosities being defined by HP minerals (garnet, glaucophane and/or jadeite, phengite and rutile)*

5.6.2.2. *Evidence for brittle behavior at peak HP-UHP*

The strength of the continental rocks during eclogite-facies metamorphism is a key element in numerical models, which can use a range of rheological laws, from "weak" (Wassmann and Stöckhert 2013) to "strong" behaviors. In support of the latter type of model, two lines of evidence for a rigid to brittle behavior have been deduced from field observations.

Brittle behavior of garnet has been observed in many rock types, including "weak" lithologies like micaschists. Indeed, the increasing use of BSE and X-ray maps has allowed imaging fractures inside pre-Alpine garnet grains that have been sealed by a new Ca-rich garnet growing at or close to peak HP conditions (Pennacchioni 1996; Le Bayon et al. 2006; Manzotti et al. 2012) (Figure 5.17). For some authors, the sealed fractures reveal that the rocks containing garnet were already fractured before the Alpine subduction (Le Bayon et al. 2006). For other researchers, garnet fractures developed during the early stages of subduction, indicating high strain rates that may be due to seismic failure (Giuntoli et al. 2018b; Engi et al. 2018).

Localized deformation may also provide evidence for rigid behavior at HP-UHP. Narrow shear zones with HP mineral assemblages are known to develop in undeformed or poorly-strained, partially transformed, granitic bodies (Monte Mucrone, Figure 5.18(a)) and granulite bodies (Emilius (Pennacchioni 1996)). In the latter case, eclogite-facies veins are associated with the shear zones, suggesting that the progress of dehydration reactions before or close to peak P conditions was responsible for an increase of the fluid pressure, leading to the opening and growth of fractures filled with HP minerals (Pennacchioni 1996; Hertgen et al. 2017). Localized shearing also sometimes occurred in pervasively transformed orthogneisses (Figure 5.18(b)), and in poorly-strained granitic orthogneisses (Gran Paradiso (Menegon and Pennachioni 2010)), where it is thought to develop in pre-existing brittle joints or faults.

Numerous veins filled with HP minerals (e.g. rutile, omphacite, garnet) have been reported (Figure 5.18(d)). Although HP veins are much more abundant in the oceanic units (e.g. Bearth 1967; Phillipot et al. 1991; Lefeuvre et al. 2020), they are also present in the continental units, where they have been searched by collectors for several decades (Piccoli et al. 2007). Some nice examples can be found in the internal Sesia Zone (Figure 5.18(d)), in the Monte Rosa (Lapen et al. 2007) and in the Emilius (Angiboust et al. 2017; Hertgen et al. 2017). Although the mode of growth of the HP minerals in these veins is not fully understood (e.g. because crack-seal fibers have not been found), they imply that the fluid pressure was equal to or even slightly exceeding the lithostatic pressure. The larger amount of HP veins in the oceanic units compared to the continental units is in itself an indication of the amount of fluid present during the HP metamorphism. The oceanic crust has been extensively hydrothermally altered during the "oceanic metamorphism", and therefore dehydration reactions were a major component of their evolution during the Alpine subduction. On the contrary, most of the Paleozoic basement has been dehydrated during the Variscan evolution, and therefore the amount of available fluid during the Alpine subduction was limited.

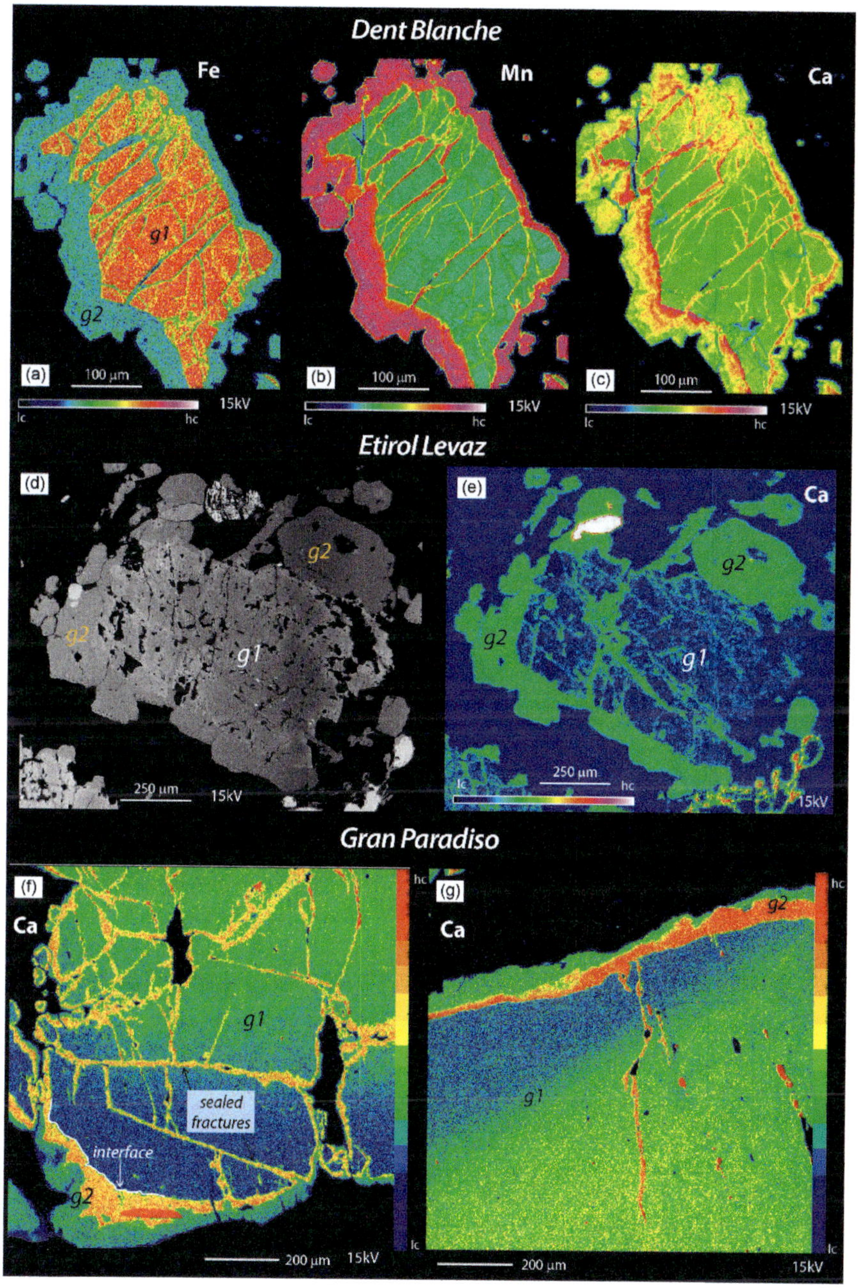

Figure 5.17. *Images of pre-Alpine garnets from polycyclic units*

COMMENTS ON FIGURE 5.17.– *(a–c) Dent Blanche, Roisan-Cignana Shear Zone (see Manzotti et al. (2012) for further details). The early garnet generation (g1) displays fractures sealed by the new garnet generation (g2), in this case, enriched in the Mn component due to the low temperature of the Alpine metamorphism. (d and e) Etirol Levaz: The early garnet generation (g1) is surrounded by a second garnet generation (g2). The latter also fills fractures within g1. Image courtesy of Regis. (f and g) Gran Paradiso. (f) Le Bayon et al. (2006), (g) courtesy of Le Bayon, same sample. The early garnet generation (g1) preserves a zonation and is overgrown by a narrow rim of a second garnet generation (g2). The interface between the two garnet generations (white line) has a lobate shape, evidence of dissolution of g1 before the growth of g2. Fractures penetrating the garnet are filled with a garnet that has the same composition as the earliest g2 and hence is interpreted to have existed before the Alpine cycle (at a time when they were filled with breakdown products from the g1, like chlorite or biotite). The fractures connect the pre-Alpine inclusions to the matrix and have allowed them to react during the Alpine cycle.*

Evidence for brittle faulting at HP-UHP is best documented when pseudotachylites are found, provided that the minerals growing during the cooling of the melt indicate HP conditions. Such evidence has been reported in eclogite-facies mantle rocks from Corsica (Austrheim and Andersen 2004) and Lanzo (Scambelluri et al. 2017; Pennacchioni et al. 2020). In subducted continental rocks, most pseudotachylites have been attributed to deformation episodes taking place when the HP-UHP rocks were already exhumed (Dora-Maira: Zechmeister et al. 2007), and they are sometimes aligned along steeply-dipping, brittle faults cutting through the nappe stack (Dent Blanche: Dal Piaz et al. 2015). Other authors have proposed a potential link with seismic cycles occurring along with the subduction interface (Menegon et al. 2007; Menant et al. 2018). Note that the pseudotachylites found up until now were observed in mantle rocks made of olivine and pyroxene aggregates (Corsica, Lanzo), or in granulites (Valpelline), both of which are notoriously "competent". Other lithologies (Figure 5.18(c)), like the mafic eclogites and the micaschists, do not show evidence of pseudotachylites at eclogite-facies conditions.

5.6.2.3. *Pervasive ductile deformation during decompression*

After continental subduction has taken place, the subducted material came back to the surface. During exhumation, the subducted material may behave as a rigid body, at scales ranging from a km to tens of km (in which case the HP-UHP structures are nicely preserved). Alternatively, a significant amount of ductile deformation may have taken place on the way back to the surface, at greenschist-facies conditions (e.g. in the external part of the Sesia Zone) or at albite-epidote amphibolite facies conditions (e.g. in the Monte Rosa, Gran Paradiso and

Dora-Maira nappes). This late pervasive deformation obscures the rheological behavior of the subducted continental crust at peak *P*, i.e. at their maximum burial.

Figure 5.18. *Localized, narrow, shear zones in incompletely a) Monte Mucrone to completely b) Pont Saint Martin transformed granitoid bodies (Sesia Zone). The shear zones display typical HP mineral assemblages. Pervasive ductile foliations in mafic rocks c) Quincinetto are frequently associated with veins displaying HP minerals (i.e. garnet, jadeite, omphacite and rutile), in this case, quartz and omphacite d) Perloz. Note that all these examples are found in the best-preserved eclogite-facies continental unit of the Western Alps, i.e. the internal Sesia Zone*

In the internal Sesia Zone, most of the HP rocks escape any significant deformation during decompression, at least under greenschist-facies conditions. A weak to moderate amount of strain may have occurred earlier, under blueschist-facies conditions (Vuichard 1987; Babist et al. 2006). In contrast, the external Sesia Zone, and large parts of the Pillonet and Arolla units, are mainly made of orthogneisses displaying a pervasive ductile deformation at greenschist facies conditions. Although precise P-T conditions are difficult to establish for this episode, the low *P* overprint is so pervasive that relicts of prior, high-*P*, parageneses are becoming very rare, or even absent.

In the Internal Crystalline Massifs (Dora-Maira – Gran Paradiso – Monte Rosa), the main ductile deformation has taken place at albite-epidote amphibolite facies and is characterized by a near-horizontal foliation with E-W stretching lineation. The strain pattern, and a few finite strain measurements, indicate that some domains record finite constrictional strains (Pinerolo (Mertz and Siddans 1985); Money (Kassem and Ring 2004)), while others record finite flattening strains (Kassem and Ring 2004). Shear criteria associated with this deformation consistently indicate top-to-the West shearing (Monte Rosa (Lacassin and Mattauer 1985; Lacassin 1987); Gran Paradiso (Le Bayon and Ballèvre 2006)) or remain ambiguous with respect to the main shearing sense (Dora Maira (Philippot 1990; Wheeler 1991; Michard et al. 1993)).

5.7. Metamorphic history of the subducted crust: P-T paths and timing

The displacement of rocks in the Earth's crust and mantle is recorded by changes in P and T, i.e. the P-T path (Figure 5.19). P-T paths provide important information about processes at convergent plate boundaries. Increasing P is associated with burial, while decreasing P records exhumation. Determining the chemical composition of coexisting minerals and calculating P-T conditions using the formalism of thermodynamic equilibrium is a necessary step for quantifying the P-T path. However, a careful examination of the thin sections is required before any calculation, to assess the extent of equilibrium between minerals and their relative timing of growth. This is because the rock record is always much more incomplete than the entire P-T path followed by a single specimen during the burial-exhumation cycle.

5.7.1. *Incompleteness of the rock record*

In order to establish the metamorphic history of the subducted continental rocks, petrologists are akin to sample three different rock types. First, Mesozoic-Cenozoic sediments should only record the Alpine subduction-collision history and are therefore prime targets for petrological purposes. However, they represent only a very minor part of the continental crust involved in the subduction. In addition, most of the Meso-Cenozoic sediments were detached from their basement during the early stages of the convergence, then stacked in the frontal part of the orogen, and therefore escaped the subduction metamorphism.

Second, in the Paleozoic basement, a few monocyclic metasediments (Pinerolo, Money) of the Carboniferous and Permian ages are known. These metasediments are always completely recrystallized, except for a few detrital minerals like garnet (Manzotti and Ballèvre 2013) and zircon (Manzotti et al. 2015a; Manzotti et al. 2016). The other rock types display contrasting behaviors. The Permian gabbros and granites have been partially recrystallized, leaving small to large volumes that escaped significant re-equilibration during the Alpine orogen, with magmatic textures (and sometimes minerals) preserved (Lardeaux and Spalla 1991). In the polycyclic basement, some samples are fully recrystallized, except for the persistence of some pre-Alpine minerals (garnet, biotite, muscovite). Other samples are incompletely recrystallized, like some hornfelses associated with the Permian intrusions (Gabudianu Radulescu et al. 2011), or not recrystallized at all like the high-grade amphibolite to granulite-facies rocks from the Permian lower crust (IIDK, Valpelline Series).

Potential causes for this contrasting behavior are still debated. However, a central role is played by the fluid phase, essentially consisting of H_2O. When water saturation is achieved since the start, the rocks are fully recrystallized. If the water saturation is achieved by some dehydration reactions during the prograde path, the rocks will be partially to fully recrystallized. Conversely, if water saturation is never achieved at any moment during the Alpine history, the rocks will remain essentially unreacted.

Mechanisms allowing water saturation in the basement rocks may be:

– initial water content in sediments, a hypothesis restricted to the Carboniferous-Permian sediments (like the Money and Pinerolo), or to those parts of the Paleozoic basement which did not enjoy Variscan metamorphism;

– alteration of granitoids (fluid-rock interaction) shortly after the following intrusion, as proposed for the whiteschists of the Monte Rosa (Pawlig and Baumgartner 2001; Marger et al. 2019; Luisier et al. 2021);

– fluid infiltration during the Jurassic extension along brittle shear zones penetrating deep into the crust;

– fluid flow associated with hydrothermal seafloor alteration, post-dating Jurassic rifting, well before the onset of Alpine subduction (e.g. the Dora-Maira whiteschists according to Gauthier-Putallaz et al. (2016));

– fluid flow expelled from the downgoing oceanic slab and penetrating the overlying continental crust during the Alpine history (e.g. the whiteschists of Dora-Maira) (Ferrando 2012; Chen et al. 2017).

During decompression, most rocks will be under-saturated with respect to water, except if water enters the system, e.g. in localized, narrow shear zones. Preservation of the "peak" assemblages is therefore dependent upon the availability of water (e.g. Guiraud et al. 2001; Le Bayon et al. 2006; Manzotti et al. 2015b).

In summary, a rather complete record of the prograde history is expected in monocyclic sediments, but an incomplete record (or event lack of record) is expected in the other cases. As a result, different rock types from coherent units will deliver different P-T estimates. Assuming that these *P* differences are larger than the uncertainties on *P* estimation (a few hundreds of MPa), they may be interpreted in two ways:

– a record of different segments along the same P-T path, reflecting a complex history in terms of availability of H_2O and the associated reaction kinetics (e.g. Rubie 1986; Proyer 2003);

– a record of differences in *P* due to mineral- to outcrop-scale gradients in differential stress during metamorphism (e.g. Luisier et al. 2019; Vaughan-Hammon et al. 2021).

In the following, we will consider that the kinetic effects associated with the fluid availability far exceed those associated with gradients in differential stress. Therefore, we will use *P* estimates to record the depth at which the continental crust has been subducted.

5.7.2. *Types of P-T paths of the subducted crust in the Alps*

In the Alps, several types of P-T paths have been reconstructed, differing (i) by the peak P-T conditions achieved (Table 5.4), (ii) by the slope of the prograde path in the P-T space, (iii) by the shape and slope of the decompression path (Figure 5.13). Most units display a clockwise P-T path (i.e. higher *T* during decompression than during burial; e.g. paths B–D in Figure 5.19), with few exceptions (path A in Figure 5.11, Vanoise cover (Goffé and Velde 1984); Southern Sesia Zone (Pognante 1989)). Maximum temperatures attained during the Alpine metamorphism have allowed the preservation of the growth zoning in garnet, with few exceptions like the Alpe Arami eclogite and garnet peridotite (Sandmann et al. 2014) and the Pohorje Mountains (Slovenia) (e.g. Miller et al. 2005; Vrabec et al. 2012). For this reason, the prograde, up-*P* path cannot be reconstructed in the latter units. With respect to simple, continuous clockwise P-T paths, two types of additional complexities have been proposed (Figure 5.19).

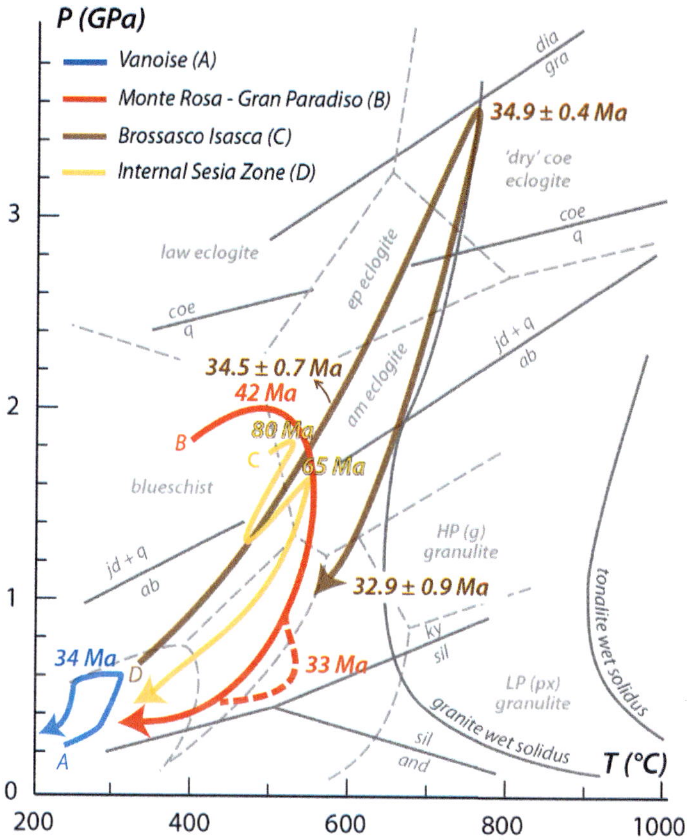

Figure 5.19. *Typical examples of P-T paths of the subducted continental crust in the Alps during the Alpine orogeny (see text for further details). Ages for the Vanoise are deduced from stratigraphic relationships. Ages for the Monte Rosa and Gran Paradiso are taken from Lapen et al. (2007) and Manzotti et al. (2018b), respectively. The P-T path from the Brossasco-Isasca Unit is taken from Groppo et al. (2019), with ages from Gauthiez-Putallaz et al. (2016). The internal Sesia Zone, illustrating two burial-exhumation cycles, is taken from Rubatto et al. (2011) and Regis et al. (2014). The P-T path for the Gran Paradiso is from Manzotti et al. (2015b), with ages from Manzotti et al. (2018b). The increase in T during decompression for the Monte Rosa is taken from Borghi et al. (1996). The anticlockwise P-T path of the Vanoise is from Goffé and Velde (1984)*

	Type of P-T path	P-T conditions of Alpine metamorphism	Isotopic age	Key references
Adula	B	Eclogite facies (metamorphic field gradient from 1.7 GPa, 640°C in the north to 2.5 GPa, 750°C in the south)	Younger than 38 Ma	Heinrich 1982; Dale and Holland 2003; Herwartz et al. 2011; Sandmann et al. 2014
Tauern	B	1.9–2.2, 630°C 2.0–2.5 GPa, 600°C	31.5 ± 0.7 Ma 32.8 ± 0.5 Ma	Hoschek 2001 Glodny et al. 2005; Nagel et al. 2013
Vanoise	A	Blueschist facies (0.6 GPa, 300°C)		Goffé and Velde 1984
Ambin, Gd St Bernard	B	Blueschist facies (1.4 GPa, 450°C)		Ganne et al. 2003; Bucher and Bousquet 2007
Monte Rosa	B, D	Eclogite facies (2.0 GPa, 500–550 °C)	44-42 Ma	Dal Piaz and Lombardo 1986; Lapen et al. 2007; Hauke et al. 2019b
Gran Paradiso	B	Eclogite facies (1.8–2.0 GPa, 500–550 °C)	42 Ma	Dal Piaz and Lombardo 1986; Le Bayon et al. 2006; Manzotti et al. 2018b
Dora Maira (UHP)	E	Coesite-eclogite facies (4.0 GPa, 700 °C)	35 Ma	Chopin 1984; Rubatto and Hermann 2001; Compagnoni and Rolfo 2003; Groppo et al. 2019
Dora Maira (HP)	B	Eclogite facies (2.0–2.4 GPa, 500–520 °C)		Pognante and Sandrone 1989; Groppo et al. 2019
Valosio	B	Eclogite facies (1.2–2.0 GPa, 530°C)		Messiga et al. 1992b

	Type of P-T path	P-T conditions of Alpine metamorphism	Isotopic age	Key references
Theodul-Glacier	B	Eclogite facies (2.0 ± 0.1 GPa, 580 ± 65°C)	56 and 58 Ma	Weber and Bucher 2015; Weber et al. 2015
Etirol-Levaz	B	Eclogite facies (2.0–2.5 GPa, 550–600°C)	46–48 Ma 53 and 62 Ma	Beltrando et al. 2010a; Fassmer et al. 2016
Emilius	B	Eclogite facies (2.1–2.4 GPa, 500–550°C)	53 Ma 56 and 58 Ma	Dal Piaz et al.1983; Angiboust et al. 2017; Hauke et al. 2019b
Dent Blanche (Arolla)	B	Greenschist- to blueschist-facies transition (0.8 GPa, 400–450°C)	ca. 50–42 Ma	Diehl et al. 1952; Angiboust et al. 2014; Manzotti et al. 2020
Sesia (Internal Zone)	B, C	Eclogite facies (2.0 GPa, 550–600°C)	80–60 Ma	Compagnoni 1977; Regis et al. 2014
Ötztal	B	Eclogite facies		
Schobergruppe	B	Eclogite facies	Max 97 Ma	Hauke et al. 2019a
Saualpe and Koralpe	B	Eclogite facies		Thöni 2006
Pohorje	E? (growth zoning not preserved)	Coesite-eclogite facies (4 GPa, >700°C)	90 Ma; 91 to 96 Ma	Janák et al. 2004, 2006; Miller et al. 2005; Vrabec et al. 2012; Janák et al. 2015; Sandmann et al. 2016

Table 5.4. *Alpine HP to ultra-HP metamorphism of the continental crust in the Alps. Units in white belong to the distal European palaeomargin, in green to the external Briançonnais, in yellow to the internal Briançonnais, in pink to the Sesia-Dent Blanche nappes and in orange to the Austro-Alpine nappes. A choice of representative units is made in this table. The letters A–E refer to the types of P-T path drawn in Figure 5.19*

– In some units of the Western Alps (part of the oceanic units (Beltrando et al. 2008); part of the internal Sesia Zone (Rubatto et al. 2011)), a complex P-T path has been proposed during the early stages of decompression, with an initial decompression (of about 0.5 GPa) associated with cooling (of about 50°C), followed by compression and heating of nearly the same value (path C in Figure 5.19). This type of P-T path (called "yo-yo subduction") has been interpreted as recording two short-lived subduction-exhumation cycles, in line with an orogenic model where repeated tectonic switches (from compression to extension) occur during the convergent history (e.g. Lister and Foster 2009).

– During the later stages of the decompression history (at ca. 0.4–0.6 GPa), some authors argued for the existence of a slight (ca. 50°C in the Dora-Maira, Gran Paradiso and Monte Rosa (Borghi et al. 1996; Gasco et al. 2010, 2011)) to moderate increase in T (Lepontine dome (Wiederkehr et al. 2008)) (path B dashed line in Figure 5.19). The implications of this record are quite important because heating at low pressure in already exhumed rocks requires a significant source of heat, which most authors would like to attribute to the detachment of the lithospheric root and its replacement by warmer asthenospheric mantle (Brouwer et al. 2004; Sizova et al. 2019).

5.7.3. *Metamorphic gradient at the scale of the Alpine belt*

Metamorphic maps of the Alps have been drawn with increasing sophistication through the years (Ernst 1973; Frey et al. 1974; Bousquet et al. 2004, 2012; Oberhänsli 2004; Beltrando et al. 2010b; Lardeaux 2014). These maps display the distribution of peak mineral assemblages in terms of metamorphic facies (Figures 5.20 and 5.21).

The first-order message given by the metamorphic maps is an increasing gradient of P-T conditions from the outer (W, NW or N) to the inner (E, SE or S) part of the Alpine belt, valid for both oceanic and continental units (Figure 5.20). This gradient abruptly ends along the Insubric Line, i.e. along the boundary between the "lower plate units" and the "upper plate units" (Schmid et al. 2017). This regional-scale gradient has been interpreted as reflecting the former dip of the subduction zone (Dal Piaz et al. 1972; Ernst 1973), a conclusion still accepted today.

A second message taken from the metamorphic maps is that crustal-scale, coherent bodies displaying a gradient in P-T conditions have been rarely found. One notable exception is the Adula Nappe, where P-T conditions are estimated at about 1.8 GPa, 650°C in the north and increase to 2.5 GPa, 750 °C in the south, along a length of ca. 50 km (Heinrich 1982; Dale and Holland 2003; Sandmann et al. 2014).

This "metamorphic field gradient" is consistent with a southward-dipping subduction zone. However, it cannot be used to decipher the geothermal gradient prevailing at the time of maximum burial of the Adula Nappe, because the latter – although remaining coherent – has been strongly deformed following peak *P* conditions (Nagel 2008; Cavargna-Sani et al. 2014): the actual distance between the two extreme values is not representative of their distance at the time of the HP event.

The third message of the metamorphic map is that major discontinuities of P-T conditions (i.e. max P-T conditions in individual rock types) occur along shear zones bounding the major units, reflecting displacement after peak-*P* conditions. This indicates that the original structure of the downgoing slab has been modified by slicing an originally continuous body during the tectonics associated with its exhumation. In line with this observation, and despite some attempts, mineral isograds cannot be drawn for the early, HP history.

This results from the fact that (i) temperature gradients were very small to small during the subduction history (in the range of 5–8°C/km), and (ii) the potential metamorphic field gradients associated with the subduction history have always been modified during nappe emplacement and post-nappe deformation.

It is generally agreed that the larger the displacement was along the shear zone, the larger the discontinuity is in terms of *P*, *T* and time! In this respect, two types of shear zones have received attention.

In the first case, tectonic boundaries may separate units with higher-grade rocks (or older ages) in the hanging wall compared to lower-grade rocks (or younger ages) in the footwall. One of the best examples of such a situation is observed in the Dora-Maira Massif, where the blueschist-facies Pinerolo Unit is overthrust by the coesite-bearing Brossasco-Isasca unit (Avigad et al. 2003; Groppo et al. 2019). A similar situation is known in the Gran Paradiso Massif, where the Money Unit is overthrust by the eclogite-facies Gran Paradiso Unit (Le Bayon et al. 2006; Manzotti et al. 2015b), both units having attained peak P-T conditions at about 42 Ma (Manzotti et al. 2018b).

In a second case, tectonic boundaries separate units with lower-grade rocks in the hanging wall compared to the footwall. A spectacular example of this is found in the NW Alps, where the coesite-bearing eclogites of the oceanic Zermatt Zone are separated from the greenschist-to-blueschist facies continental Dent-Blanche Nappe by a greenschist-facies shear zone (Combin Fault (Ballèvre and Merle 1986)). This boundary, also taking into account the difference in age of the HP episode, is interpreted either as an out-of-sequence thrust (Ring 1995) or as an extensional ductile fault, allowing the exhumation of the eclogite-facies nappe stack in the footwall (Reddy et al. 1999; Manzotti et al. 2018b).

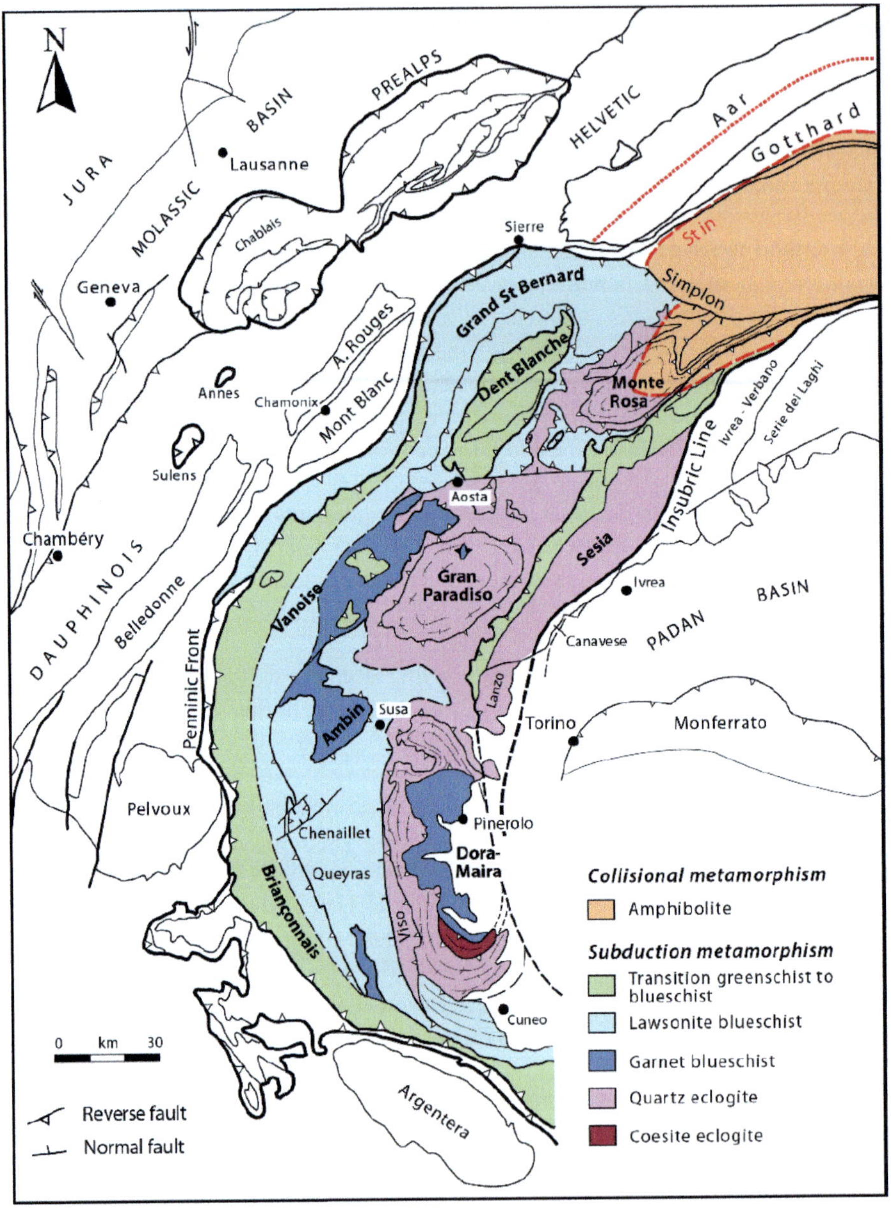

Figure 5.20. *A simplified metamorphic map of the Western Alps. According to petrological studies, peak T has been achieved in most units at the same time as peak P (see Figure 5.19). Therefore, this map displays the peak P-T conditions in the different Alpine units, in terms of metamorphic facies*

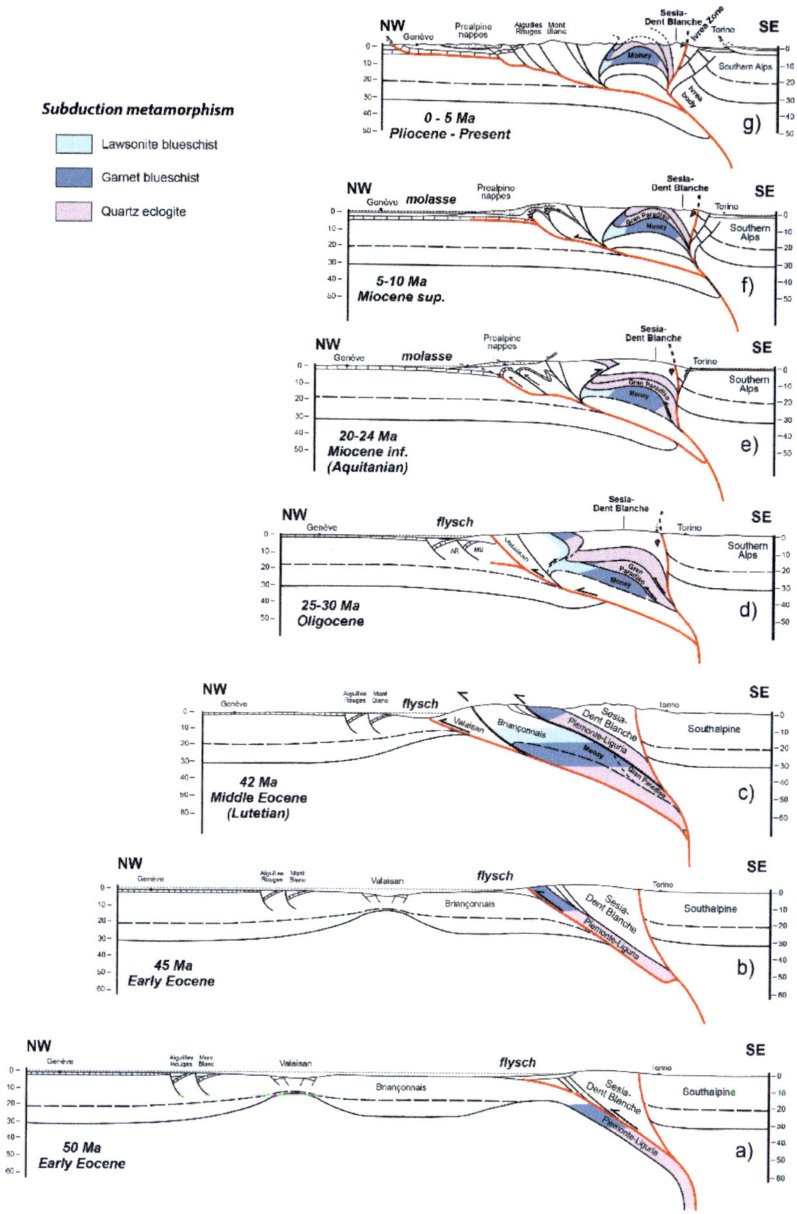

Figure 5.21. *A schematic tectonic evolution of the Western Alps, displaying how the metamorphic (baric) structure of the downgoing plate (the Piemonte-Liguria ocean during the Early Eocene, or the Briançonnais basement in the Upper Eocene) is destructed when the oceanic/continental slices detach from the downgoing plate*

Large-scale folding and/or shear zones post-dating the peak P-T conditions associated with continental subduction may also occur after nappe emplacement, and therefore overprint the metamorphic pattern inherited from the early stages of exhumation of the UHP-HP rocks. It has long been recognized that the internal part of the Briançonnais Zone has been significantly modified by back-folding/back-thrusting (e.g. Argand 1911; Michard 1959, 1977; Caby 1996; Escher and Beaumont 1997; Bucher et al. 2003; Scheiber et al. 2013). A nice example of ductile shear zones post-dating nappe stacking is provided by the Central Alps, on both sides of the Lepontine structural dome. To the west of the dome, the Simplon Fault displays a ductile to brittle normal-sense shear allowing exhumation of the high-temperature nappes in its footwall during the Miocene (Mancktelow 1990; Campani et al. 2014).

On the east flank of the Lepontine dome, the Forcola Fault, on top of the Adula nappe, accommodates at least 3 km of vertical displacement during the Miocene (Meyre et al. 1998). Other examples in the Eastern Alps are the Brenner and Katschberg Faults, which develop during north-south shortening following nappe emplacement (for a review, see Rosenberg et al. (2018)).

5.7.4. *Timing of the continental subduction in the Alps*

The timing of the continental subduction in the Alps has been constrained using different radiometric methods. Increasing knowledge of the behavior of the isotopic systems and increasingly sophisticated instruments have opened new avenues for understanding the ages and their significance. Two main strategies have been used.

The first strategy, used to date the HP metamorphism, considered rocks where HP parageneses are well equilibrated and well preserved, and where the HP minerals can be related to a specific fabric.

^{40}Ar-^{39}Ar ages on white mica have long been considered a primary tool because the timing of growth of white mica can be easily determined based on its microstructural position and chemical composition. This method has encountered major difficulties for two main reasons. First, most grains are chemically zoned, and the interpretation of this zoning is itself delicate. It may be due to partial re-equilibration during exhumation or to one or several fluid pulses occurring during decompression (e.g. Sesia (Halama et al. 2014, 2018)). Secondly, white mica has been shown to have the capacity of retaining Ar (the excess or inherited component), leading to older ages without geological significance. This problem is especially acute in continentally-derived basement rocks (e.g. Dora-Maira (Scaillet 1996; Di Vincenzo et al. 2006; Schertl and Hamerschmidt 2016); Sesia (Ruffet et al. 1997)). For these reasons, we excluded ^{40}Ar-^{39}Ar ages from our compilation, except when no other ages were available. The Rb-Sr method has been combined with the ^{40}Ar-^{39}Ar dating, in order to evaluate the excess Ar (e.g. Ruffet et al. 1997; Di Vincenzo et al. 2006; Halama et al. 2018).

Lu-Hf and Sm-Nd garnet geochronology have been used to assess the prograde and peak metamorphic history in both continental and oceanic units (e.g. Thöni and Miller 1996; Duchêne et al. 1997; Hauke et al. 2019a). This is because petrological modeling shows that garnet grows at increasing pressure and temperature. Provided growth zoning is preserved and rare earth element rich inclusions are absent, this is a fair assumption, providing robust ages. However, one major pitfall of these two methods is the possibility of preserving older, i.e. pre-Alpine, garnet cores that have not been fully dissolved or re-equilibrated at peak P-T conditions, and may therefore have a large impact on the obtained ages. Without detailed petrology, garnet geochronology may be difficult to interpret, especially in subducted pre-Alpine basements.

The second strategy is to date accessory phases, like zircon (e.g. Rubatto et al. 1999), titanite (e.g. Rubatto and Hermann 2001; Castelli et al. 2002), rutile (Lapen et al. 2007), allanite (e.g. Regis et al. 2014), and monazite (e.g. Manzotti et al. 2018b) using the U-Th-Pb method. Amongst these accessory phases, zircon is the most largely dated for constraining the age of the HP metamorphism (Figures 5.22 and 5.23). The main difficulty in this second strategy is to assess the timing of growth of the accessory phases with respect to the main, HP, paragenesis. To achieve this, we can consider (i) their textural position, (ii) the nature of their mineral inclusions, and (iii) their chemical composition, assessing whether they were in equilibrium with other, major, silicates grown during the HP event.

It is now possible to date accessory minerals using three different techniques. The first one is to extract a significant quantity of grains and date them after dissolution (ID-TIMS). A second technique is to date individual grains, after mounting them in an epoxy resin, allowing us to resolve core to rim differences in age ("in situ" dating, using SIMS or LA-ICP-MS). A third technique consists of dating grains in thin sections, therefore ascertaining their structural position, inclusions and chemistry ("in situ and in context" dating). It is therefore advisable to enter into the details of the methods used before comparing ages obtained in a single unit. The progress has been so huge and rewarding during the last decade that many older studies are now out of date.

The main results of the geochronological studies are as follows (Figures 5.22 and 5.23 and references therein):

– The oldest HP rocks are found in the Austroalpine nappes, where they cluster around 90–95 Ma.

– In the Central and Western Alps, an age gradient is observed across the belt. The oldest HP rocks are found in the Sesia Zone (ca. 80-60 Ma), then the Emilius-type continental slivers (ca. 50 Ma), succeeded by the eclogite-facies oceanic units (ca. 45 Ma), and immediately followed by the eclogite-facies continental units (Monte Rosa and Gran Paradiso: 42 Ma). The iconic Dora-Maira coesite eclogites (i.e. the Brossasco-Isasca Unit) are dated at around 35–33 Ma and are therefore the youngest eclogites in the Western Alps.

– Few estimates of burial rates have been made in the continental units (Figure 5.24). One exception is the Brossasco-Isasca Unit, where Gauthier-Putallaz et al. (2016) estimated burial rates >2 cm/year. This estimation is higher than those obtained in the oceanic units (of the order of 5 mm/year (de Meyer et al. 2014; Dragovic et al. 2020)).

– Estimation of exhumation rates varies considerably (Figure 5.24), with maximum values around 1–3 cm/year for the Brossasco-Isasca Unit (Gebauer et al. 1997; Rubatto and Hermann 2001) and much lower values for the Gran Paradiso Unit (of the order of 2–5 mm/year (Manzotti et al. 2018b)), similar to values obtained in the oceanic units (1-2 mm/year (Agard et al. 2002)).

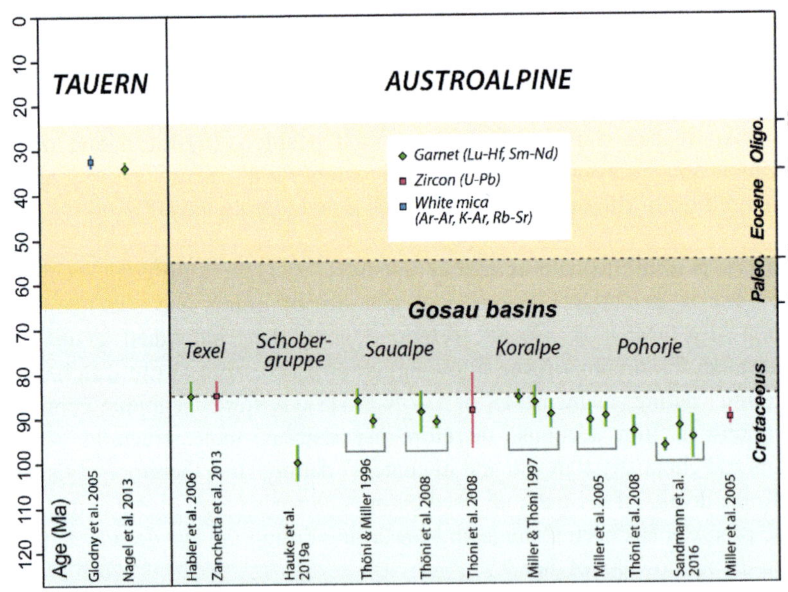

Figure 5.22. Geochronological data on the Alpine HP metamorphism in the Eastern Alps. The timing of sedimentation in the Gosau basins (Eastern Alps) is shown for comparison, to emphasize the major difference in evolution between Austroalpine units in the Eastern Alps and the Sesia–Dent Blanche nappes (see Figure 5.23) (modified after Manzotti et al. (2014a))

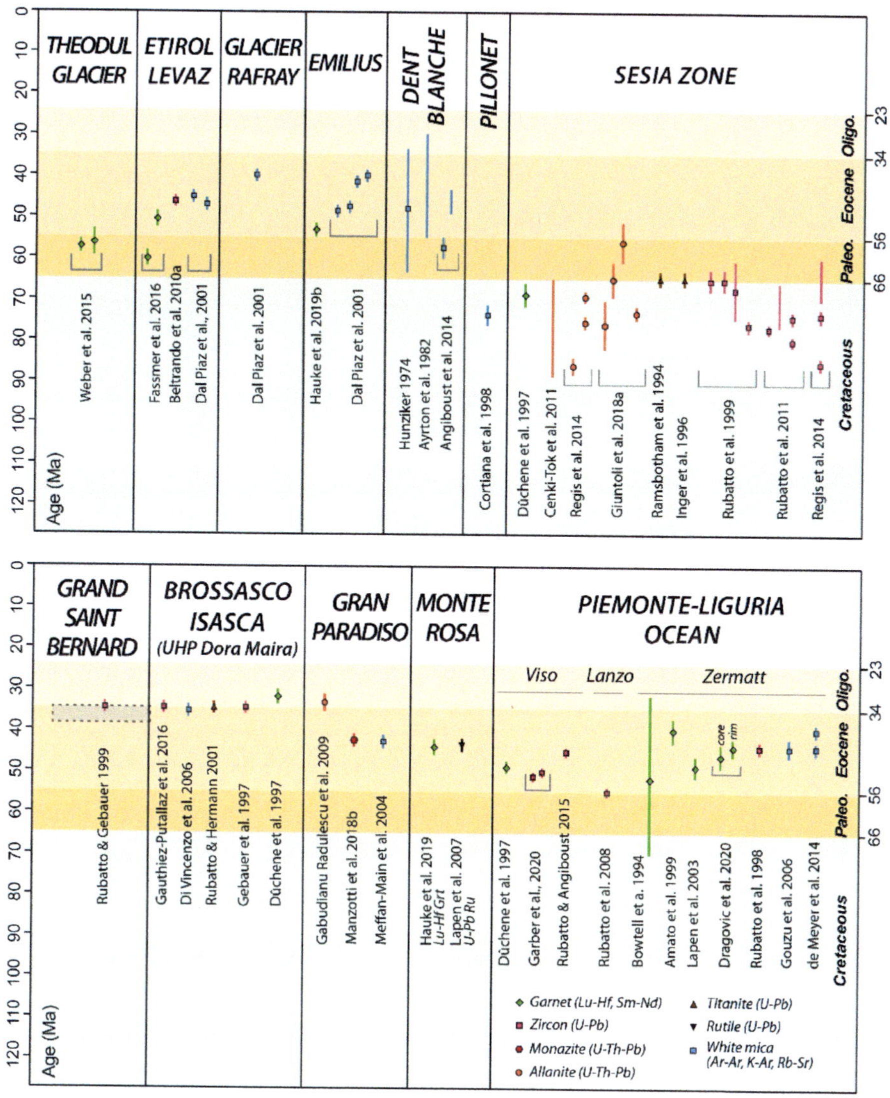

Figure 5.23. *Geochronological data on the Alpine HP metamorphism in the Western Alps (modified after Manzotti et al. (2014a))*

5.7.5. The exhumation record in syn- to post-orogenic sediments

Exhumation of HP rocks may also be constrained using data from the sediments deposited in the foreland basins surrounding the Alpine belt (Figure 5.24).

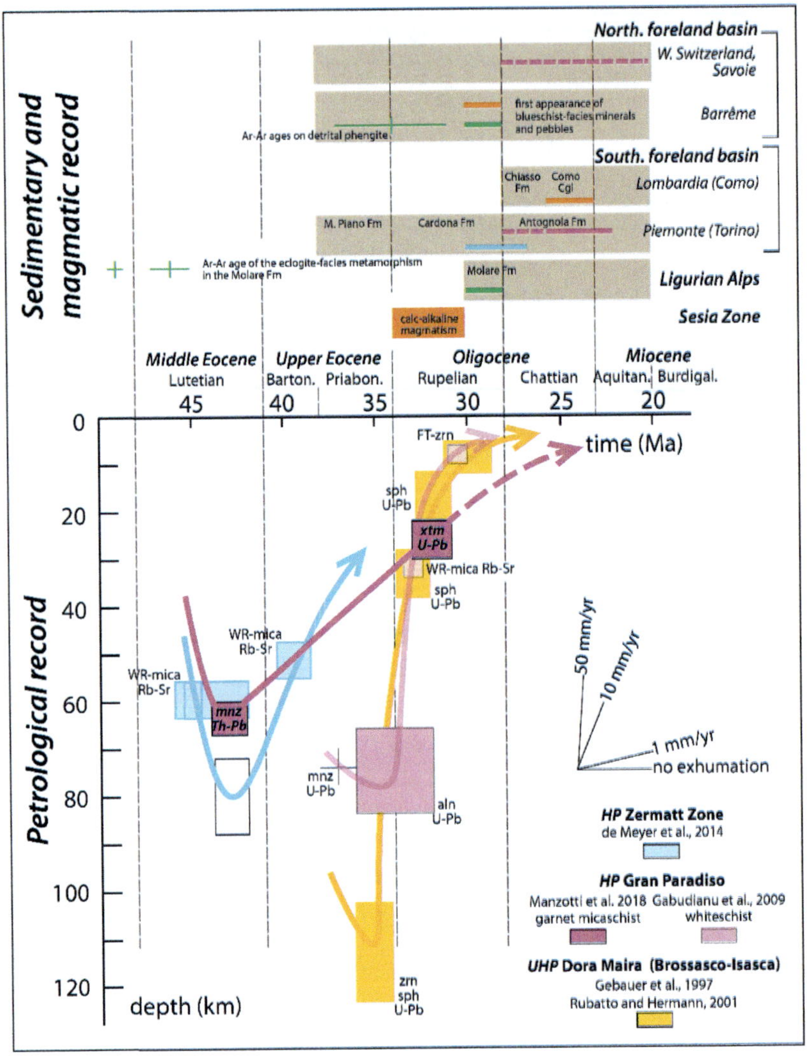

Figure 5.24. *Sedimentary, magmatic and metamorphic records of the Alpine evolution. Depth vs. time paths from which the mean exhumation rates are calculated. The conversion from pressure to depth was calculated by assuming a 20-km-thick upper crust and a 10-km-thick lower crust followed by an upper mantle with densities of 2.7, 3.0, and 3.3 g/cm3, respectively. In this model, a minimum crustal thickness of 30 km for the upper plate is assumed, because the subduction of the Gran Paradiso Unit occurred before the continental collision. Any increase in crustal thickness would lead to a slight increase in subduction depth and hence to higher exhumation rates (Manzotti et al. 2018b)*

The northern foreland basin, extending from Lyon (France) across Switzerland to Vienna (Austria), is characterized by a transition from flysch- to molasse-type sediments (e.g. Allen et al. 1991; Sinclair 1997a; Garefalakis and Schlunegger 2019). In the flysch successions, the Taveyannaz Formation, of the Rupelian age, is characterized by the abundance of volcanic andesitic clasts derived from a near-contemporaneous source (e.g. Ruffini et al. 1997; Boyet et al. 2001; Di Capua and Groppelli 2016). Although non-volcanic clasts have been encountered in the Taveyannaz sandstones, evidence for HP clasts is lacking. By contrast, in the molassic sediments, clasts of HP rocks and minerals (glaucophane, lawsonite) have been reported since the Chattian from boreholes, as well as outcrops located essentially in the westernmost part of the foreland basin (e.g. von Moos 1935; Vatan 1949; Vernet 1958, 1964; Mange-Rajetsky and Oberhänsli 1982; von Eynatten 2003).

A similar situation is found in the Barrême area (SW France), where part of a top-wedge basin is well preserved (Evans and Elliott 1999; Ford et al. 1999). There, the sedimentation covers a long time range, from Upper Eocene to the earliest Miocene. The record of exhumation in this basin ranges from pebbles of material derived from the oceanic units (radiolarian cherts (Cordey et al. 2012); serpentinites (Schwartz et al. 2012)) to detrital minerals and clasts including HP minerals such as glaucophane and phengite (de Graciansky et al. 1971; Evans and Mange-Rajetsky 1991). The first appearance of HP rocks and minerals occurred during the late Rupelian (i.e. 30–28 Ma), and the Ar-Ar ages on detrital white micas indicate blueschist-facies metamorphism at 34 ± 3 Ma in the source rocks (Morag et al. 2008).

In the southern foreland basin, the Oligo-Miocene deposits crop out in the "Gonfolite Lombarda Group" (Bernoulli et al. 1989; Tremolada et al. 2010), where the main source of detrital material is represented by metamorphic and plutonic rocks (Carrapa and Giulio 2001), including granitoids of the Bergell intrusion dated at 33 Ma. This reflects the proximal (northern) source of detrital material.

Further south, in the Torino area (Elter et al. 1966; Polino et al. 1991; Festa et al. 2003), the Oligo-Miocene sediments record the progressive erosion of deeper structural levels. Sesia-derived HP rocks are found in the late Oligocene conglomerates, where they are associated with clasts derived from the South-Alpine and Piemonte-Liguria units. From the latter, blueschist-facies clasts are known in the Burdigalian, while eclogite-facies clasts appear in the Aquitanian, and become more abundant in the Langhian. Finally, clasts similar to lithologies found in the Dora-Maira and Gran Paradiso massifs become recognizable in the Langhian.

Direct evidence for the timing of denudation of HP rocks is found in two localities, namely in the Ligurian Alps and the Sesia Zone.

– In the Ligurian Alps, blueschist- to eclogite-facies oceanic rocks of the Montenotte and Voltri Groups are unconformably covered by fluviatile to shallow marine sediments of Oligocene age (Lorenz 1969, 1979), representing the oldest deposits in the Tertiary Piemonte Basin (Rossi et al. 2009). The Oligocene deposits are faulted or gently tilted, showing that the post-Oligocene deformation in the Ligurian Alps is negligible. HP clasts derived from blueschist- or eclogite-facies oceanic units are found in numerous localities (Federico et al. 2004, 2005; Smye et al. 2021), at stratigraphic levels ranging from the late Rupelian to the Chattian. Detrital material sourced in continentally-derived, eclogite-facies, units have not been identified in the Ligurian Alps.

– Along the eastern margin of the Sesia Zone, late Oligocene, coarse-grained, alluvial deposits containing locally-derived clasts of the HP Internal Sesia Zone and calc-alkaline volcanics (dated at 30.5 ± 0.5 Ma (Kapferer et al. 2012)) rest unconformably on top of the "eclogitic micaschists" (Scheuring et al. 1974; Callegari et al. 2004; Kapferer et al. 2011). This formation is a unique witness of the exhumation of the Sesia Zone and is tilted, faulted and affected by cataclastic deformation against the brittle Canavese Line, an offset of the Insubric Line.

Overall, the studies of the syn- to post-orogenic sediments indicate that (i) there is no evidence of erosion of HP rocks before the late Rupelian (ca. 30 Ma); (ii) a drastic change of tectonic setting occurred during or slightly after the episode of calc-alkaline magmatism, dated at ca. 30–33 Ma), potentially associated with a slab breakoff (von Blankenberg and Davies 1995; Sinclair 1997b); and (iii) progressive unroofing of deeper structural levels is recorded in the Western Alps from the late Oligocene to the Miocene.

5.8. A dynamic view of continental subduction in the Western Alps

5.8.1. *Kinematics of the subduction and collision*

As described above, the Alpine belt displays numerous examples of subducted continental material. Structural considerations and timing data show that two distinct, unrelated, episodes of continental subduction have taken place (e.g. Neubauer et al. 2000; Schmid et al. 2004; Handy et al. 2010; Angrand and Mouthereau 2021).

5.8.1.1. *Cretaceous continental subduction in the Austro-Alpine domain (Eastern Alps)*

A first, short-lived, episode of continental subduction has been achieved during the early Upper Cretaceous (ca. 90 Ma: Figure 5.22), whose record is only preserved in the Austroalpine units in the Eastern Alps. There, the eclogite-facies units were exhumed shortly after peak *P*, before or during the deposition of the Gosau basins, of the middle to late Upper Cretaceous age (e.g. the Kainach basin, east of Graz (Ebner and Rantitsch 2000)).

The geodynamic setting of the eclogite-facies metamorphism in the Austroalpine is still uncertain. It may be unrelated to the closure of the Triassic Meliata Ocean because this closure occurred by the end of the Jurassic. Therefore, the eclogite-facies metamorphism may have resulted from intra-continental subduction, which had been initiated in a Permian rifted (thinned) crust located inside the Austroalpine Domain (Janàk et al. 2004, 2009, 2015; Sandmann et al. 2016).

5.8.1.2. *Late Cretaceous to Oligocene continental subduction in the Penninic Domain (from the Western to the Eastern Alps)*

A second, long-lived (from the Late Cretaceous to the Oligocene), although episodic, continental subduction is mainly recorded in the Western Alps, but also in the Central (Adula Nappe) and Eastern Alps (Eclogite Zone). This episode is associated with the subduction of the Jurassic Piemonte-Liguria Ocean. In the Eastern Alps, the two eclogite belts are recognized, but the eclogites from the first episode were part, during the second episode, of the upper plate (they were already exhumed). The peculiar geometry of the pre-orogenic setting (i.e. of the Jurassic extensional system) is characterized by narrow oceanic basins separated by ribbons of thinned continental crust and extensional allochthons. Southwestward convergence involved the sequential subduction of oceanic basins.

Because of the arcuate shape of the Alpine belt, and taking into account the generally accepted “internal” origin of higher nappes with respect to lower ones, it is impossible to build a kinematic model with a unique transport direction throughout the belt (e.g. Goguel 1963; Ricou and Siddans 1986). The link between ductile deformation at the sample to outcrop scale and plate displacement has been discussed in the frame of the entire belt (Malavieille et al. 1984; Choukroune et al. 1986; Platt et al. 1989; Schmid et al. 2017). In the Western Alps, superposed deformations are related to an early episode of top-to-the-north or top-to-the NNW shearing, followed by a later episode of top-to-the-west shearing (Figure 5.25; Schmid et al. 2017).

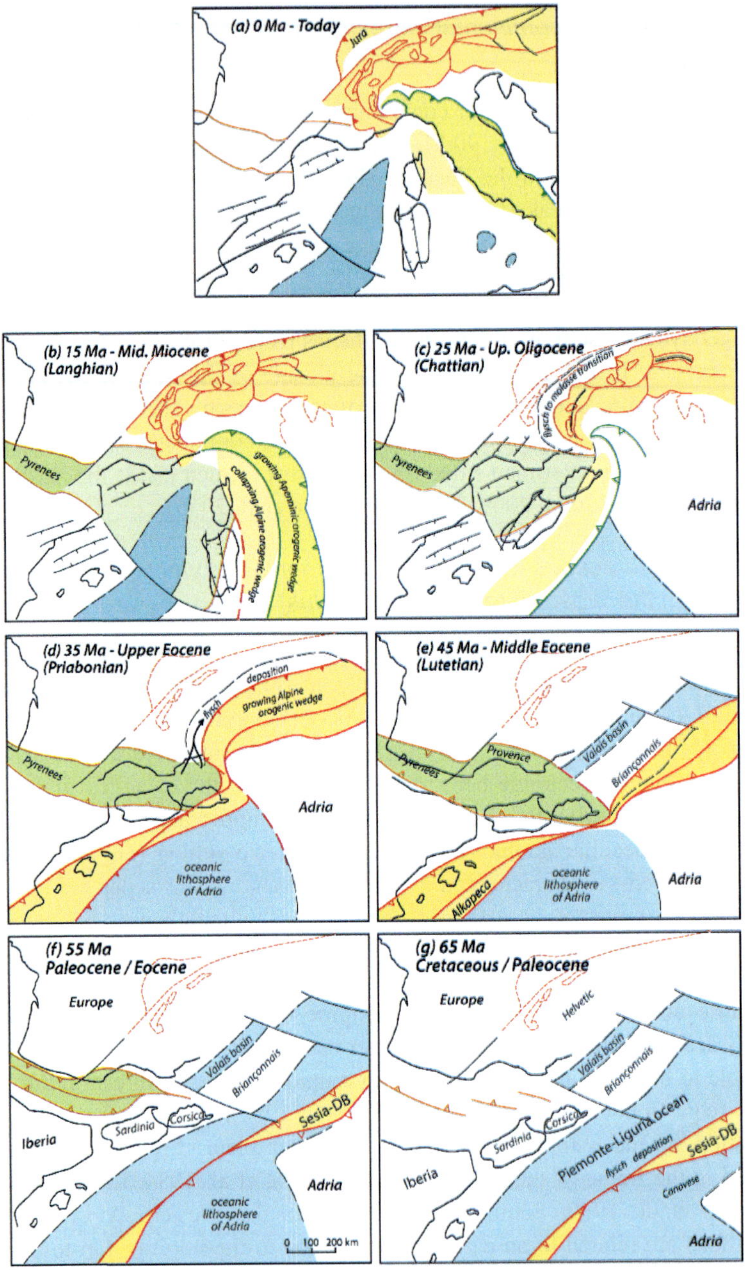

Figure 5.25. *A kinematic reconstruction of the subduction to collision history in the Western Alps (modified from Schmid et al. (2017))*

The first continental stripe involved in the subduction zone was the Sesia-Dent Blanche. This stage followed the consumption of a piece of Jurassic ocean, now poorly represented in the Canavese Zone (Elter et al. 1966; Ferrando et al. 2004). It is therefore probable that no or few ophiolites returned to the surface along this suture zone (McCarthy et al. 2020; Le Breton et al. 2021).

Alternatively, the Sesia-Dent Blanche nappes may have been accreted to the upper plate by subduction erosion (Dal Piaz et al. 1972; Polino et al. 1990; Angiboust et al. 2014), although this model ignores the ophiolite slices in the Canavese Zone and is not supported by independent data on the migration towards the former trench of an unrecognized Cretaceous-Palaeocene volcanic arc.

The Briançonnais microplate was entering the subduction zone during the late Middle to Upper Eocene, at a time when most of the Piemonte-Liguria ocean was already consumed. During this process, the Briançonnais crust was sliced into small pieces, progressively detached from the downgoing slab, and thrusted on top of each other, building the thickened Alpine crust in the internal zones of the Western Alps.

The entrance of the Valaisan basin into the subduction zone (i.e. below the frontal part of the Briançonnais thickening wedge) was a short-lived episode, occurring close to the Eocene-Oligocene boundary.

A drastic change in the geodynamic setting occurred during the early Oligocene (e.g. Ford et al. 2006; Dumont et al. 2012; Romagny et al. 2020; Jolivet et al. 2021): subduction has gone, and collision is now on. This "Oligocene revolution" is best displayed by the widespread occurrence of calc-alkaline magmas (mainly intrusions, but also lava flows and volcaniclastic material) related to partial melting in the asthenosphere. A popular and valuable model for explaining this short-lived magmatism is the slab break-off (von Blanckenburg and Davies 1995). The occurrence of the Early Oligocene calc-alkaline magmatism coincides with the latest evidence for continental subduction, meanwhile, crustal shortening continues in the External Zones, marked by propagating thrusts at progressively higher crustal levels, and by a change from an underfilled (i.e. flysch) to overfilled (i.e. molasse) sedimentation in the northern foreland basin (e.g. Sinclair 1997a, 1997b).

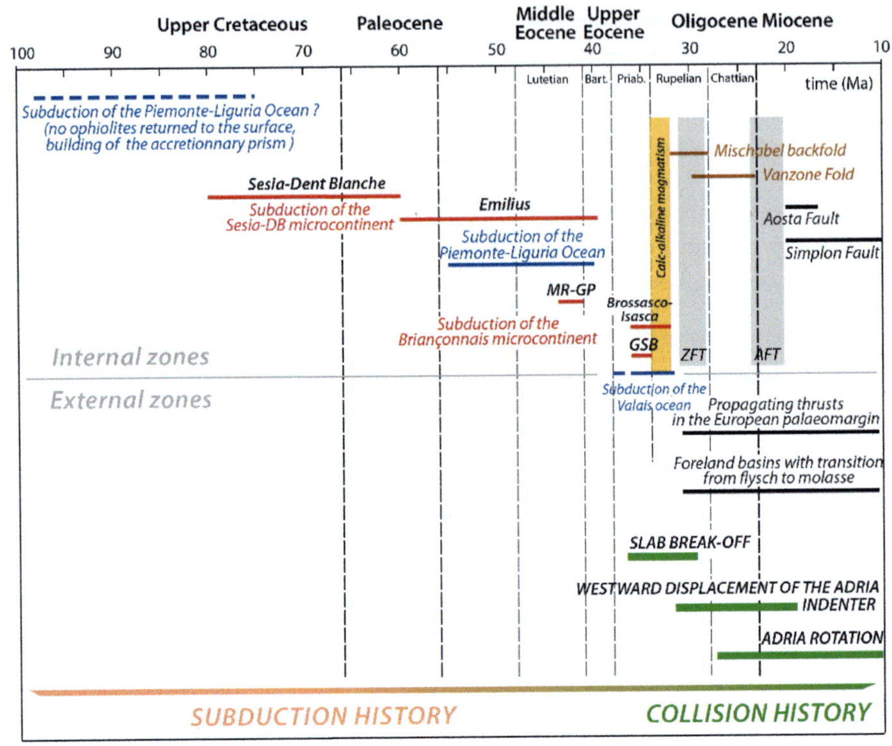

Figure 5.26. *A schematic overview of the subduction history in the Western Alps*

5.8.2. *Exhumation mechanisms of UHP-HP rocks*

The long-standing debate about the exhumation mechanisms of UHP-HP rocks is not a specific feature of the Alpine belt (e.g. Andersen et al. 1991; Platt 1993; Jolivet et al. 1994, 2003; Vignaroli et al. 2008; Warren 2013; Schenker et al. 2015), but is also relevant for other mountain belts. This debate focuses on the driving forces for exhumation and needs to integrate data from a large wealth of disciplines, including geological mapping, structural data, P-T paths and timing constraints. In addition, numerical models have been developed in order to better understand the dynamics of exhumation (Gerya et al. 2008; Butler et al. 2013, 2014; Burov et al. 2014; Schmalholz and Schenker 2016).

– Erosion alone cannot account for exhuming HP-UHP rocks, because the amount of material deposited in the foreland basins is far too small compared to the size and depths to which continental (and oceanic) units have been subducted. Accordingly, the driving force for exhumation would be the density difference between the subducted material (especially so in the case of the continental crust) and its surrounding environment (such as the mantle rocks from the upper plate). Tectonic processes acting at the scale of the belt must have taken place contemporaneously with erosion, whatever its location and rate.

– There is an orderly sequence of deformation events propagating from the most internal to the most external zones (or higher units towards lower units), typified by the Late Cretaceous ages in the internal Sesia Zone, to the Eocene ages in the Monte Rosa-Gran Paradiso nappes (e.g. Duchêne et al. 1997; Rubatto et al. 1998; Manzotti et al. 2018b). This precludes models that involve "tectonic mélange" in a subduction channel as the main process to obtain the actual geometry of the Alpine belt (e.g. Polino et al. 1990; Roda et al. 2020, 2021).

– Extensional, ductile to brittle, faults have been recognized at various stages in the evolution of the Alpine belt. Orogen-parallel extensional faults, like the Simplon (Mancktelow 1990), Forcola, Brenner (Selverstone 1988) and Katschberg (Genser and Neubauer 1989; Scharf et al. 2013) Faults, have played a significant role in the late (Miocene to Present) stages of exhumation of the HP-UHP rocks. At the belt scale, the importance of orogen-parallel, extensional, faults, increases from the Western Alps to the Eastern Alps, where they are associated with strike-slip faults, significantly contributing to the "lateral extrusion" of the Alpine belt towards the Carpathians-Pannonian system (e.g. Ratschbacher et al. 1991; Rosenberg et al. 2018).

– Orogen-parallel extension during convergence is not, however, a viable mechanism for explaining the exhumation of HP-UHP rocks shortly following peak-P conditions, as documented in most units from the Western Alps. In this area, the extent and significance of orogen-perpendicular extensional faults, like the West Viso detachment, also called the Susa Shear Zone (Ballèvre et al. 1990; Ghignone et al. 2020) and the Combin Shear Zone (Ballèvre and Merle 1993; Reddy et al. 1999; Manzotti et al. 2018b) has not been widely acknowledged. These ductile faults were acting during the early stages of exhumation, immediately following the burial at peak P depths. This favors a model of an extrusion wedge, bounded by a "reverse fault" at the base (i.e. a thrust) and a "normal fault" at its top (e.g. Chemenda et al. 1996).

– This extrusion wedge model – our favorite one – may imply horizontal, plate-scale extension, at least transiently (Malusà et al. 2011), but may also occur during continuing convergence, when slab-roll back has taken place. This process is generally associated with horizontal, crustal-scale, extension in the upper plate (e.g. Jolivet et al. 1994, 2003; Brun and Faccenna 2008), a feature that is not observed from the Upper Cretaceous to present in the Southern Alps (Zanchetta et al. 2015). However, if the upper plate is converging at about the same rate as the retreat of the subducting plate, shortening will be maintained during the entire history (Schlunegger and Kissling 2015; Kissling and Schlunegger 2018; Manzotti et al. 2018).

5.9. Prospects for the future

After more than a century of study on HP-UHP continental rocks in the Alps, there is still ample space for future research. Three main topics deserve attention.

– Duration and rates of metamorphic processes have been investigated in a small number of places. Specifically, differences in peak *P* ages, like 35 Ma in Dora-Maira (Gebauer et al. 1997; Rubatto and Hermann 2001; Gauthiez-Putallaz et al. 2016) and 42 Ma in Monte Rosa (Lapen et al. 2007) and Gran Paradiso (Manzotti et al. 2018b), need to be assessed. Similarly, the large spread of ages in the internal Sesia Zone, from about 80–60 Ma (Rubatto et al. 2011; Regis et al. 2014; Giuntoli et al. 2018a), is still poorly understood. It can be ascribed to differences in stacked subunits, or to episodic fluid flow in the eclogite-facies crust during its subsequent exhumation.

– The amount and timing of fluid infiltration into the subducting crust have just been explored (Konrad-Schmolke et al. 2011a, 2011b; Gauthiez-Putallaz et al. 2016; Engi et al. 2018). A better understanding of the role of fluids is crucial for the progress of metamorphic reactions, and therefore the changes in volume, density and mechanical properties of the subducting crust.

– The rheology of the subducted continental crust should be better constrained, based on accurate and detailed fieldwork, with implications for numerical models aimed at understanding the dynamics of mountain belts.

– A better understanding of the geophysical data on the deep structure of the Alps is necessary in order to accurately describe the geometry and nature of the lithospheric slabs below and around the mountain belt. Numerous pending questions may be resolved (e.g. whether the European lithosphere is affected or not by the slab break-off) thanks to international cooperative programs (Zhao et al. 2015; Scarponi et al. 2020; Handy et al. 2021; Malusà et al. 2021).

5.10. Acknowledgments

After more than a century of study, writing this synthesis represented a challenge given the large number of papers published and the diverse array of opinions/models. Many thanks are due to the reviewers S. Duchêne, N. Froitzheim, and L. Jolivet for their comments on the early draft of this paper, and to the editors N. Bellahsen and C.L. Rosenberg for their invitation to write this review.

5.11. References

Agard, P., Monié, P., Jolivet, L., Goffé, B. (2002). Exhumation of the Schistes Lustrés complex: In situ laser probe $^{40}Ar/^{39}Ar$ constraints and implications for the Western Alps. *J. Metamorph. Geol.*, 20, 599–618.

Allen, P.A., Crampton, S.L., Sinclair, H.D. (1991). The inception and early evolution of the North Alpine Foreland Basin, Switzerland. *Basin Research*, 3, 143–163.

Amato, J.M., Johnson, C.M., Baumgarner, L.P., Beard, B.L. (1999). Rapid exhumation of the Zermatt-Saas ophiolite deduced from high-precision Sm-Nd and Rb-Sr geochronology. *Earth Planet. Sci. Lett.*, 171, 425–438.

Andersen, T.B., Jamveit, B., Dewey, J.F., Swensson, E. (1991). Subduction and eduction of continental crust: Major mechanisms during continent-continent collision and orogenic extensional collapse, a model based on the south Norwegian Caledonides. *Terra Nova*, 3, 303–310.

Angiboust, S., Glodny, J., Oncken, O., Chopin, C. (2014). In search of transient subduction interfaces in the Dent Blanche-Sesia Tectonic System (W. Alps). *Lithos*, 205, 298–321.

Angiboust, S., Yamato, P., Hertgen, S., Hyppolito, T., Bebout, G.E., Morales, L. (2017). Fluid pathways and high-P metasomatism in a subducted continental slice (Mt. Emilius klippe, W. Alps). *J. Metamorph. Geol.*, 35, 471–492.

Angrand, P. and Mouthereau, F. (2021). Evolution of the Alpine orogenic belts in the Western Mediterranean region as resolved by the kinematics of the Europe-Adria diffuse plate boundary. *Bull. Soc. Geol. Fr.*, 192, 42.

Argand, E. (1911). Sur les plissements en retour et la structure en éventail dans les Alpes occidentales. *Bull. Soc. Vaud. Sci. Nat.*, 47, 33–36.

Austrheim, H. and Andersen, T.B. (2004). Pseudotachylites from Corsica: Fossil earthquakes from a subduction complex. *Terra Nova*, 16, 193–197.

Avigad, D., Chopin, C., Le Bayon, R. (2003). Thrusting and extension in the Southern Dora-Maira ultra-high-pressure massif (Western Alps): View from below the coesite-bearing unit. *J. Geol.*, 111, 57–70.

Ayrton, S., Bugnon, C., Haarpaintner, T., Weidmann, M., Frank, E. (1982). Géologie du front de la nappe de la Dent-Blanche dans la region des Monts-Dolins, Valais. *Eclogae Geol. Helv.*, 75, 269–286.

Babist, J., Handy, M.R., Konrad-Schmolke, M., Hammerschmidt, K. (2006). Precollisional, multistage exhumation of subducted continental crust: The Sesia Zone, Western Alps. *Tectonics,* 25, TC6008.

Ballèvre, M. and Merle, O. (1993). The Combin Fault: Compressional reactivation of a Late Cretaceous-early Tertiary detachment fault in the Western Alps. *Schweiz. Mineral. Petrogr. Mitt.,* 73, 205–227.

Ballèvre, M., Lagabrielle, Y., Merle, O. (1990). Tertiary ductile normal faulting as a consequence of lithospheric stacking in the Western Alps. *Soc. Géol. Fr. Mémoires,* 156, 227–236.

Ballèvre, M., Manzotti, P., Dal Piaz, G.V. (2018). Pre-Alpine (Variscan) inheritance: A key for the location of the future Valaisan Basin (Western Alps). *Tectonics,* 37, 786–817.

Ballèvre, M., Camonin, A., Manzotti, P., Poujol, M. (2020). A step towards unravelling the paleogeographic attribution of pre-Mesozoic basement complexes in the Western Alps based on U-Pb geochronology of Permian magmatism. *Swiss J. Geosci.,* 113, 12.

Bearth, P. (1952). Geologie und petrographie des Monte Rosa. *Beit. Geol. Karte Schweiz, Neue Folge,* 96, 118.

Bearth, P. (1967). Die ophiolithe der Zone von Zermatt-Saas Fee. *Beit. Geol. Karte Schweiz, Neue Folge,* 132, 130.

Bellahsen, N., Mouthereau, F., Boutoux, A., Bellanger, M., Lacombe, O., Jolivet, L., Rolland, Y. (2014). Collision kinematics in the western external Alps. *Tectonics,* 33, 1055–1088.

Beltrando, M., Lister, G.S., Hermann, J., Forster, M., Compagnoni, R. (2008). Deformation mode switches in the Penninic units of the Urtier Valley Western Alps: Evidence for a dynamic orogeny. *J. Struct. Geol.*, 30, 194–219.

Beltrando, M., Rubatto, D., Manatschal, G. (2010a). From passive margins to orogens: The link between ocean-continent transition zones and (ultra)high-pressure metamorphism. *Geology*, 38, 559–562.

Beltrando, M., Compagnoni, R., Lombardo, B. (2010b). (Ultra)High-pressure metamorphism and orogenesis: An Alpine perspective. *Gondwana Res.*, 18, 147–166.

Bernoulli, G., Bertotti, G., Zingg, A. (1989). Northward thrusting of the Gonfolite Lombarda (South-Alpine molasse) onto the Mesozoic sequence of the Lombardian Alps: Implications for the deformation history of the Southern Alps. *Eclogae Geol. Helv.*, 82, 841–856.

Bertotti, G. (2001). Subsidence, deformation, thermal and mechanical evolution of the Mesozoic South Alpine rifted margin: An analogue for Atlantic-type margins. *Geol. Soc. London Spec. Publ.*, 187, 125–141.

Bertotti, G., Picotti, V., Bernoulli, D., Castellarin, A. (1993). From rifting to drifting: Tectonic evolution of the South-Alpine upper crust from the Triassic to the Early Cretaceous. *Sediment. Geol.,* 86, 53–76.

Bertrand, J.-M. (1968). Etude structurale du versant occidental du Massif du Grand Paradis Alpes Graies. *Géol. Alpine*, 44, 55–87.

Bertrand, J.-M., Paquette, J.-L., Guillot, F. (2005). Permian zircon U-Pb ages in the Gran Paradiso massif: Revisiting post-Variscan events in the Western Alps. *Schweiz. Mineral. Petrogr. Mitt.*, 85, 15–29.

Borghi, A., Compagnoni, R., Sandrone, R. (1996). Composite P–T paths in the internal Penninic massifs of the Western Alps: Petrological constrains to their thermo-mechanical evolution. *Eclogae Geol. Helv.*, 89, 345–367.

Boriani, A., Burlini, L., Sacchi, R. (1990). The Cossato-Mergozzo-Brissago Line and the Pogallo Line (Southern Alps, Northern Italy) and their relationships with the late Hercynian magmatic and metamorphic events. *Tectonophysics*, 182, 91–102.

Bousquet, R., Engi, M., Gosso, G., Oberhänsli, R., Berger, A., Spalla, M.I., Zucali, M., Goffé B. (2004). Explanatory notes to the map: Metamorphic structure of the Alps transition from the Western to the Central Alps. *Mitt. Öst. Mineral. Gesell.*, 149, 145–156.

Bousquet, R., Oberhänsli, R., Schmid, S.M., Berger, A., Wiederkehr, M., Robert, C., Möller, A., Rosenberg, C.L., Zeilinger, G., Molli, G. et al. (2012). Metamorphic framework of the Alps. *Commission for the Geological Map of the World; Subcommission for Magmatic and Metamorphic Maps.* IUGS and IUGG, Paris.

Bowtell, S.A., Cliff, R.A., Barnicoat, C. (1994). Sm-Nd isotopic evidence on the age of eclogitization in the Zermatt-Saas ophiolite. *J. Metamorph. Geol.*, 12, 187–196.

Boyet, M., Lapierre, H., Tardy, M., Bosch, D., Maury, R. (2001). Nature des sources des composants andésitiques des Grès du Champsaur et des Grès de Taveyannaz. Implications dans l'évolution des Alpes occidentales au Paléogène. *Bull. Soc. Géol. Fr.*, 172, 487–501.

Brouwer, F.M., van de Zedde, D.M.A., Wortel, M.J.R., Vissers, R.L.M. (2004). Late-orogenic heating during exhumation: Alpine PTt trajectories and thermomechanical models. *Earth Planet. Sci. Lett.*, 220, 185–199.

Brun, J.-P. and Beslier, M.-O. (1996). Mantle exhumation at passive margins. *Earth Planet. Sci. Lett.*, 142, 161–173.

Brun, J.-P. and Faccenna, C. (2008). Exhumation of high-pressure rocks driven by slab rollback. *Earth Planet. Sci. Lett.*, 272, 1–7.

Bucher, S. and Bousquet, R. (2007). Metamorphic evolution of the Briançonnais units along the ECORS-CROP profile (Western Alps): New data on metasedimentary rocks. *Swiss J. Geosci.*, 100, 227–242.

Bucher, S., Schmid, S.M., Bousquet, R., Fügenschuh, B. (2003). Late-stage deformation in a collisional orogen (Western Alps): Nappe refolding, back-thrusting or normal faulting? *Terra Nova,* 15, 109–117.

Bucher, K., Fazis, Y., De Capitani, C., Grapes, R. (2005). Blueschists, eclogites, and decompression assemblages of the Zermatt-Saas ophiolite: High-pressure metamorphism of subducted Tethys lithosphere. *Am. Mineral.,* 90, 821–835.

Bürgmann, R. and Dresen, G. (2008). Rheology of the lower crust and upper mantle: Evidence from rock mechanics, geodesy, and field observations. *Ann. Rev. Earth Planet. Sci.*, 36, 531–567.

Burov, E., Francois, T., Agard, P., Le Pourhiet, L., Meyer, B., Tirel, C., Lebedev, S., Yamato, P., Brun, J.-P. (2014). Rheological and geodynamic controls on the mechanisms of subduction and HP/UHP exhumation of crustal rocks during continental collision: Insights from numerical models. *Tectonophysics*, 631, 212–250.

Bussy, F. and Cadoppi, P. (1996). U-Pb zircon dating of granitoids from the Dora-Maira massif (Western Italian Alps). *Schweiz. Mineral. Petrogr. Mitt.*, 76, 217–233.

Butler, J.P., Beaumont, C., Jamieson, R.A. (2013). The Alps 1: A working geodynamic model for burial and exhumation of (ultra)high-pressure rocks in Alpine-type orogens. *Earth Planet. Sci. Lett.*, 377–378, 114–131.

Butler, J.P., Beaumont, C., Jamieson, R.A. (2014). The Alps 2: Controls on crustal subduction and (ultra)high-pressure rock exhumation in Alpine-type orogens. *J. Geophy. Res. Solid Earth*, 119, 5987–6022.

Buzzi, L. and Gaggero, L. (2008). Petrogenesis of post-orogenic Lower Permian andesites in Southern Europe: Insights into the collapse of the Variscan range. *Geodin. Acta*, 21, 273–290.

Caby, R. (1996). Low-angle extrusion of high-pressure rocks and the balance between outward and inward displacements of Middle Pennines units in the Western Alps. *Eclogae Geol. Helv.,* 89, 229–267.

Callegari, E., Compagnoni, R., Dal Piaz, G.V. (1969). Relitti di strutture intrusive erciniche e scisti a sillimanite nel Massiccio del Gran Paradiso. *Boll. Soc. Geol. Ital.*, 88, 59–69.

Callegari, E., Cigolini, C., Medeot, O., d'Antonio, M. (2004). Petrogenesis of calc-alkaline and shoshonitic post-collisional Oligocene volcanics of the cover series of the Sesia Zone, Western Italian Alps. *Geodin. Acta*, 17, 1–29.

Campani. M., Mancktelow. N., Courrioux. G. (2014). The 3D interplay between folding and faulting in a syn-orogenic extensional system: The Simplon Fault Zone in the Central Alps (Switzerland and Italy). *Swiss J. Geosci.*, 107, 251–271.

Carrapa, B. and Di Giulio, A. (2001). The sedimentary record of the exhumation of a granitic intrusion into a collisional setting: The lower Gonfolite Group, Southern Alps, Italy. *Sediment. Geol.*, 139, 217–228.

Carraro, F., Dal Piaz, G.V., Sacchi, R. (1970). Serie di Valpelline e II Zona Diorito-Kinzigitca sono i relitti di un ricoprimento proveniente dalla zona Ivrea-Verbano. *Mem. Soc. Geol. Ital.*, 9, 197–224.

Castelli, D. (1991). Eclogitic metamorphism in carbonate rocks: The example of impure marbles from the Sesia-Lanzo Zone, Italia Western Alps. *J. Metamorph. Geol.*, 9, 61–77.

Castelli, D. and Rubatto, D. (2002). Stability of Al- and F-rich titanite in metacarbonate: Petrologic and isotopic constraints from a polymetamorphic eclogitic marble of the internal Sesia Zone (Western Alps). *Contrib. Mineral. Petrol.*, 142, 627–639.

Cavargna-Sani, M., Epard, J.-L., Steck A. (2014). Structure, geometry and kinematics of the Northern Adula nappe (Central Alps). *Swiss J. Geosci.,* 107, 135–156.

Cenki-Tok, B., Oliot, E., Rubatto, D., Berger, A., Engi, M., Janots, E., Thomsen, T.B., Manzotti, P., Regis, D., Spandler, C. et al. (2011). Preservation of Permian allanite within an Alpine eclogite facies shear zone at Mt Mucrone, Italy: Mechanical and chemical behaviour of allanite during mylonitization. *Lithos*, 125, 40–50.

Chemenda, A.I., Mattauer, M., Bokun, A.N. (1996). Continental subduction and a mechanism for exhumation of high-pressure metamorphic rocks: New modelling and field data from Oman. *Earth Planet. Sci. Lett.*, 143, 173–182.

Chen, Y.-X., Zhou, K., Zheng, Y.-F., Schertl, H.-P. (2017). Zircon geochemical constraints on the protolith nature and metasomatic process of the Mg-rich whiteschists from the Western Alps. *Chem. Geol.,* 467, 177–195.

Chopin, C. (1984). Coesite and pure pyrope in high-grade blueschist of the Western Alps: A first record and some consequences. *Contrib. Mineral. Petrol.*, 86, 107–118.

Chopin, C., Goffé, B., Ungaretti, L., Oberti, R. (2003). Magnesiostaurolite and zincostaurolite: Mineral description with a petrogenetic and crystal-chemical update. *Eur. J. Mineral.,* 15, 167–176.

Choukroune, P., Ballèvre, M., Cobbold, P.R., Gautier, Y., Merle, O., Vuichard, J.-P. (1986). Deformation and motion in the Western Alpine arc. *Tectonics,* 5, 215–226.

Compagnoni, R. (1977). The Sesia-Lanzo Zone: High pressure-low temperature metamorphism in the Austroalpine continental margin. *Rend. Soc. Ital. Mineral. Petrol.*, 33, 335–378.

Compagnoni, R. and Maffeo, B. (1973). Jadeite-bearing metagranites l.s. and related rocks in the Mount Mucrone area (Sesia-Lanzo Zone, Western Italian Alps). *Schweiz. Mineral. Petrogr. Mitt.,* 53, 355–378.

Compagnoni, R. and Rolfo, F. (2003). UHPM units in the Western Alps. *Eur. Mineral. Union Notes Mineral.,* 5, 13–49.

Compagnoni, R., Elter, G., Lombardo, B. (1974). Eterogeneità stratigrafica del complesso degli "gneiss minuti" nel massicio cristallino del Gran Paradiso. *Mem. Soc. Geol. Ital.,* 13, 227–239.

Cordey, F., Tricart, P., Guillot, S., Schwartz, S. (2012). Dating the Tethyan ocean in the Western Alps with radiolarite pebbles from synorogenic Oligocene molasses basins (southeast France). *Swiss J. Geosci.,*105, 39–48.

Cortiana, G., Dal Piaz, G.V., Del Moro, A., Hunziker, J.C., Martin, S. (1998). ^{40}Ar-^{39}Ar and Rb-Sr dating of the Pillonet klippe and Sesia-Lanzo basal slice in the Ayas Valley and evolution of the Austroalpine-Piedmont nappe stack. *Mem. Sci. Geol.*, 50, 177–194.

Dal Piaz, G.V. (1971). Nuovi ritrovamenti di cianite alpina nel cristallino antico del Monte Rosa. *Rend. Soc. Ital. Mineral. Petrol.*, 27, 437–477.

Dal Piaz, G.V. (2001). Geology of the Monte Rosa massif: Historical review and personal comments. *Schweiz. Mineral. Petrogr. Mitt.*, 81, 275–303.

Dal Piaz, G.V. and Lombardo, B. (1986). Early Alpine eclogite metamorphism in the Penninic Monte Rosa-Gran Paradiso basement of the Northwestern Alps. In *Blueschists and Eclogites*, Evans B.W. and Brown E.H. (eds). Geological Society of America, Boulder, CO.

Dal Piaz, G.V., Hunziker, J.C., Martinotti, G. (1972). La zona Sesia-Lanzo e l'evoluzione tettonico-metamorfica delle Alpi nordoccidentali interne. *Mem. Soc. Geol. Ital.*, 11, 433–466.

Dal Piaz, G.V., Lombardo, B., Gosso, G. (1983). Metamorphic evolution of the Mt. Emilius klippe, Dent Blanche nappe, Western Alps. *Am. J. Sci.*, 283-A, 438–458.

Dal Piaz, G.V., Cortiana, G., Del Moro, A., Martin, S., Pennacchioni, G., Tartarotti, P. (2001). Tertiary age and paleostructural inferences of the eclogitic imprint in the Austroalpine outliers and Zermatt-Saas ophiolite, Western Alps. *Int. J. Earth Sci.*, 90, 668–684.

Dal Piaz, G.V., Bistacchi, A., Gianotti, F., Monopoli, B., Passeri, L., Schiavo, A. (2015). Note illustrative della carta Geologica d'Italia alla scala 1:50.000. Foglio 070, Monte Cervino. *ISPRA, Servizio Geologico d'Italia*, 070, 1–431.

Dale, J. and Holland, T.J.B. (2003). Geothermobarometry, P-T paths and metamorphic field gradients of high-pressure rocks from the Adula Nappe, Central Alps. *J. Metamorph. Geol.*, 21, 813–829.

De Graciansky, P.-C., Lemoine, M., Saliot, P. (1971). Remarques sur la présence de minéraux et de paragenèses du métamorphisme alpin dans les galets des conglomérats oligocènes du synclinal de Barrême (Alpes de Haute-Provence). *C. R. Acad. Sci. Paris,* D 272, 3243–3245.

De Meyer, C., Baumgartner, L.P., Beard, B.L., Johnson, C.M. (2014). Rb-Sr ages from phengite inclusions in garnets from high pressure rocks of the Swiss Western Alps. *Earth Planet. Sci. Lett.*, 395, 205–216.

Dewey, J.F. and Bird, J.M. (1970). Mountain belts and the new global tectonics. *J. Geophy. Res.*, 75, 2625–2647.

Di Capua, A. and Groppelli, G. (2016). Application of actualistic models to unravel primary volcanic control on sedimentation (Taveyanne Sandstones, Oligocene Northalpine Foreland Basin). *Sediment. Geol.*, 36, 147–160.

Di Vincenzo, G., Tonarini, S.M., Lombardo, B., Castelli, D., Ottolini, L. (2006). Comparison of ^{40}Ar-^{39}Ar and Rb-Sr data on phengites from the UHP Brossasco-Isasca Unit (Dora Maira Massif, Italy): Implications for dating white mica. *J. Petrol.,* 47, 1439–1465.

Diehl, E.A., Masson, R., Stutz, A.H. (1952). Contributo alla conoscenza del ricoprimento della Dent Blanche. *Mem. Ist. Geol. Mineral. Univ. Padova,* 17, 1–52.

Dragovic, B., Angiboust, S., Tappa, M.J. (2020). Petrochronological close-up on the thermal structure of a paelo-subduction zone (W. Alps). *Earth Planet. Sci. Lett.,* 547, 116446.

Duchêne, S., Blichert-Toft, J., Luais, B., Télouk, P., Lardeaux, J.-M., Albarède, F. (1997). The Lu-Hf dating of garnets and the ages of the Alpine high-pressure metamorphism. *Nature,* 387, 586–589.

Dumont, T., Schwartz, S., Guillot, S., Simon-Labric, T., Tricart, P., Jourdan, S. (2012). Structural and sedimentary records of the Oligocene revolution in the Western Alpine arc. *J. Geodyn.*, 56–57, 18–38.

Ebner, F. and Rantitsch, G. (2000). Das Gosaubecken von Kainach – ein Überblick. *Mitt. Gesell. Geol.- und Bergbaustud. Öst.,* 44, 157–172.

Ellenberger, F. (1952). Sur l'extension des facies briançonnais en Suisse, dans les Préalpes medianes et les Pennides. *Eclogae Geol. Helv.*, 45, 285–286.

Ellenberger, F. (1960). Sur une paragenèse éphémère à lawsonite et glaucophane dans le métamorphisme alpin en Haute Maurienne. *Bull. Soc. Géol. Fr.*, (7)II, 190–194.

Elter, G., Elter, P., Sturani, C., Weidmann, M. (1966). Sur la prolongation du domaine ligure de l'Apennin dans le Monferrat et les Alpes et sur l'origine de la nappe de la Simme s.l. des Préalpes romandes et chablaisiennes. *Arch. Sci. Genève*, 90, 279–378.

Engi, M., Giuntoli, F., Lanari, P., Burn, M., Kunz, B., Bouvier, A.-S. (2018). Pervasive eclogitization due to brittle deformation and rehydration of subducted basement: Effects on continental recycling? *Geochemistry Geophysics Geosystems,* 19, 865–881.

Ernst, W. (1971). Metamorphic zonations on presumably subducted lithospheric plates from Japan, California and the Alps. *Contrib. Mineral. Petrol.,* 34, 43–59.

Ernst, W. (1973). Interpretative summary of metamorphism in the Alps. *Geol. Soc. Am. Bull.,* 84, 2053–2078.

Ernst, W. and Liou, J.C. (2008). High- and ultrahigh-pressure metamorphism: Past results and future prospects. *Am. Mineral.,* 93, 1771–1786.

Escher, A. and Beaumont, C. (1997). Formation, burial and exhumation of basement nappes at crustal scale: A geometric model based on the Western Swiss-Italian Alps. *J. Struct. Geol.,* 19, 995–974.

Eskola, P. (1920). The mineral facies of rocks. *Norsk Geol. Tid.,* 6, 143–194.

Evans, M.J. and Elliott, T. (1999). Evolution of a thrust-sheet-top basin: The Tertiary Barrême basin, Alpes-de-Haute-Provence, France. *Geol. Soc. Am. Bull.*, 111, 1617–1643.

Evans, M.J. and Mange-Rajetzky, M.A. (1991). The provenance of sediments in the Barrême thrust-top basin, Haute-Provence, France. *Geol. Soc. London Spec. Publ.*, 57, 323–342.

Faryad, S.W., Melcher, F., Hoinkes, G., Puhl, J., Meisel, T., Frank, W. (2002). Relics of eclogite facies metamorphism in the Austroalpine basement, Hochgrössen (Speik complex), Austria. *Mineral. Petrol.*, 74, 49–73.

Fassmer, K., Obermüller, G., Nagel, T.J., Kirst, F., Froitzheim, N., Sandmann, S., Miladinova, I., Fonseca, R.O.C., Münker, C. (2016). High-pressure metamorphic age and significance of eclogite-facies continental fragments associated with oceanic lithosphere in the Western Alps (Etirol-Levaz Slice, Valtournenche, Italy). *Lithos*, 252–253, 145–159.

Federico, L., Capponi, G., Crispini, L., Scambelluri, M. (2004). Exhumation of alpine high-pressure rocks: Insights from petrology of eclogite clasts in the Tertiary Piedmontese basin (Ligurian Alps, Italy). *Lithos,* 74, 21–40.

Federico, L., Capponi, G., Crispini, L., Scambelluri, M., Villa, I.M. (2005). ^{39}Ar/^{40}Ar dating of high-pressure rocks from the Ligurian Alps: Evidence for a continuous subduction-exhumation cycle. *Earth Planet. Sci. Lett.,* 240, 668–680.

Ferrando, S. (2012). Mg-metasomatism of metagranitoids from the Alps: Genesis and possible tectonic scenarios. *Terra Nova*, 24, 423–436.

Ferrando, S., Bernoulli, D., Compagnoni, R. (2004). The Canavese zone (internal Western Alps): A distal margin of Adria. *Schweiz. Mineral. Petrogr. Mitt.*, 84, 237–256.

Festa, A., Dela Pierre, F., Irace, A., Piana, F., Fioraso, G., Lucchesi, S., Boano, P., Forno, M.G. (2003). Note illustrative della carta geologica d'Italia alla scala 1:50.000, Foglio 156 Torino Est. *APAT – Servizio Geologico d'Italia*, 141.

Ford, M., Lickorish, W.H., Kusznir, N.J. (1999). Tertiary foreland sedimentation in the southern subalpine chains, SE France: A geodynamic appraisal. *Basin Research,* 11, 315–336.

Ford, M., Duchêne, S., Gasquet, D., Vanderhaegue, O. (2006). Two-phase orogenic convergence in the external and internal SW Alps. *J. Geol. Soc. London*, 163, 815–826.

Fountain, D.M. (1976). The Ivrea-Verbano and Strona-Ceneri Zones, northern Italy: A cross-section of the continental crust – New evidence from seismic velocities of rock samples. *Tectonophysics,* 33, 145–165.

Franz, L. and Romer, R.L. (2007). Caledonian high-pressure metamorphism in the Strona-Ceneri Zone (Southern Alps of southern Switzerland and northern Italy). *Swiss J. Geosci.,* 100, 457–467.

Frey, M., Hunziker, J.C., Frank, W., Bocquet, J., Dal Piaz, G.V., Jäger, E., Niggli, E. (1974). Alpine metamorphism of the Alps. A review. *Schweiz. Mineral. Petrogr. Mitt.,* 54, 247–290.

Froitzheim, N. and Manatschal, G. (1996). Kinematics of Jurassic rifting, mantle exhumation, and passive-margin formation in the Austroalpine and Penninic nappes (Eastern Switzerland). *Geol. Soc. Am. Bull.,* 9, 1120–1133

Gabudianu Radulescu, I., Rubatto, D., Gregory, C., Compagnoni, R. (2009). The age of HP metamorphism in the Gran Paradiso Massif, Western Alps: A petrological and geochronological study of "silvery micaschists". *Lithos,* 110, 95–108.

Gabudianu Radulescu, I., Compagnoni, R., Lombardo, B. (2011). Polymetamorphic history of a relict Permian hornfels from the central Gran Paradiso Massif (Western Alps, Italy): A microstructural and thermodynamic modelling study. *J. Metamorph. Geol.,* 29, 851–874.

Gaidies, F., Krenn, E., De Capitani, C., Abart, R. (2008). Coupling forward modelling of garnet growth with monazite geochronology: An application to the Rappold Complex (Austroalpine crystalline basement). *J. Metamorph. Geol.,* 26, 775–793.

Galli, A., Le Bayon, B., Schmidt, M.W., Burg, J.-P., Reusser, E., Sergeev, S.A., Larionov, A. (2012). U-Pb zircon dating of the Gruf Complex: Disclosing the late Variscan granulitic lower crust of Europe stranded in the Central Alps. *Contrib. Mineral. Petrol.*, 163, 353–378.

Galli, A., Le Bayon, B., Schmidt, M.W., Burg, J.-P., Reusser, E. (2013). Tectonometamorphic history of the Gruf Complex (Central Alps): Exhumation of a granulite-migmatite complex with the Bergell pluton. *Swiss J. Geosci.*, 106, 33–62.

Ganne, J., Bussy, F., Vidal, O. (2003). Multi-stage garnet in the internal Briançonnais basement (Ambin Massif, Savoy): New petrological constraints on the blueschist-facies metamorphism in the Western Alps and tectonic implications. *J. Petrol.*, 44, 1281–1308.

Garber, J.M., Smye, A.J., Feineman, M.D., Kylander-Clark, A.R.C., Matthews, S. (2020). Decoupling of zircon U-Pb and trace-element systematics driven by U diffusion in eclogite-facies zircon (Monviso meta-ophiolite, W. Alps). *Contrib. Mineral. Petrol.,* 175, 55.

Gardien, V., Reusser, E., Marquer, D. (1994). Pre-Alpine metamorphic evolution of the gneisses from Valpelline series (Western Alps, Italy). *Schweiz. Mineral. Petrogr. Mitt.*, 74, 489–502.

Garefalakis, P. and Schlunegger, F. (2019). Tectonic processes, variations in sediment flux, and eustatic sea level recorded in the 20 Myr old Burdigalian transgression in the Swiss Molasse Basin. *Solid Earth,* 10, 2045–2072.

Gasco, I., Borghi, A., Gattiglio, M. (2011). P-T Alpine metamorphic evolution of the Monte Rosa nappe along the Piedmont Zone boundary (Gressoney Valley, NW Italy). *Lithos*, 127, 336–353.

Gasco, I., Borghi, A., Gattiglio, M. (2012). Metamorphic evolution of the Gran Paradiso Massif: A case study of an eclogitic metagabbro and a polymetamorphic glaucophane-garnet micaschist. *Lithos*, 115, 101–120.

Gauthiez-Putallaz, L., Rubatto, D., Hermann, J. (2016). Dating prograde fluid pulses during subduction by in situ U-Pb and oxygen isotope analysis. *Contrib. Mineral. Petrol.*, 171, 15.

Gebauer, D., Schertl, H.P., Brix, M., Schreyer, W. (1997). 35 Ma old ultra-high-pressure metamorphism and evidence for very rapid exhumation in the Dora Maira Massif, Western Alps. *Lithos,* 41, 5–24.

Genser, J. and Neubauer, F. (1989). Low-angle normal faults at the eastern margin of the Tauern Window (Eastern Alps). *Mitt. Öst. Geol. Gesell.,* 81, 233–243.

Gerya, T.V., Perchuk, L.L., Burg, J.P. (2008). Transient hot channels: Perpetrating and regurgitating ultrahigh-pressure, high-temperature crust-mantle associations in collision belts. *Lithos,* 103, 236–256.

Ghignone, S., Gattiglio, M., Balestro, G., Borghi, A. (2020). Geology of the Susa Shear Zone (Susa Valley, Western Alps). *J. Maps,* 16, 79–86.

Gillet, P. and Goffé, B. (1988). On the significance of aragonite occurrences in the Western Alps). *Contrib. Mineral. Petrol.,* 99, 70–81.

Giorgis, D., Thélin, P., Stampfli, G., Bussy, F. (1999). The Mont-Mort metapelites: Variscan metamorphism and geodynamic context (Briançonnais basement, Western Alps, Switzerland). *Schweiz. Mineral. Petrogr. Mitt.,* 79, 381–398.

Giuntoli, F. and Engi, M. (2016). Internal geometry of the central Sesia Zone (Aosta Valley, Italy): HP tectonic assembly of continental slices. *Swiss J. Geosci.,* 109, 445–471.

Giuntoli, F., Lanari, P., Burn, M., Kunz, B.E., Engi, M. (2018a). Deeply subducted continental fragments – Part II: Insight from petrochronology in the central Sesia Zone (Western Italian Alps). *Solid Earth,* 9, 191–222.

Giuntoli, F., Lanari, P., Engi, M. (2018b). Deeply subducted continental fragments – Part 1: Fracturing, dissolution-precipitation, and diffusion processes recorded by garnet textures of the central Sesia Zone (Western Italian Alps). *Solid Earth*, 9, 167–189.

Glodny, J., Ring, U., Kühn, A., Gleissner, P., Franz, G. (2005). Crystallization and very rapid exhumation of the youngest Alpine eclogites (Tauern Window, Eastern Alps) from Rb/Sr mineral assemblage analysis. *Contrib. Mineral. Petrol.*, 149, 699–712.

Goffé, B. (1979). La lawsonite et les associations à pyrophyllite-calcite dans les sédiments alumineux du Briançonnais. Premières occurrences. *C. R. Acad. Sci. Paris*, D 289, 813–816.

Goffé, B. (1980). Magnésiocarpholite, cookéite et euclase dans le niveaux continentaux métamorphiques de la zone briançonnaise. Données minéralogiques et nouvelles occurrences. *Bull. Minéral.*, 103, 297–302.

Goffé, B. (1984). Le faciès à carpholite-chloritoide dans la couverture briançonnaise des Alpes ligures : un témoin de l'histoire tectono-métamorphique régionale. *Mem. Soc. Geol. Ital.*, 28, 461–479.

Goffé, B. and Chopin, C. (1986). High-pressure metamorphism in the Western Alps: Zoneography of metapelites, chronology and consequences. *Schweiz. Mineral. Petrogr. Mitt.,* 66, 41–52.

Goffé, B. and Saliot, P. (1977). Les association minéralogiques des roches hyperalumineuses du Dogger de Vanoise. Leur significations dans le métamorphisme régional. *Bull. Soc. Fr. Minéral. Cristallogr.*, 100, 302–309.

Goffé, B. and Velde, B. (1984). Contrasted metamorphic evolutions in thrusted cover units of the Briançonnais zone (French Alps): A model for the conservation of HP-LT metamorphic mineral assemblages. *Earth Planet. Sci. Lett.,* 68, 351–360.

Goffé, B., Schwartz, S., Lardeaux, J.M., Bousquet, R. (2004). Explanatory notes to the map: Metamorphic structure of the Alps (Western and Ligurian Alps). *Mitt. Öst. Mineral. Gesell.*, 149, 125–144.

Goguel, J. (1963). Sur l'arc des Alpes occidentales. *Bull. Soc. Géol. Fr.*, 7, 20–33.

Gosso, G. (1977). Metamorphic evolution and fold history in the eclogitic micaschists of the Upper Gressoney Valley (Sesia-Lanzo Zone, Western Alps). *Rend. Soc. Ital. Mineral. Petrol.*, 33, 389–407.

Gosso, G., Messiga, B., Rebay, G., Spalla, M.I. (2010). Interplay between deformation and metamorphism during eclogitization of amphibolites in the Sesia-Lanzo Zone of the Western Alps. *Int. Geol. Rev.,* 52, 1193–1219.

Gouzu, C., Itaya, T., Hyodo, H., Matsuda, T. (2006). Excess ^{40}Ar-free phengite in ultrahigh-pressure metamorphic rocks from the Lago di Cignana area, Western Alps. *Lithos,* 92, 418–430.

Groppo, C., Ferrando, S., Gilio, M., Botta, S., Nosenzo, F., Balestro, G., Festa, A., Rolfo, F. (2019). What's in the sandwich? New P-T constraints for the (U)HP nappe stack of southern Dora-Maira Massif (Western Alps). *Eur. J. Mineral.,* 31, 665–683.

Guiraud, M., Powell, R., Rebay, G. (2001). H_2O in metamorphism and unexpected behaviour in the preservation of metamorphic mineral assemblages. *J. Metamorph. Geol.,* 19, 445–454.

Habler, G., Thöni, M., Sölva, H. (2006). Tracing the high-pressure stage in the polymetamorphic Texel Complex (Austroalpine basement unit, Eastern Alps): P-T-t-d constraints. *Mineral. Petrol.,* 88, 269–296.

Hacker, B.R., Rubie, D.C., Kirby, S.H., Bohlen, S.R. (2005). The calcite-aragonite transformation in low-Mg marble: Equilibrium relations, transformation mechanisms, and rates. *J. Geophy. Res. Solid Earth*, 110, B03205.

Halama, R., Glodny, J., Konrad-Schmolke, M., Sudo, M., Marschall, H.R., Wiedenbeck, M. (2014). Effects of fluid-rock interaction on ^{40}Ar/^{39}Ar geochronology in high-pressure rocks (Sesia-Lanzo Zone, Western Alps). *Geochim. Cosmochim. Acta,* 126, 475–494.

Halama, R., Glodny, J., Konrad-Schmolke, M., Sudo, M. (2018). Rb-Sr and in situ $^{40}Ar/^{39}Ar$ dating of exhumation-related shearing and fluid-induced recrystallization in the Sesia Zone (Western Alps, Italy). *Geosphere,* 14, 1–26.

Handy, M.R., Franz, L., Heller, F., Janott, B., Zurbriggen, R. (1999). Multistage accretion and exhumation of the continental crust (Ivrea crustal section, Italy and Switzerland). *Tectonics*, 18, 1154–1177.

Handy, M.R., Schmid, S.M., Bousquet, R., Kissling, E., Bernoulli, D. (2010). Reconciling plate-tectonic reconstructions of Alpine Tethys with the geological-geophysical record of spreading and subduction in the Alps. *Earth-Sci. Rev.*, 102, 121–158.

Handy, M.R., Schmid, S.M., Paffrath, M., Friederich, W., AlpArray Working Group (2021). European tectosphere and slabs beneath the greater Alpine area – Interpretation of mantle structure in the Alps-Apennines-Pannonian region from teleseismic V_p studies. *Solid Earth*. doi: 10.5194/se-2021-49.

Hauke, M., Froitzheim, N., Nagel, T.J., Miladinova, I., Fassmer, K., Fonseca, R.O.C., Sprung, P., Münker, C. (2019a). Two high-pressure metamorphic events, Variscan and Alpine, dated by Lu-Hf in an eclogite complex of the Austroalpine nappes (Schobergruppe, Austria). *Int. J. Earth Sci.,* 108, 1317–1331.

Hauke, M., Froitzheim, N., Weber, S., Münker, C. (2019b). Ages for high-pressure metamorphism in the Western Alps obtained by Lu-Hf garnet dating. *14th EGU Emile Argand Conference on Alpine Geological Studies*, Sion.

Heinrich, C.A. (1982). Kyanite-eclogite to amphibolite facies evolution of hydrous mafic and politic rocks, Adula Nappe, Central Alps. *Contrib. Mineral. Petrol.,* 81, 30–38.

Hertgen, S., Yamato, P., Morales, L.F.G., Angiboust, S. (2017). Evidence for brittle deformation events at eclogite-facies P-T conditions (example of the Emilius klippe, Western Alps). *Tectonophysics,* 706–707, 1–13.

Herwartz, D., Nagel, T.J., Münker, C., Scherer, E.E., Froitzheim, N. (2011). Tracing two orogenic cycles in one eclogite sample by Lu-Hf garnet chronometry. *Nature Geoscience,* 4, 178–183.

Holland, T.J.B. and Powell, R. (1998). An internally consistent thermodynamic data set for phases of petrological interest. *J. Metamorph. Geol.,* 16, 309–343.

Hoschek, G. (2001). Thermobarometry of metasediments and metabasites from the Eclogite zone of the Hohe Tauern, Eastern Alps, Austria. *Lithos,* 59, 127–150.

Hunziker, J.C. (1974). Rb-Sr and K-Ar age determination and the Alpine tectonic history of the Western Alps. *Mem. Ist. Geol. Mineral. Univ. Padova,* 31, 3–45.

Ildefonse, B., Lardeaux, J.-M., Caron, J.-M. (1990). The behavior of shape preferred orientations in metamorphic rocks: Amphiboles and jadeites from the Monte Mucrone area (Sesia-Lanzo Zone, Italian Western Alps). *J. Struct. Geol.,* 12, 1005–1011.

Inger, S., Ramsbotham, W., Cliff, R.A., Rex, D.C. (1996). Metamorphic evolution of the Sesia-Lanzo Zone, Western Alps: Time constraints from multi-system geochronology. *Contrib. Mineral. Petrol.,* 126, 152–168.

Janák, M., Froitzheim, N., Luptak, B., Vrabec, M., Krogh Ravna, E.J. (2004). First evidence of ultrahigh-pressure metamorphism of eclogites in Pohorje, Slovenia: Tracing deep continental subduction in the Eastern Alps. *Tectonics,* 23, TC5014.

Janák, M., Froitzheim, N., Vrabec, M., Krogh Ravna, E.J., De Hoog, J.C.M. (2006). Ultrahigh-pressure metamorphism and exhumation of garnet peridotite in Pohorje, Eastern Alps. *J. Metamorphic Geol.,* 24, 19–31.

Janák, M., Cornell, D., Froitzheim, N., De Hoog, J.C.M., Broska, I., Vrabec, M., Hurai, V. (2009). Eclogite-hosting metapelites from the Pohorje Mountains (Eastern Alps): P-T evolution, zircon geochronology and tectonic implications. *Eur. J. Mineral.,* 21, 1191–1212.

Janák, M., Froitzheim, N., Yoshida, K., Sasinkková, M., Nosko, T., Kobayashi, T., Hirajima, T., Vrabec, M. (2015). Diamond in metasedimentary crustal rocks from Pohorje Eastern Alps: A window to deep continental subduction. *J. Metamorph. Geol.,* 33, 495–512.

Jolivet, L., Daniel, J.M., Truffert, C., Goffé, B. (1994). Exhumation of deep crustal metamorphic rocks and crustal extension in arc and back-arc regions. *Lithos,* 33, 3–30.

Jolivet, L., Faccenna, C., Goffé, B., Burov, E., Agard, P. (2003). Subduction tectonics and exhumation of high-pressure metamorphic rocks in the Mediterranean orogens. *Amer. J. Sci.,* 303, 353–409.

Jolivet, L., Baudin, T., Calassou, S., Chevrot, S., Ford, M., Issautier, B., Lasseur, E., Masini, E., Manatschal, G., Mouthereau, F., Thinon, I., Vidal, O. (2021). Geodynamic evolution of a wide plate boundary in the Western Mediterranean, near-field versus far-field interactions. *Bull. Soc. Géol. Fr.*, 192, 48.

Kapferer, N., Mercolli, I., Berger, A. (2011). The composition and evolution of an Oligocene regolith on top of the Sesia-Lanzo Zone (Western Alps). *Int. J. Earth Sci.,* 100, 1115–1127.

Kapferer, N., Mercolli, I., Berger, A. (2012). Dating emplacement and evolution of the orogenic magmatism in the internal Western Alps 2: The Biella Volcanic Suite. *Swiss J. Geosci.,* 105, 67–84.

Karakas, O., Wotzlaw, J.-F., Guillong, M., Ulmer, P., Brack, P., Economos, R., Bergantz, G.W., Sinigoi, S., Bachmann, O. (2019). The pace of crustal-scale magma accretion and differentiation beneath silicic caldera volcanoes. *Geology*, 47, 719–723.

Kassem, O.K. and Ring, U. (2004). Underplating-related finite-strain patterns in the Gran Paradiso massif, Western Alps, Italy: Heterogeneous ductile strain superimposed on a nappe stack. *J. Geol. Soc. London,* 161, 875–884.

Kienast, J.-R. and Nicot, E. (1971). Présence d'une paragenèse à disthène et chloritoïde (d'âge alpin probable) dans les gneiss à sillimanite, grenat et cordiérite de Valpelline (Val d'Aoste, Italie). *C.R. Acad. Sci. Paris,* 272, 1836–1839.

Kissling E. and Schlunegger F. (2018). Rollback orogeny model for the evolution of the Swiss Alps. *Tectonics,* 37, 1097–1115.

Konrad-Schmolke, M., O'Brien, P.J., Zack, T. (2011a). Fluid migration above a subducted slab – Constraints on amount, pathways and major element mobility from partially overprinted eclogite-facies rocks (Sesia Zone, Western Alps). *J. Petrol.,* 52, 457–486.

Konrad-Schmolke, M., Zack, T., O'Brien, P.J., Barth, M. (2011b). Fluid migration above a subducted slab – Thermodynamic and trace element modelling of fluid-rock interaction in partially overprinted eclogite-facies rocks (Sesia Zone, Western Alps). *Earth Planet. Sci. Lett.,* 311, 287–298.

Kunz, B.E., Manzotti, P., von Niederhäusern, B., Engi, M., Darling, J.R., Giuntoli, F., Lanari, P. (2018). Permian high-temperature metamorphism in the Western Alps (NW Italy). *Int. J. Earth Sci.,* 107, 203–229.

Lacassin, R. (1987). Kinematics of ductile shearing from outcrop to crustal-scale in the Monte Rosa nappe, Western Alps. *Tectonics,* 6, 69–88.

Lacassin, R. and Mattauer, M. (1985). Kilometer-scale sheath fold at Mattmark and implications for transport direction in the Alps. *Nature,* 315, 739–742.

Ladenhauf, C., Armstrong, R., Konzett, J., Miller, C. (2001). The timing of pre-Alpine high-pressure metamorphism in the Eastern Alps: Constraints from U-Pb SHRIMP dating of eclogite zircons from the Austroalpine Silvretta nappe. *J. Conf. Abs. (EUG XI),* 6, 600.

Lagabrielle, Y. and Cannat, M. (1990). The Alpine Jurassic ophiolites resemble the modern Central Atlantic basement. *Geology,* 18, 319–322.

Lapen, T.J., Johnson, C.M., Baumgartner, L.P., Mahlen, N.J., Beard, B.L., Amato, J.M. (2003). Burial rates during prograde metamorphism of an ultra-high-pressure terrane: An example from Lago di Cignana, western Alps, Italy. *Earth Planet. Sci. Lett.,* 215, 57–72.

Lapen, T.J., Johnson, C.M., Baumgartner, L.P., Dal Piaz, G.V., Skora, S., Beard, B.L. (2007). Coupling of oceanic and continental crust during Eocene eclogite-facies metamorphism: Evidence from the Monte Rosa nappe, Western Alps. *Contrib. Mineral. Petrol.,* 153, 139–157.

Lardeaux, J.-M. (2014). Deciphering orogeny: A metamorphic perspective. Examples from European and Variscan belts. Part I: Alpine metamorphism in the Western Alps. A review. *Bull. Soc. Géol. Fr.,* 185, 93–114.

Lardeaux, J.-M. and Spalla, M.I. (1991). From granulites to eclogites in the Sesia zone (Italian Western Alps): A record of the opening and closure of the Piedmont ocean. *J. Metamorph. Geol.,* 9, 35–59.

Lardeaux, J.-M., Gosso, G., Kiénast, J.-R., Lombardo, B. (1982). Relations entre le métamorphisme et la déformation dans la zone Sesia-Lanzo (Alpes Occidentales) et le problème de l'éclogitisation de la croûte continentale. *Bull. Soc. Géol. Fr.*, (7) XXIV, 793–800.

Laubscher, H.P. (1971). The large scale kinematics of the Western Alps and the Northern Apennines and its palinspastic implications. *Amer. J. Sci.*, 271, 193–226.

Le Bayon, B. and Ballèvre, M. (2006). Deformation history of a subducted continental crust (Gran Paradiso, Western Alps): Continuing crustal shortening during exhumation. *J. Struct. Geol.*, 28, 793–815.

Le Bayon, B., Pitra, P., Ballèvre, M., Bohn, M. (2006). Reconstructing P-T paths during continental collision using multi-stage garnet (Gran Paradiso nappe, Western Alps). *J. Metamorph. Geol.*, 24, 477–496.

Le Breton, N. and Thompson, A.L. (1988). Fluid-absent (dehydration) melting of biotite in metapelites in the early stages of crustal anatexis. *Contrib. Mineral. Petrol.*, 99, 226–237.

Le Breton, E., Brune, S., Ustaszewski, K., Zahirovic, S., Seton, M., Müller, R.D. (2021). Kinematics and extent of the Piemont-Liguria Basin – Implications for subduction processes in the Alps. *Solid Earth,* 12, 885–913.

Le Guernic, J. (1967). La zone du Roure : contribution à l'étude du Briançonnais interne et du Piémontais en haute Ubaye. *Trav. Lab. Géol. Fac. Sci. Grenoble*, 43, 95–127.

Lefeuvre, B., Agard, P., Verlaguet, A., Dubacq, B., Plunder, A. (2020). Massive formation of lawsonite in subducted sediments from the Schistes Lustrés (W. Alps): Implications for mass transfer and decarbonation in cold subduction zones. *Lithos,* 370–371, 105629.

Lefèvre, R. and Michard, A. (1976). Les nappes briançonnaises internes et ultrabriançonnaises de la bande d'Acceglio (Alpes franco-italiennes). Une étude structurale et pétrographique dans le faciès des schistes bleus à jadeite. *Sci. Géol. Bull.*, 29, 183–222.

Lemoine, M. (1960). Découverte d'une microfaune du Crétacé supérieur au col du Longet (sources de l'Ubaye, Basses-Alpes) ; conséquences tectoniques et paléogéographiques. *C.R. sommaire séances Soc. Géol. Fr.*, 1960, 234–236.

Lemoine, M. and Trümpy, R. (1987). Pre-oceanic rifting in the Alps. *Tectonophysics,* 133, 305–320.

Lemoine, M., Bas, T., Arnaud-Vanneau, A., Arnaud, H., Dumont, T., Gidon, M., Bourbon, M., de Gracianskyet, P.-C., Rudkiewicz, J.-L., Megard-Galli, J. et al. (1986). The continental margin of the Mesozoic Tethys in the Western Alps. *Marine Petroleum Geol.*, 3, 179–199.

Lemoine, M., Tricart, P., Boillot, G. (1987). Pre-oceanic rifting in the Alps. *Geology,* 15, 622–625.

Lister, G. and Foster, M. (2009). Tectonic mode switches and the nature of orogenesis. *Lithos,* 113, 274–291.

Lorenz, C. (1969). Contribution à l'étude stratigraphique de l'Oligocène et du Miocène inférieur des confins liguro-piémontais. *Atti Ist. Geol. Univ. Genova*, 6, 273–889.

Lorenz, C. (1979). L'Oligo-Miocène ligure ; un exemple de transgression. *Bull. Soc. Géol. Fr.,* 4, 375–378.

Luisier, C., Baumgartner, L., Schmalholz, S.M., Siron, G., Vennemann, T. (2019). Metamorphic pressure variation in a coherent Alpine nappe challenges lithostatic pressure paradigm. *Nature Com.,* 10, 4734.

Luisier, C., Baumgartner, L., Putlitz, B., Vennemann, T. (2021). Whiteschist genesis through metasomatism and metamorphism in the Monte Rosa nappe (Western Alps). *Contrib. Mineral. Petrol.*, 176, 10.

Malavieille, J., Lacassin, R., Mattauer, M. (1984). Signification tectonique des linéations d'allongement dans les Alpes occidentales. *Bull. Soc. Géol. Fr.*, 26, 895–906.

Malusà, M.G., Faccenna, C., Garzanti, E., Polino, R. (2011). Divergence in subduction zones and exhumation of high pressure rocks (Eocene Western Alps). *Earth Planet. Sci. Lett.,* 310, 21–32.

Malusà, M.G., Guillot, S., Zhao, L., Paul, A., Solarino, S., Dumont, T., Schwartz, S., Aubert, C., Baccheschi, P., Eva, E. et al. (2021). The deep structure of the Alps based on the CIFALPS seismic experiment: A synthesis. *Geochemistry Geophysics Geosystems,* 22, e2020GC009466.

Manatschal, G. (2004). New models for evolution of magma-poor rifted margins based on a review of data and concepts from West Iberia and the Alps. *Int. J. Earth Sci.,* 93, 432–466.

Manatschal, G. and Bernoulli, D. (1998). Rifting and early evolution of ancient ocean basins: The record of the Mesozoic Tethys and of the Galicia-Newfoundland margins. *Marine Geophysical Researchers,* 20, 371–381.

Mancktelow, N. (1990). The Simplon Fault Zone. *Beit. Geol. Karte Schweiz*, 163, 74.

Mange-Rajetzky, M. and Oberhänsli, R. (1982). Detrital lawsonite and blue sodic amphibole in the molasses of Savoy, France and their significance in assessing Alpine evolution. *Schweiz. Mineral. Petrogr. Mitt.*, 62, 415–436.

Manzotti, P. and Ballèvre, M. (2013). Multi-stage garnet in high-pressure metasediments: Alpine overgrowths on Variscan detrital grains. *Geology,* 41, 151–1154.

Manzotti, P. and Ballèvre, M. (2017). Tectonic history of the Dent Blanche. *Geological Field Trips,* 9, 1–73.

Manzotti, P. and Zucali, M. (2013). The pre-Alpine tectonic history of the Austroalpine continental basement in the Valpelline unit (Western Italian Alps). *Geol. Mag.*, 150, 153–172.

Manzotti, P., Rubatto, D., Darling, J., Zucali, M., Cenki-Tok, B., Engi, M. (2012). From Permo-Triassic lithospheric thinning to Jurassic rifting at the Adriatic margin: Petrological and geochronological record in Valtournenche (Western Italian Alps). *Lithos,* 146–147, 276–292.

Manzotti, P., Ballèvre, M., Zucali, M., Robyr, M., Engi, M. (2014a). The tectonometamorphic evolution of the Sesia-Dent Blanche nappes (internal Western Alps): Review and synthesis. *Swiss J. Geosci.,* 107, 309–336.

Manzotti, P., Zucali, M., Ballèvre, M., Robyr, M., Engi, M. (2014b). Geometry and kinematics of the Roisan-Cignana Shear Zone, and the orogenic evolution of the Dent Blanche Tectonic System. *Swiss J. Geosci.,* 107, 23–47.

Manzotti, P., Le Carlier de Veslud, C., Le Bayon, B., Ballèvre, M. (2014c). Petro-structural map of the Money Unit (Gran Paradiso Massif, Valnontey valley, Western Alps). *J. Maps,* 10, 324–340.

Manzotti, P., Poujol, M., Ballèvre, M. (2015a). Detrital zircon in blueschist-facies-metaconglomerates from the Western Alps: Implications for the late Carboniferous to early Permian palaeogeography. *Int. J. Earth Sci.,* 104, 703–731.

Manzotti, P., Pitra, P., Langlade, J., Ballèvre, M. (2015b). Constraining P-T conditions during thrusting of a higher pressure unit over a lower pressure one (Gran Paradiso, Western Alps). *J. metamorphic Geol.,* 33, 981–1002.

Manzotti, P., Ballèvre, M., Poujol, M. (2016). Detrital zircon geochronology in the Dora-Maira and Zone Houillère: A record of sediment travel paths in the Carboniferous. *Terra Nova*, 28, 279–288.

Manzotti, P., Ballèvre, M., Dal Piaz, G.V. (2017). Continental gabbros in the Dent Blanche Tectonic System (Western Alps): From the pre-Alpine crustal structure of the Adriatic palaeo-margin to the geometry of an alleged subduction interface. *J. Geol. Soc. London,* 174, 541–556.

Manzotti, P., Rubatto, D., Zucali, M., El Korh, A., Cenki-Tok, B., Engi, M. (2018a). Permian magmatism and metamorphism in the Dent Blanche nappe: Constraints from field observations and geochronology. *Swiss J. Geosci.,* 111, 79–97.

Manzotti, P., Bosse, V., Pitra, P., Robyr, M., Schiavi, F., Ballèvre, M. (2018b). Exhumation rates in the Gran Paradiso (Western Alps) constrained b in situ U-Th-Pb dating of accessory phases (monazite, allanite and xenotime). *Contrib. Mineral. Petrol.*, 173, 24.

Manzotti, P., Ballèvre, M., Pitra, P., Robyr, M., Putlitz, B., Müntener, O. (2020). The growth of sodic amphibole at the greenschist-to blueschist-facies transition (Dent-Blanche, Western Alps): Bulk-rock chemical control and thermodynamic modelling. *J. Petrol.*, 61, egaa044.

Manzotti, P., Ballèvre, M., Pitra, P., Schiavi, F. (2021). Missing lawsonite and aragonite found: P-T and fluid composition in meta-marls from the Combin Zone (Western Alps). *Contrib. Mineral. Petrol.,* 176, 60.

Marger, K., Luisier, C., Baumgartner, L.P., Putlitz, B., Dutrow, B.L., Bouvier, A.-S., Dini, A. (2019). Origin of Monte Rosa whiteschist from in-situ tourmaline and quartz oxygen isotope analysis by SIMS using new tourmaline reference materials. *Am. Mineral.*, 104, 1503–1520.

Marquer, D., Challandes, N., Baudin, T. (1996). Shear zone patterns and strain distribution at the scale of a Penninic nappe: The Suretta nappe (Eastern Swiss Alps). *J. Struct. Geol.,* 18, 753–764.

Masson, R. (1938). Geologisch-Petrographische Untersuchungen im unteren Valpelline, Provinz Aosta (Italien). *Schweiz. Mineral. Petrogr. Mitt.,* 18, 54–213.

McCarthy, A., Tugend, J., Mohn, G., Candioti, L., Chelle-Michou, C., Arculus, R., Schmalholz, S.M., Müntener, O. (2020). A case of Ampferer-type subduction and consequences for the Alps and the Pyrenees. *Am. J. Sci.,* 320, 313–372.

Meffan-Main, S., Cliff, R.A., Barnicoat, A.C., Lombardo, B., Compagnoni, R. (2004). A tertiary age for Alpine high-pressure metamorphism in the Gran Paradiso massif, Western Alps: A Rb-St microsampling study. *J. Metamorph. Geol.,* 22, 267–281.

Mehnert, K.R. (1975). The Ivrea Zone: A model of the deep crust. *N. Jb. Mineral. Abh.,* 125, 156–199.

Melcher F. and Meisel T. (2004). A metamorphosed Early Cambrian crust-mantle transition in the Eastern Alps, Austria. *J. Petrol.,* 48, 1689–1723.

Menant, A., Angiboust, S., Monié, P., Oncken, O., Guigner, J.-M. (2018). Brittle deformation during Alpine basal accretion and the origin of seismicity nets above the subduction interface. *Earth Planet. Sci. Lett.*, 487, 84–93.

Menegon, L. and Pennacchioni, G. (2010). Local shear zone pattern and bulk deformation in the Gran Paradiso metagranite (NW Italian Alps). *Int. J. Earth Sci.*, 99, 1805–1825.

Menegon, L., Di Toro, G., Pennacchioni, G. (2007). Produzione ciclica di pseudotachiliti in ambiente duttile: Evidenza di una mega asperità lungo un thrust alpino (Falda Dent Blanche, Alpi Nord-Occidentali). *Rend. Soc. Geol. Ital., Nuova Serie,* 4, 249–250.

Mertz, J.D. and Siddans, A.W.B. (1985). Finite strain states, quartz textures and the significance of lineation in Permo-Carboniferous metasediments of the Dora Maira Massif, Val Germanasca, Italy. *Tectonophysics,* 118, 61–73.

Messiga, B., Tribuzio, R., Caucia, F. (1992a). Amphibole evolution in Variscan eclogite-amphibolites from the Savona crystalline massif (Western Ligurian Alps, Italy): Controls on the decompressional P–T–t path. *Lithos,* 27, 215–230.

Messiga, B., Tribuzio, R., Scambelluri, M. (1992b). Mafic eclogites from the Valosio crystalline massif (Ligurian Alps, Italy). *Schweiz. Mineral. Petrogr. Mitt.,* 72, 363–375.

Meyre, C., Marquer, D., Schmid, S.M., Ciancaleoni, L. (1998). Syn-orogenic extension along the Forcola fault: Correlation of Alpine deformations in the Tambo and Adua nappes (Eastern Penninic Alps). *Eclogae Geol. Helv.,* 91, 409–420.

Michard, A. (1959). Contribution à l'étude géologique de la zone d'Acceglio-Longet dans la haute Varaita (Alpes cottiennes, Italie). *Bull. Soc. Géol. Fr.*, (7)I, 52–61.

Michard, A. (1967). *Etudes géologiques dans les zones internes des Alpes cottiennes*. CNRS éditions, Paris.

Michard, A. (1977). Charriages et métamorphisme haute pression dans les Alpes cottiennes méridionales : à propos des schistes à jadéïte de la bande d'Acceglio. *Bull. Soc. Géol. Fr.*, (7) XIX, 883–892.

Michard, A., Chopin, C., Henry, C. (1993). Compression versus extension in the exhumation of the Dora-Maira coesite-bearing unit, Western Alps, Italy. *Tectonophysics,* 221, 173–193.

Michard, A., Avigad, D., Goffé, B., Chopin, C. (2004). The high-pressure metamorphic front of the south Western Alps (Ubaye-Maira transect, France, Italy). *Schweiz. Mineral. Petrogr. Mitt.,* 84, 215–235.

Miller, C. and Thöni, M. (1995). Origin of eclogites from the Austroalpine Ötztal basement (Tirol, Austria): Geochemistry and Sm-Nd vs. Rb–Sr isotope systematics. *Chem. Geol.*, 122, 199–225.

Miller, C. and Thöni, M. (1997). Eo-Alpine eclogitisation of Permian MORB-type gabbros in the Koralpe (Eastern Alps, Austria): New geochronological, geochemical and petrological data. *Chem. Geol.,* 137, 283–310.

Miller, C., Mundil, R., Thöni, M., Konzett, J. (2005). Refining the timing of eclogite metamorphism: A geochemical, petrological, Sm-Nd and U-Pb case study from the Pohorje Mountains, Slovenia (Eastern Alps). *Contrib. Mineral. Petrol.,* 150, 70–84.

Morag, N., Avigad, D., Harlavan, Y., McWilliams, M.O., Michard, A. (2008). Rapid exhumation and mountain building in the Western Alps: Petrology and $^{40}Ar/^{39}Ar$ geochronology of detritus from Tertiary basins of Southeastern France. *Tectonics*, 27, TC2004.

Nagel, T.J. (2008). Tertiary subduction, collision and exhumation recorded in the Adula nappe, central Alps. In *Tectonic Aspects of the Alpine–Dinaride–Carpathian System*, Siegesmund, S., Fügenschuh, B., Froitzheim, N. (eds). Geological Society, London.

Nagel, T.J., Herwarts, D., Rexroth, S., Münker, C., Froitzheim, N., Kurz, W. (2013). Lu-Hf dating, petrography, and tectonic implications of the youngest Alpine eclogites (Tauern Window, Austria). *Lithos,* 170–171, 179–190.

Neubauer, F., Dallmeyer, R.D., Dunkl, I., Schirnik, D. (1995). Late Cretaceous exhumation of the metamorphic Gleinalm dome, Eastern Alps: Kinematics, cooling history and sedimentary response in a sinistral wrench corridor. *Tectonophysics,* 242, 79–98.

Nicollet, C., Bosse, V., Spalla, M.I., Schiavi, F. (2018). Eocene ultra high temperature (UHT) metamorphism in the Gruf Complex (Central Alps): Constraints by LA-ICPMS zircon and monazite dating in petrographic context. *J. Geol. Soc. London,* 175, 774–787.

Nosenzo, F., Manzotti, P., Poujol, M., Ballèvre, M., Langlade, J. (2022). A window into an older orogenic cycle: P-T conditions and timing of the pre-Alpine history of the Dora-Maira Massif (Western Alps). *J. Metamorph. Geol.,* 40, 789–821.

Novarese, V. (1898). *I giacimenti di grafite delle Alpi Cozie.* Tipografia Nazionale di G. Bertero EC, Rome.

Nussbaum, C., Marqueur, D., Biino, G. (1998). Two subduction events in a polycyclic basement: Alpine and pre-Alpine high-pressure metamorphism in the Suretta nappe, Swiss Eastern Alps. *J. Metamorph. Geol.*, 16, 591–605.

Oberhänsli, R. (2004). Metamorphic structure of the Alps 1:1,000,000. *Commission for the Geological Map of the World*, International Union of Geological Sciences, Paris.

Pantet, A., Epard, J.-L., Masson, H. (2020). Mimicking Alpine thrusts by passive deformation of synsedimentary normal faults: A record of the Jurassic extension of the European margin (Mont Fort nappe, Pennine Alps). *Swiss J. Geosci.,* 113, 2020.

Pawlig, S. (2001). Geological evolution of the Monte Rosa: Constraints from geochronology and geochemistry of a tal-kyanite-chloritoid shear zone within the Monte Rosa Granite (Monte Rosa Nappe, Italian Western Alps). PhD Dissertation, Mainz.

Pawlig, S. and Baumgartner, L.P. (2001). Geochemistry of a talc-kyanite-chloritoid shear zone within the Monte Rosa granite, Val d'Ayas, Italy. *Schweiz. Mineral. Petrogr. Mitt.,* 81, 329–346.

Pennacchioni, G. (1996). Progressive eclogitization under fluid-present conditions of pre-Alpine mafic granulites in the Austroalpine Mt Emilius Klippe (Italian Western Alps). *J. Struct. Geol.*, 18, 549–561.

Pennacchioni, G. and Guermani, A. (1993). The mylonites of the Austroalpine Dent Blanche nappe along the northwestern side of the Valpelline Valley (Italian Western Alps). *Mem. Sci. Geol.*, 45, 37–55.

Pennacchioni, G., Scambelluri, M., Bestmann, M., Notini, L., Nimis, P., Plümper, O., Faccenda, M., Nestola, F. (2020). Record of intermediate-depth subduction seismicity in a dry slab from an exhumed ophiolite. *Earth Planet. Sci. Lett.*, 548, 116490.

Petri, B., Mohn, G., Štípská, P., Schulman, K., Manatschal, G. (2016). The Sondalo gabbro contact aureole (Campo unit, Eastern Alps): Implications for mid-crustal mafic magma emplacement. *Contrib. Mineral. Petrol.*, 171, 52.

Petri, B., Duretz, T., Mohn, G., Schmalholz, S., Karner, G., Müntener, O. (2019). Thinning mechanisms of heterogeneous continental lithosphere. *Earth Planet. Sci. Lett.,* 512, 147–162.

Pfiffner, O.A. (2014). *Geology of the Alps.* Wiley-Blackwell, Oxford.

Philippot, P. (1990). Opposite vergence of nappes and crustal extension in the French-Italian Western Alps. *Tectonics*, 9, 1143–1164.

Philippot, P. and Selverstone, J. (1991). Trace-element-rich brines in eclogitic veins – Implications for fluid composition and transport during subduction. *Contrib. Mineral. Petrol.*, 106, 417–430.

Piccoli, G.C., Maletto, G., Bosio, P., Lombardo, B. (2007). Minerali del Piemonte e della Valle d'Aosta (Alba). *Piccoli GC (ed) Associazione Amici del Museo "F. Eusebio", Alba (Cuneo)*, 607.

Platt, J.P. (1993). Exhumation of high-pressure rocks: A review of concepts and processes. *Terra Nova,* 5, 119–133.

Platt, J.P., Behrmann, J.H., Cunningham, P.C., Dewey, J.F., Helman, M., Parish, M., Shepley, M.G., Wallis, S., Weston, P.J. (1989). Kinematics of the Alpine arc and the motion history of Adria. *Nature,* 337, 158–161.

Pognante, U. (1989). Tectonic implications of lawsonite formation in the Sesia Zone (Western Alps). *Tectonophysics,* 162, 219–227.

Pognante, U. and Sandrone R. (1989). Eclogites in the Northern Dora-Maira Nappe (Western Alps, Italy). *Mineral. Petrol.,* 40, 57–71.

Polino, R., Dal Piaz, G.V., Gosso, G. (1990). Tectonic erosion at the Adria margin and accretionary processes for the Cretaceous orogeny of the Alps. *Soc. Géol. Fr. Mém.,* 156, 345–367.

Polino, R., Ruffini, R., Ricci, B. (1991). Le molasse terziare della collina di Torino: Relazioni con la cinematica alpina. *Atti Tic. Sci. Terra,* 34, 85–95.

Proyer, A. (2003). The preservation of high-pressure rocks during exhumation: Metagranites and metapelites. *Lithos,* 70, 183–194.

Quick, J.E., Sinigoi, S., Peressini, G., Demarchi, G., Wooden, J.L., Sbisá, A. (2009). Magmatic plumbing of a large Permian caldera exposed to a depth of 25 km. *Geology,* 37, 603–606.

Raguin, E. (1925). Découverte d'une faune de Foraminifères, très probablement crétacés, dans les calcaires hautement métamorphiques du vallon du Paquier, près de la Grande-Motte (Savoie). *C. R. Acad. Sci. Paris,* 181, 726–727.

Rahn, M. (1991). Eclogites from the Minugrat, Siviez-Mischabel nappe (Valais, Switzerland). *Schweiz. Mineral. Petrogr. Mitt.,* 71, 415–426.

Ramsbotham, W., Inger, S., Cliff, R.A., Rex, D., Barnicoat, A. (1994). Time constraints on the metamorphic and structural evolution of the southern Sesia-Lanzo Zone, Western Italian Alps. *Mineral. Mag.,* 58A, 758–759.

Ratschbacher, L., Frisch, W., Linzer, H.-G. (1991). Lateral extrusion in the Eastern Alps: Part II. Structural analysis. *Tectonics,* 10, 257–271.

Reddy, S.M., Wheeler, J., Cliff, R.A. (1999). The geometry and timing of orogenic extension: An example from the Western Italian Alps. *J. Metamorph. Geol.,* 17, 573–589.

Regis, D., Rubatto, D., Darling, J., Cenki-Tok, B., Zucali, M., Engi, M. (2014). Multiple metamorphic stages within an eclogite-facies terrane (Sesia Zone, Western Alps) revealed by Th-U-Pb petrochronology. *J. Petrol.*, 55, 1429–1456.

Regis, D., Venturini, G., Engi, M. (2016). Geology of the Scalaro valley – Sesia Zone (Italian Western Alps. *J. Maps,* 12, 621–629.

Ricou, L.E. and Siddans, A.W. (1986). Collision tectonics in the Western Alps. In *Collision Tectonics*, Cowards, M.P. and Ries, A.C. (eds). Geological Society, London.

Ring, U. (1995). Horizontal contraction or horizontal extension? Heterogeneous Late Eocene and Early Oligocene general shearing during blueschist and greenschist facies metamorphism at the Pennine-Austroalpine boundary of the Western Alps. *Geol. Rundsch.,* 84, 843–859.

Ring, U., Collins, A.S., Kassem, O.K. (2005). U-Pb SHRIMP data on the crystallization age of the Gran Paradiso augengneiss, Italian Western Alps: Further evidence for Permian magmatic activity in the Alps during break-up of Pangea. *Eclogae Geol. Helv.,* 98, 363–370.

Roda, M., Zucali, M., Regorda, A., Spalla, M.I. (2020). Formation and evolution of a subduction-related mélange: The example of the Rocca Canavese Thrust Sheets (Western Alps). *Geol. Soc. Am. Bull.,* 132, 884–896.

Roda, M., Zucali, M., Corti, L., Visalli, R., Ortolano, G., Spalla, M.I. (2021). Blueschist mylonitic zones accomodating syn-subduction exhumation of deeply buried continental crust: The example of the Rocca Canavese Thrust Sheets Unit (Sesia-Lanzo Zone, Italian Western Alps). *Swiss J. Geosci.,* 114, 6.

Romagny, A., Jolivet, L., Menant, A., Bessière, E., Maillard, A., Canva, A., Gorini, C., Augier, R. (2020). Detailed tectonic reconstructions of the Western Mediterranean region for the last 35 Ma, insights on driving mechanisms. *Bull. Soc. Géol. Fr.*, 191, 37.

Romer, R.L. and Franz, L. (1998). Ordovician Barrow-type metamorphism in the Strona-Ceneri Zone (Northern Italy) dated by U-Pb on staurolite. *Schweiz. Mineral. Petrogr. Mitt.,* 78, 383–395.

Rosenberg, C.L., Schneider, S., Scharf, A., Bertrand, A., Hammerschmidt, K., Rabaute, A., Brun, J.-P. (2018). Relating collisional kinematics to exhumation processes in the Eastern Alps. *Earth-Sci. Rev.,* 176, 311–344.

Rossi, M., Mosca, P., Polino, R., Rogledi, S., Biffi, U. (2009). New outcrop and subsurface data in the Tertiary Piedmont Basin (NW-Italy); Unconformity-bounded stratigraphic units and their relationships with basin-modification phases. *Riv. Ital. Pal. Strati.,* 115, 305–335.

Rubatto, D. and Angiboust, S. (2015). Oxygen isotope record of oceanic and high-pressure metasomatism: A P-T-time-fluid path for the Monviso eclogites (Italy). *Contrib. Mineral. Petrol.,* 170, 44.

Rubatto, D. and Gebauer, D. (1999). Eo/Oligocene (35 Ma) high-pressure metamorphism in the Gornergrat Zone (Monte Rosa, Western Alps): Implications for paleogeography. *Schweiz. Mineral. Petrogr. Mitt.,* 79, 353–362.

Rubatto, D. and Hermann, J. (2001). Exhumation as fast as subduction? *Geology,* 29, 3–6.

Rubatto, D., Gebauer, D., Fanning, M. (1998). Jurassic formation and Eocene subduction of the Zermatt-Saas-Fe ophiolites: Implications for the geodynamic evolution of the Central and Western Alps. *Contrib. Mineral. Petrol.,* 132, 269–287.

Rubatto, D., Gebauer, D., Compagnoni, R. (1999). Dating of eclogite-facies zircons: The age of Alpine metamorphism in the Sesia-Lanzo Zone (Western Alps). *Earth Planet. Sci. Lett.,* 167, 141–158.

Rubatto, D., Müntener, O., Barnhoorn, A., Gregory, C. (2008). Dissolution-reprecipitation of zircon at low-temperature, high-pressure conditions (Lanzo Massif, Italy). *Am. Mineral.,* 93, 1519–1529.

Rubatto, D., Regis, D., Hermann, J., Boston, K., Engi, M., Beltrando, M., McAlpine, S.R.B. (2011). Yo-yo subduction recorded by accessory minerals in the Italian Western Alps. *Nature Geoscience*, 4, 338–342.

Rubie, D.C. (1986). The catalysis of mineral reactions by water and restrictions on the presence of aqueous fluid during metamorphism. *Mineral. Mag.,* 50, 399–415.

Ruffet, G., Gruau, G., Ballèvre, M., Féraud, G., Philippot, P. (1997). Rb-Sr and ^{40}Ar-^{39}Ar laser probe dating of high-pressure phengites from the Sesia Zone (Western Alps): Underscoring of excess argon and new age constraints on the high-pressure metamorphism. *Chem. Geol.,* 141, 1–18.

Ruffini, R., Polino, R., Callegari, E. (1997). Volcanic clast-rich turbidites of the Taveyanne sandstones from the Thônes syncline (Savoie, France): Records of a Tertiary postcollisional volcanism. *Schweiz. Mineral. Petrogr. Mitt.,* 77, 161–174.

Sandmann, S., Nagel, T.J., Herwartz, D., Fonseca, R.O.C., Kurzawski, R.M., Münker, C., Froitzheim, N. (2014). Lu-Hf garnet systematics of a polymetamorphic basement unit: New evidence for coherent exhumation of the Adula Nappe (Central Alps) from eclogite-facies conditions. *Contrib. Mineral. Petrol.,* 168, 1075.

Sandmann, S., Herwatz, D., Kirst, F., Froitzheim, N., Nagel, T.J., Fonseca, R.O.C., Münker, C., Janák, M. (2016). Timing of eclogite facies metamorphism of mafic and ultramafic rocks from the Pohorje Mountains (Eastern Alps, Slovenia) based on Lu-Hf garnet geochronometry. *Lithos,* 262, 576–585.

Sartori, M. (1990). L'Unité du Barrhorn (Zone Pennique, Valais, Suisse), un lien entre les Préalpes médianes rigides et leur socle paléozoïque. *Mém. Géol. Lausanne,* 6, 156.

Scaillet, S. (1996). Excess ^{40}Ar transport scale and mechanism in high-pressure phengites: A case study from an eclogitized metabasite of the Dora-Maira nappe, Western Alps. *Geochim. Cosmochim. Acta,* 60, 1075–1090.

Scambelluri, M., Pennacchioni, G., Gilio, M., Bestmann, M., Plümper, O., Nestola, F. (2017). Fossil intermediate-depth earthquakes in subducting slabs linked to differential stress release. *Nature Geoscience*, 10, 960.

Scarponi, M., Hetényi, G., Berthet, T., Baron, L., Manzotti, P., Petri, B., Pistone, M., Müntener, O. (2020). New gravity data and 3-D density model constraints on the Ivrea Geophysical Body (Western Alps). *Geophysical J. Int.*, 222, 1977–1991.

Scharf, A., Handy, M.R., Favaro, S., Schmid, S.M., Bertrand, A. (2013). Modes of orogen-parallel stretching and extensional exhumation of thickening orogenic crust in response to microplate indentation and slab roll-back (Tauern Window, Eastern Alps). *Int. J. Earth Sci.*, 102, 1627–1654.

Scheiber, T., Pfiffner, O.A., Schreurs, G. (2013). Upper crustal deformation in continent-continent collision: A case study from the central Bernard nappe complex (Valais, Switzerland). *Tectonics*, 32, 1320–1342.

Schenker, F.L., Schmalholz, S.M., Moulas, E., Pleuger, J., Baumgartner, L.P., Podladchikov, Y., Vrijmoed, J., Buchs, N., Müntener, O. (2015). Current challenges for explaining (ultra)high-pressure tectonism in the Pennine domain of the Central and Western Alps. *J. Metamorph. Geol.*, 33, 869–886.

Schertl, H.P. and Hammerschmidt K. (2016). Tracking the incidence of excess argon in white mica Ar-Ar data from UHP conditions to upper crustal levels in the Dora-Maira Massif. *Eur. J. Mineral.*, 28, 1255–1275.

Scheuring, B., Ahrendt, H., Hunziker, J.C., Zingg, A. (1974). Palaeobotanical and geochronological evidence for the Alpine age of the metamorphism in the Sesia Zone. *Geol. Rundsch.*, 62, 305–326.

Schmalholz, S.M. and Schenker, F.L. (2015). Exhumation of the Dora Maira ultrahigh-pressure unit by buoyant uprise within a low-viscosity mantle oblique-slip shear zone. *Terra Nova*, 28, 348–355.

Schmid, S.M., Fügenschuh, B., Kissling, E., Schuster, R. (2004). Tectonic map and overall architecture of the Alpine orogeny. *Eclogae Geol. Helv.*, 97, 93–117.

Schmid, S.M., Kissling, E., Diehl, T., van Hinsberger D.J.J., Molli, G. (2017). Ivrea mantle wedge, arc of the Western Alps, and kinematic evolution of the Alps-Apennines orogenic system. *Swiss J. Geosci.*, 110, 581–612.

Schwartz, S., Guillot, S., Tricart, P., Bernet, M., Jourdan, S., Dumont, T., Montagnac, G. (2012). Source tracing of detrital serpentinite in the Oligocene molasse deposits from the Western Alps (Barrême basin): Implications for relief formation in the internal zone. *Geol. Mag.*, 149, 841–856.

Selverstone, J. (1988). Evidence for east-west crustal extension in the Eastern Alps: Implications for the unroofing history of the Tauern Window. *Tectonics*, 7, 87–105.

Sinclair, H.D. (1997a). Tectonostratigraphic model of underfilled peripheral foreland basins: An Alpine perspective. *Geol. Soc. Am. Bull.*, 109, 323–347.

Sinclair, H.D. (1997b). Flysch to molasse transition in peripheral foreland basins: The role of the passive margin versus slab breakoff. *Geology,* 25, 1123–1126.

Sizova, E., Hauzenberger, C., Fritz, H., Faryad, S.W., Gerya, T. (2019). Late orogenic heating of (ultra)high pressure rocks: Slab rollback vs slab breakoff. *Geosciences,* 9, 499.

Smye, A.J., Seman, S.M., Scambelluri, M., Starr, P.G., Federico, L. (2021). Exhumation dynamics of high-pressure metamorphic rocks from the Voltri Unit, Western Alps: Constraints from phengite Rb-Sr geochronology. *Contrib. Mineral. Petrol.*, 176, 14.

Spear, F.S., Kohn, M.J., Cheney, J.T. (1999). P-T paths from anatectic pelites. *Contrib. Mineral. Petrol.,* 134, 17–32.

Steck, A., Masson, H., Robyr, M. (2015). Tectonics of the Monte Rosa and surrounding nappes (Switzerland and Italy): Tertiary phases of subduction, thrusting and folding in the Pennine Alps. *Swiss J. Geosci.,* 108, 3–34.

Stella, A. (1894). Relazione sul rilevamento eseguito nell'anno 1893 sulle Alpi occidentali (valli dell'Orco e della Soana). *Boll. Reale Com. Geol. Italia,* XXV, 343–371.

Thélin, P., Sartori, M., Lengeler, R., Schaerer, J.-P. (1990). Eclogites of Palaeozoic or early Alpine age in the basement of the Penninic Siviez-Mischabel nappe, Wallis, Switzerland. *Lithos,* 25, 71–88.

Thöni, M. (2006). Dating eclogite-facies metamorphism in the Eastern Alps – Approaches, results, interpretations: A review. *Mineral. Petrol.,* 88, 123–148.

Thöni, M. and Miller, C. (1996). Garnet Sm-Nd data from the Saualpe and the Koralpe (Eastern Alps, Austria): Chronological and P-T constraints on the thermal and tectonic history. *J. Metamorph. Geol.,* 14, 453–466.

Thöni, M., Miller, C., Blichert-Toft, J., Whitehouse, M.J., Konzett, J., Zanetti, A. (2008). Timing of high-pressure metamorphism and exhumation of the eclogite type-locality (Kupplerbrunn-Prickler Halt, Saualpe, southeastern Austria): Constraints from correlations of the Sm-Nd, Lu-Hf, U-Pb and Rb-Sr isotopic systems. *J. Metamorph. Geol.,* 26, 561–581.

Tremolada, F., Guasti, E., Scardia, G., Carcano, C., Rogledi, S., Sciunnach, D. (2010). Reassessing the biostratigraphy and the paleobathymetry of the Gonfolite Lombarda Group in the Como area (Northern Italy). *Riv. Ital. Paleont. Strati.,* 116, 35–49.

Trepmann, C.A. and Stöckhert, B. (2001). Mechanical twinning of jadeite – An indication of synseismic loading beneath the brittle-plastic transition. *Int. J. Earth Sci.,* 90, 4–13.

Trepmann, C.A. and Stöckhert, B. (2002). Cataclastic deformation of garnet: A record of synseismic loading and postseismic creep. *J. Struct. Geol.,* 4, 1845–1856.

Tumiati, S., Thöni, M., Nimis, P., Martin, S., Mair, V. (2003). Mantle-crust interactions during Variscan subduction in the Eastern Alps (Nonsberg-Ulten zone): Geochronology and new petrological constraints. *Earth Planet. Sci. Lett.,* 210, 509–526.

Turner, F.J. (1981). *Metamorphic Petrology: Mineralogical, Field and Tectonic Aspects.* McGraw Hill, Washington, New York.

Vatan, A. (1949). La sédimentation détritique dans la zone subalpine et le Jura méridional, au Crétacé et au Tertiaire. *C.R. Som. Séances Soc. Géol. Fr.*, 102–104.

Vaughan-Hammon, J.D., Luisier, C., Baumgartner, L.P., Schmalholz, S.M. (2021). Peak Alpine metamorphic conditions from staurolite-bearing metapelites in the Monte Rosa nappe (Central European Alps) and geodynamic implications. *J. Metamorph. Geol.*, 39. doi: 10.1111/jmg.12595.

Venturini, G., Martinotti. G., Armando, G., Barbero, M., Hunziker, J.C. (1994). The Central Sesia-Lanzo Zone (Western Italian Alps): New field observations and lithostratigraphic subdivisions. *Schweiz. Mineral. Petrogr. Mitt.*, 74, 115–125.

Vernet, J.-P. (1958). Les minéraux lourds d'une série chattienne de la molasse du Plateau Suisse. *Bull. Soc. Vaud. Sci. Nat.*, 67, 93–101.

Vernet, J.-P. (1964). Pétrographie sédimentaire dans la molasse de la région d'Yverdon. *Bull. Soc. Vaud. Sci. Nat.*, 68, 443–454.

Vignaroli, G., Faccenna, C., Jolivet, L., Piromallo, C., Rossetti, F. (2008). Subduction polarity at the junction between the Western Alps and the Northern Apennines, Italy. *Tectonophysics,* 450, 34-50.

Von Blanckenburg, F. and Davies, J.H. (1995). Slab breakoff: A model for syncollisional magmatism and tectonics in the Alps. *Tectonics,* 14, 120–131.

Von Eynatten, H. (2003). Petrography and chemistry of sandstones from the Swiss Molasse Basin: An archive of the Oligocene to Miocene evolution of the Central Alps. *Sedimentology,* 50, 703–724.

Von Moos, A. (1935). Sedimentpetrographische Untersuchungen an Molassesandsteinen. *Schweiz. Mineral. Petrogr. Mitt.,* 15, 169–265.

Vrabec, M., Janak, M., Froitzheim, N., de Hoog, J.C.M. (2012). Phase relations during peak metamorphism and decompression of the UHP kyanite eclogites, Pohorje Mountains (Eastern Alps, Slovenia). *Lithos,* 144/145, 40–55.

Vuichard, J.-P. (1987). Conditions P-T du métamorphisme anté-alpin dans la "seconde zone diorito-kinzigitique" (Zone Sesia-Lanzo, Alpes occidentales). *Schweiz. Mineral. Petrogr. Mitt.,* 67, 257–271.

Warren, C.J. (2013). Exhumation of (ultra-)high-pressure terranes: Concepts and mechanisms. *Solid Earth,* 4, 75–92.

Wassmann, S. and Stöckhert, B. (2013). Rheology of the plate interface – Dissolution precipitation creep in high pressure metamorphic rocks. *Tectonophysics,* 608, 1–29.

Weber, S. and Bucher, K. (2015). An eclogite-bearing continental tectonic slice in the Zermatt-Saas high-pressure ophiolites at Trockener Steg (Zermatt, Swiss Western Alps). *Lithos,* 232, 336–359.

Weber, S., Sandmann, S., Miladinova, I., Fonseca, R.O.C., Froitzheim, N., Münker, C., Bucher, K. (2015). Dating the initiation of Piemonte-Liguria Ocean subduction: Lu-Hf garnet chronometry of eclogites from the Theodul Glacier Unit (Zermatt-Saas zone, Switzerland). *Swiss J. Geosci.,* 108, 183–199.

Wheeler, J. (1991). Structural evolution of a subducted continental sliver: The northern Dora-Maira massif, Italian Alps. *J. Geol. Soc. London,* 148, 1101–1113.

Wiederkehr, M., Bousquet, R., Schmid, S., Berger, A. (2008). From subduction to collision: Thermal overprint of HP/LT meta-sediments in the north-eastern Lepontine Dome (Swiss Alps) and consequences regarding the tectono-metamorphic evolution of the Alpine orogenic wedge. *Swiss J. Geosci.,* 101, S127–S155.

Zanchetta, S., Poli, S., Rubatto, D., Zanchi, A., Bove, G.M. (2013). Evidence for deep subduction of the Austroalpine crust (Texel Complex, NE Italy). *Rend. Lincei Sci. Fisiche Naturali,* 24, 163–176.

Zanchetta, S., Malusá, M.G., Zanchi, A. (2015). Precollisional development and Cenozoic evolution of the Southalpine retrobelt (European Alps). *Lithosphere,* 7, 662–681.

Zechmeister, M.S., Ferré, E.C., Cosca, M.A., Geissman, J.W. (2007). Slow and fast deformation in the Dora Maira Massif, Italian Alps: Pseudotachylites and inferences on exhumation history. *J. Struct. Geol.,* 29, 1114–1130.

Zhao, L., Paul, A., Guillot, S., Solarino, S., Malusà, M.G., Zheng, T., Aubert, C., Salimbeni, S., Dumont, T., Schwartz, S., Zhu, R., Wang, Q. (2015). First seismic evidence for continental subduction beneath the Western Alps. *Geology,* 43, 815–818.

Zucali, M., Spalla, M.I., Gosso, G. (2002). Strain partitioning and fabric evolution as a correlation tool: The example of the Eclogitic Micaschist Complex in the Sesia Lanzo Zone (Monte Mucrone-Monte Mars, Western Alps, Italy. *Schweiz. Mineral. Petrogr. Mitt.,* 82, 429–454.

Zurbriggen, R. (2017). The Cenerian orogeny (early Paleozoic) from the perspective of the Alpine region. *Int. J. Earth Sci.,* 106, 517–529.

List of Authors

Philippe AGARD
Sorbonne Université
CNRS-INSU
Institut des Sciences de la Terre Paris
ISTeP UMR 7193
France

Michel BALLÈVRE
Géosciences Rennes-UMR 6118
Université de Rennes 1
France

Nicolas BELLAHSEN
ISTeP
Sorbonne Université
UMR 7193
CNRS
Paris
France

Michel FAURE
Institut des Sciences de la Terre d'Orléans
Université d'Orléans
CNRS
France

Stéphane GUILLOT
Université Savoie Mont Blanc
CNRS
IRD
UGE
ISTerre
Université Grenoble Alpes
France

Jean-Baptiste JACOB
Université Savoie Mont Blanc
CNRS
IRD
UGE
ISTerre
Université Grenoble Alpes
France

Yves LAGABRIELLE
CNRS
Géosciences Rennes
Université de Rennes 1
France

Paola MANZOTTI
Department of Geological Sciences
Stockholm University
Sweden

Didier MARQUER
UMR 6249 Chrono-environnement
Université de Bourgogne-Franche-Comté
Besançon
France

Anders MCCARTHY
CY Cergy Paris Université
France
and
Institute of Geochemistry and Petrology
ETH Zürich
Switzerland

Othmar MÜNTENER
Institute of Earth Sciences
University of Lausanne
Switzerland

Claudio L. ROSENBERG
ISTeP
Sorbonne Université
UMR 7193
CNRS
Paris
France

Denis THIÉBLEMONT
Bureau de Recherches Géologiques
et Minières
Orléans
France

Pierre TRAP
UMR 6249 Chrono-environnement
Université de Bourgogne-Franche-Comté
Besançon
France

Jonas VANARDOIS
UMR 6249 Chrono-environnement
Université de Bourgogne-Franche-Comté
Besançon
France

Index

M, N

O

P

R, S

U, V, W

Summary of Volume 1

Chapter 4. Late-Orogenic Extension and Current Tectonics in the Western and Central Alps: A Review

Christian SUE

Part 3. Regional Alpine Studies from Extension to Collision

Chapter 5. Tectonic Evolution of the Eastern Alps

Walter KURZ, Harald FRITZ, Andreas WÖLFLER

List of Authors

Index

Summary of Volume 3

Chapter 3. A Review of How Our Understanding on the Relationships between the Alps and the North Alpine Foreland Basin has Evolved

Fritz SCHLUNEGGER and Edi KISSLING

Chapter 4. The Detrital Thermochronological Record of Exhumation of the European Alps since the Early Oligocene

Matthias BERNET, Peter VAN DER BEEK, Pierre G. VALLA and Christoph GLOTZBACH

Printed and bound by CPI Group (UK) Ltd, Croydon, CR0 4YY

13/09/2026

14961000-0003